THE UNITED BROTHERHOOD OF CARPENTERS

Wertheim Publications in Industrial Relations

Established in 1923 by the family
of the late Jacob Wertheim "for the support of original research
in the field of industrial cooperation."

THE UNITED BROTHERHOOD OF CARPENTERS

The First Hundred Years

WALTER GALENSON

Harvard University Press

Cambridge, Massachusetts
and London, England   1983

Library of Congress Cataloging in Publication Data

Galenson, Walter, 1914–
The United Brotherhood of Carpenters.

(Wertheim publications in industrial relations)
Includes bibliographical references and index.
1. United Brotherhood of Carpenters and Joiners of
America — History.   I. Title.   II. Series.
HD6515.C2U54   1983   331.88'194'0973   82–23402
ISBN 0–674–92196–8

FOREWORD

The United Brotherhood of Carpenters and Joiners of America has been at the center of the American labor movement for the past century. Both the American Federation of Labor (now the AFL-CIO) and the Carpenters date their beginnings to 1881. Peter J. McGuire, founder of the Carpenters and its leader as secretary-treasurer for the first two decades, was the major collaborator with Samuel Gompers in forming the Federation. McGuire initiated the call for the conferences of 1881 and 1886 from which the Federation emerged, he chaired the first Federation convention in 1886, and he was secretary-treasurer on a part-time basis and then its first vice-president.

The Carpenters have produced a sequence of national leaders who have played a major role in the labor movement — Peter J. McGuire, William D. Huber, Frank Duffy, James Kirby, William L. Hutcheson, M. A. Hutcheson, and William Sidell, to mention officers who completed their active service before 1981. From the outset, except for the years 1901 to 1905, the Carpenters have had a key officer on the Executive Council of the Federation; no other affiliated union can approach this record. As Professor Galenson points out, "In the course of a century, the United Brotherhood of Carpenters and Joiners of America grew from a small organization founded by a devoted band of carpenters into one of the giants of the trade union world."

Although the history of the Carpenters has not been well known, even by students of labor history and collective bargaining, this union provides a significant perspective from which to understand the debates and the trial and error process through which America's national unions have evolved. Prominent among the basic problems are the interactions and responsibilities of local and national levels of governance; the relations of individual members to their local and national organizations; the legislative and judicial systems within the unions; the conflicts among autonomous national unions in the Federation and the significance of jurisdiction; and the changing relations between Canadian members and an organization based in the United States. In their growth as the leading institutions in our labor movement, the national unions have accommodated the roles of ideology, political organization, and legislative activities; the central role and changing structure of collective bargaining and attitudes toward employers; the concerns with federal, state, and local governments as employers in peace and wartime; and policies toward pension and other union benefits. It is instructive to see policy issues emerge and be shaped by new environmental forces and challenges. It is illuminating to trace their growth within a single organization.

The Carpenters provide a particularly revealing setting for the evolving tensions and conflicts, as well as cooperation, between local community interests and national policies in times of recession and prosperity, in peace and wartime over the past century. Some processes remain staunchly local, while others have become nationalized to varying degrees in complex mixtures. The emergence of the district council comprising local units in a community has further enriched and complicated these processes.

The narrative of this volume reminds us how intense has been the hostility of many employers to the growth of labor organizations and to collective bargaining policies over the century. It also provides some needed perspective on the recent period of severe recession and stagnation or decline in union membership. But as Professor Galenson concludes, "If there is one thing to be learned from the Brotherhood experience, it is that a well-administered, well-financed, and solidly structured organization yields large returns to the working people who sustain it."

Professor Galenson is ideally qualified to write this definitive history of the Carpenters and to present on a full canvas the picture of this significant and somewhat neglected national union. Since his graduate student days, he has been interested in rival unionism, and he has written seminal work on labor history in the 1930s and 1940s. He is the author or editor of many volumes on labor movements and economic development in advanced and developing countries. He is widely acquainted with the archives and bibliography of the period dealing with labor organizations and public policy, and he has enjoyed full access to all materials in the well-kept records of the Carpenters. These extensive archives have yielded a number of documents previously unknown to scholars, including an unpublished manuscript by Frank Duffy on the history of the Brotherhood, an early diary of Peter J. J. McGuire, and unpublished testimony by McGuire before the United States Industrial Commission, records that have enhanced the review of the union's internal processes and its relations to outside organizations.

This is the fourth book by Professor Galenson published in the Wertheim Publications in Industrial Relations, for which I have served as editor since 1945. This is the fifty-first volume in the series, forty-six of which have been published since 1945.

This book should be of special interest both to scholars and to practitioners in labor, management, or government, who are concerned with understanding the labor movement and collective bargaining, the shaping of their policies and values, and with projecting their future. The Brotherhood of Carpenters is the largest union in the building and construction industry and reflects well those technological and market environments, including the special features of those labor markets, although it has also attracted members and organizations in other related sectors. Representing these diverse interests has been a challenge to organizational building and administration. The theme of the impact and limitations of public policies, which runs through this narrative and analysis, should be of wide interest. The book also discusses some of the most significant personalities of the labor movement over the past century at critical moments in their careers. It will long remain an authoritative study.

March 1983   John T. Dunlop

PREFACE

It is a pleasure to acknowledge the cooperation I received from the officers and staff of the United Brotherhood of Carpenters and Joiners in the preparation of this study. Their files were opened; they responded cheerfully to requests for information and data, and gave me the benefit of their knowledge and experience. General President William Konyha was supportive of the project throughout. I was also fortunate in being able to interview his two predecessors in that office, Maurice A. Hutcheson and William Sidell, an experience which afforded me the unusual opportunity of hearing about the accomplishments and problems of the union directly from the men who guided it over a period of thirty years.

General Secretary John S. Rogers maintained a close interest in the project from its inception to its conclusion; he was indefatigable in helping track down sources of information; and his careful reading of the manuscript gives me some confidence that if errors of fact remain, they are not egregious. Professor John T. Dunlop of Harvard University shared with me his great knowledge of the American labor movement through numerous discussions and a line-by-line commentary on an earlier draft of the study. It is no exaggeration to say that without the encouragement of these two men, completion of this volume would have been long delayed, and perhaps never achieved at all. Mary Carnell did her usual admirable job of transforming my almost illegible scrawl into a letter-perfect typed manuscript. Mary Ellen Geer, who edited the volume, contributed greatly to its readability.

Walter Galenson

CONTENTS

THE UNITED BROTHERHOOD OF CARPENTERS

THE PREDECESSORS OF THE
UNITED BROTHERHOOD

Carpenters were among the first of the American craftsmen to organize into associations for mutual benefit and protection. "It is a matter of tradition that the idea of overturning the tea in Boston Harbor was first promulgated at a meeting of the ship carpenters and calkers, and that these men before that sometimes acted together in political matters."[1] Whether or not the ship carpenters were actually the instigators of the Tea Party, some of them do appear to have been followers of Samuel Adams. At any rate, there is no doubt that organizations of carpenters were in being before the American Revolution.

THE EIGHTEENTH CENTURY

According to the standard history of American labor, "The first authenticated organization in the building trades was that of the house carpenters of Philadelphia in 1724."[2] The Carpenters' Company of Philadelphia included both masters and journeymen. A five-member committee was established to determine the price of all new types of work, and to settle controversies on the measurement and valuation of carpentry work. A uniform scale of prices for work was established, and anyone accepting less than the scale was liable to expulsion.

This "book of prices" later provided the basis for costing jobs, and was designed to prevent undue competition. What it did was to establish piece rates for each portion of the work involved in putting up a building. The rates were often kept secret by the master mechanics to prevent their employees, who were usually paid by the day, from knowing what their profits were. As one local journeymen's union complained in 1836, "Many of us would prefer working piece-work if we could know the price that would be put on our work; but we cannot know it in consequence of the employers' book of prices being secret; that it is unjust, numbers of us have learned to our cost."[3]

The Carpenters' Company of Philadelphia was involved in at least one famous historical event. In 1770 it erected its own meeting hall. When the first congress of colonial representatives assembled in Philadelphia in 1774 and was refused the use of the state capitol, the delegates met at the

Carpenters' Hall. What transpired there has been described in the following terms:

> At ten in the morning of the fifth of September, 1774, the delegates assembled at the City Tavern on Second Street. From that point they marched on foot along the street until they reached the threshhold of Carpenters' Hall. The members are soon seated and the doors are shut. A sense of the extraordinary importance and the serious responsibility of the occasion caused a profound silence to fall upon the assembly . . . That day, in Carpenters' Hall, there was a scene worthy of the painter's art. Washington was kneeling there, and Henry, and Randolph, and Rutledge, and Lee, and Jay, and by their side stood, bowed in reverence, the Puritan Patriots of New England, whose humble households were being menaced by armed soldiery.[4]

The constitutional assembly met in the Carpenters' Hall in 1788, and it was there that the Constitution of the United States was drawn up.

The first recorded association of journeymen carpenters was the Union Society of Carpenters, formed in Philadelphia in 1791. They complained about being obliged to work from sunup to sundown in the summertime, and of being put on piecework in winter. They demanded a uniform working day of 6:00 A.M. to 6:00 P.M. all year, with an hour off for dinner, and when they were refused, went on strike. As a means of putting pressure on the masters, they offered to work directly for owners at 25 percent below the current rate established by the master carpenters, and promised that the quality of the work would be high. The masters warned that such action could only injure the journeymen, using an argument that is reminiscent of the great economic work of the time, Adam Smith's *The Wealth of Nations*, published in 1776:

> The wages of all artificers must be regulated by the number of persons wanting employment; high wages induce masters to increase the number of apprentices, and journeymen to come from other places; low wages produce the contrary effect. It is not, therefore, in the power of any set of them in a free country to keep the price of labor much below, or raise it far above a certain medium for any great length of time together, although they may, by confederating together, for some time injure themselves and others of the same occupation by undertaking work at a price lower than that at which it can reasonably be performed.[5]

Many economists and employers in present-day America would still subscribe to this statement as an accurate description of how wages are determined. Some ideas persist over long periods of time, unaffected by the intrusion of the real world. As far as we know, the Philadelphia journeymen did not succeed in their objectives.

Only a few other scattered references to eighteenth century carpenter organizations survive, although there were undoubtedly more. Carlisle,

Pennsylvania, had a local union in 1795, which maintained a price list for carpentry work.[6] New York City carpenters managed to secure an increase in wages in 1795, and in the following year the Association of House Carpenters of New York was formed. Its main effort was to obtain the ten-hour day, but it disbanded a few years later when it failed to reach its objective.[7] Finally, there is a record of a Brotherhood of Carpenters of Halifax, Nova Scotia, established in 1798. It adopted death benefit and pension schemes, the latter to be paid if funds were available. Starting out with 46 members, it reached a maximum of 115 in 1815, but the total dropped to 73 in 1826, the last year for which there are data.[8]

THE EARLY NINETEENTH CENTURY

The Commons history notes that "the period to 1820 may aptly be characterized as the dormant period. A continuous and persistent effort of wage-earners is prevalent in but two industries [shoemakers and printers]. All other collective economic efforts on the part of laborers to improve their condition as wage-earners were spasmodic and but few in number."[9]

There is some evidence that the carpenters should be included among the crafts that showed persistence in collective action during these years. Both the shipwrights and the house carpenters in New York City organized unions around 1800, and the latter group remained in existence for many years. There also appears to have been continuing organization among the journeymen carpenters in Philadelphia and Boston. Other locals of a more sporadic nature involved the house carpenters of Savannah, Georgia, and the journeymen cabinet makers of New York and Philadelphia.[10] Further research into the records of other cities will undoubtedly reveal additional attempts at organizing journeymen carpenters in protest against the excessively long hours of work — as many as fourteen in the summer — that were prevalent at the time.

The economic depression that set in at the close of the Napoleonic wars and reached its nadir in 1820 acted as a brake on all labor organization. But gradual recovery thereafter led to some major movements among the carpenters, and it is the late 1820s that mark the beginning of a more familiar type of organization.

Philadelphia

Philadelphia appears to have been the first city to experience the upswing in unionism. This is not surprising, for at the time, together with New York, it was the major center for commerce and industry in the United States. On June 13, 1827, the Journeymen House Carpenters of Philadelphia met and passed a resolution calling for the institution of a ten-hour day, on the ground that "a man of common constitution is unable to perform more than ten hours faithful labor in one day, and that men in the habit of laboring from sun rise until dark, are generally subject to nervous and other complaints arising from continued hard labor and they believe that all men have a just right, derived from their Creator, to have sufficient time in each day for the cultivation of their mind and for self improvement."[11] Two days later, the master carpenters met; they deplored the formation of a society "that has a tendency to subvert good order, and coerce and mislead those who have been industriously pursuing their avocation and honestly maintaining their families"; they found that the current carpenters' wages were as high as could be afforded; and they resolved "that we will not employ any Journeyman who will not give his time and labor as usual."[12]

The Journeymen replied to the masters' argument that a ten-hour day would considerably reduce working time with an interesting bit of arithmetic:

> In the longest day in summer there are but 15 hours sun, and deducting 2 hours for meals, leaves 13 hours for work; in the shortest day there is but 9 hours sun, and of course 8 hours work, averaging 10½ throughout the year; now we propose to work 10 hours throughout the summer, and as long as we can see in the winter, taking only one hour for dinner, and we can accomplish nearly 9 hours work in this manner in the shortest day. The average is 9½ hours; thus their loss would be but one 12th part of the time, *and we maintain not any in the work.* [italics added][13]

The Journeymen decided to stop work pending negotiations with the masters, and appointed a negotiating committee of twelve. The committee was empowered to accept carpentry work to be executed on reasonable terms, and to provide assistance to those in need of it "during the stand out." The master carpenters responded by advertising in other cities for 300 to 400 workers, which led eventually to the loss of the strike.[14]

Another start was made in 1835 when the Journeymen House Carpenters Association of Philadelphia was organized. The association demanded that wages be raised to $1.50 per day from March 20 to November 20, and to $1.25 for the rest of the year. The masters responded with a blast against the new union, terming it "not of American birth," "subversive of all regularity in business, and destructive of confidence in the parties concerned," and "the mother of countless evils, and the source of no good." Their resolution is worth quoting in full as an excellent example of the attitude held by employers at the time to organization of their employees:

Resolved, that the Trades' Union is arbitrary, unjust and mischievous in its operation, inasmuch as it forcibly compels the well disposed journeymen to become members, as there is a By Law of this Association prohibiting all members from working at the same building or in the same shop with any journeyman who is not a member, thus compelling him to join the Association and contribute weekly his earnings for the support of the idle and discontented, or he will be thrown out of work himself, as his employer will be compelled to discharge him, fearing that all his other hands will strike.

Resolved, that we view the Trades' Union as a powerful engine of the levelling system; its operation is calculated to reduce the employer to the condition of a Journeyman, and to keep the well disposed and industrious Journeyman a Journeyman all the days of his life, as he is restricted from doing over work, even though he is so disposed.

Resolved, that the Trades' Union is calculated to weaken and destroy the harmony and ties of mutual interest that formerly existed between the master and apprentices; the boys have their minds poisoned by the members of the association, until they no longer consult their masters' interest, and finally look upon him as a hard task master and oppressor, instead of a friend and protector.

Resolved, that the Trades' Union is the growth of Monarchial Government, and ill adapted to our Republican Institutions.

Resolved, that we claim the right as Free Citizens, to make our contracts with the journeymen mechanics themselves, without the intervention of the Trades' Union; and that we do not recognize the right of any association or combination of men to interfere in the ordinary transaction of our business.

Resolved, that we are willing, owing to the advanced price of living, to increase the wages to good workmen, provided they furnish themselves with tools, as is the custom of other cities, and that we deem it inexpedient to name any sum, preferring that every man be at liberty to make his own bargains, and be master of his own shop or building.

Resolved, that the Master Mechanics of the city and county of Philadelphia, be invited to attend a meeting for the purpose of forming an association to be called the Anti-trades' Union Association, the object will be to protect us from the mischievous effects arising from the combination.[15]

The wage demand that occasioned this outburst came after increases of 38 percent in the price of food and 30 percent in the price of fuel from 1834 to 1836,[16] which put pressure on what were already low living standards based on the 1835 daily wage levels of $1.25 and $1.12½ during the summer and winter months, respectively. The union responded with the following description of the conditions under which its members worked:

Not more than one half of the journeymen have employment more than nine months in the year, we are very much exposed to the heat of the summer sun while roofing or framing, and in the winter time to work in cold bleak shops or in open buildings without fire, or are compelled to lose time. The cost of journeymen's tools is from ten to a hundred dollars, which they are obliged to carry very often on their backs like pack horses . . . and which they are liable to have stolen from them or destroyed by fire . . . A member of the committee that drafted resolutions for the meeting of the employers has been known to purchase fifty dollars worth of hats at a discount of thirty percent, and palm them on his journeymen in payment of their labor.[17]

We do not know precisely what the outcome of the strike was, but it is probable that wages were raised in line with the rapid increase in the cost of living. One result of the renewed militancy of the journeymen was the decision to convene a national convention of carpenters in an effort to secure the ten-hour day. The call was made by the Philadelphia carpenters, and the convention was held on October 24, 1836, the first national gathering of the trade. Delegates from Albany, Pittsburgh, Baltimore, Washington, and Philadelphia were present. A resolution was adopted calling for the ten-hour day and uniformity of wages in all cities and towns in the United States. A constitutional convention was called for April 1837, but apparently there was still not enough interest among carpenters to sustain a national body, for the convention was never held.[18]

New York City

The journeymen house carpenters of New York had already gained the ten-hour day by 1830, but rising prices impelled them to seek an increase in wages. In 1833 they demanded that wages be raised from $1.37½ to $1.50 a day, and after a month's strike they gained the higher level. *The New York Journal of Commerce*, a business paper, condemned all combinations as "wicked" and charged:

> Combinations among journeymen are usually set on foot by the dissolute, improvident, and therefore restless; and in the outset clearly sustained by the second and third rate class of hands . . . Turn-outs are always miserably profitless jobs. If they are successful they cannot in the long run benefit the class whose wages are raised; for the diminution in the quantity of occupation and the increased number of laborers drawn to the spot, will more than compensate for all the gains. If a day's work receives a higher reward, that advantage will be more than counterbalanced by days spent in idleness for want of occupation.[19]

These dire predictions notwithstanding, in March 1836 the New York carpenters put forth a demand that their wages be increased to $1.75 a day, and raised to $2.00 a day in June. When the employers refused, they went on strike, which they won after staying out for several weeks.[20]

Boston

Much the same course of events characterized the situation in Boston. The Carpenters' Union of Boston, founded in 1812, raised a demand for the ten-hour day in 1825. As usual, the master carpenters responded negatively, saying that journeymen combinations were "fraught with numerous and pernicious evils," and adding:

> We cannot believe this project to have originated with any of the faithful and industrious sons of New England, but are compelled to consider it an evil of foreign growth, and one which we hope and trust will not take root in the favored soil of Massachusetts. And especially, that our city, the early rising and industry of whose inhabitants are universally proverbial, may not be infested with the unnatural production.[21]

The masters advertised the availability of jobs in the hope of attracting journeymen from other cities, which prompted the union to take an advertisement warning carpenters that there was considerable unemployment in Boston, and that "on the present system, it is impossible for a Journeyman Housewright and House Carpenter to maintain a family at the present time, within the wages which are now usually given to the Journeymen House Carpenters in this city."[22] This is one of the first examples of the "stay away" notice that subsequently became a major feature in *The Carpenter*, the official organ of the United Brotherhood after its formation.

The Boston carpenters continued their proverbial "early rising and industry," for the masters prevailed. But another effort began in 1832, this time among the ship carpenters. The Columbian Charitable Society of Shipwrights and Caulkers of Boston asked the masters to meet with them to discuss a reduction of working hours, and were met instead with a lockout. A meeting of merchants and shipowners resolved that "we will neither employ any journeymen who, at the time, belong to [a] combination, nor will we give work to any master mechanic who shall employ them while they continue thus pledged to each other, and refuse to work during the hours that it has been and is now customary for other mechanics to work in this city."[23] The journeymen complained that their conditions of labor were very onerous: they were averaging only one dollar a day; they lost a great deal of time in cold or wet weather; they often worked several hours and received no pay if weather conditions forced work to break off; when ships were moved from one dock to another, they often had to remain aboard and received two days' pay for four days of work. They expressed a willingness to work over ten hours if they received extra pay.[24] But the master ship carpenters remained adamant, and the cause was lost.

The next organizational attempt came several years later. In 1834 mechanics of various trades met to form the Trades' Union of Boston and Vicinity. One of the leaders of this movement was Seth Luther, a house carpenter by trade. He has been described as "the first American in the anti-child-labor crusade"; he traveled extensively in New England, speaking to the factory workers, and was one of the most influential trade unionists of his time.[25] When the house carpenters struck in 1835, they had the support of masons and other crafts. They issued a statement attacking the employers for subjecting them to an "odious, cruel, unjust, and tyrannical system which compels the operative Mechanic to exhaust his physical and mental powers by excessive toil, until he has no desire but to eat and sleep." They warned all brother mechanics to beware of advertisements for jobs, and noted that they had not asked for an increase in wages, "but are willing that demand and supply should govern the price as that of all other disposable property." And they concluded:

> Mechanics of Boston — stand firm — be true to yourselves. Now is the time to enroll your names on the scroll of history as the undaunted enemies of oppression, as the enemies of mental, moral, and physical degradation, as the friends of the human race. The God of the Universe has given us time, health and strength. We utterly deny the right of any man to dictate how much of it we shall sell. Brethren in the City, Towns and Country, our cause is yours, the cause of Liberty, the cause of God.[26]

This appeal, which was probably written by Seth Luther, was circulated in a number of cities, and funds were raised to help the strikers. The strike appears to have lasted seven months, but once again it was lost. One of the obstacles was that the line between a small master and a journeyman was not sharply drawn, and many of the latter tended to identify themselves with the former.

With the exception of Boston, however, the strikes of 1835 appear to have achieved the ten-hour day objective. But it was not until 1840 that the federal government followed suit by establishing a ten-hour day for all government works, the navy yards in particular.[27]

Other Cities

Organizations of carpenters spread slowly to cities other than the main centers of Philadelphia, New York, and Boston, under the impetus of the rapid increase in the cost of living. In 1835 and 1836, house carpenters formed unions in Albany and Poughkeepsie, New York; Hartford, Connecticut; the District of Columbia; Cleveland, Ohio; Pittsburgh, Pennsylvia; and Indianapolis, Indiana, among others. Their demands were more or less the same: the ten-hour day, a wage of $1.50 per day, and a uniform wage throughout the year. There were even the beginnings of intercity mutuality, a concept that was later to play an important role in the formation of national unions. The Hartford local union, for example, stipulated in its bylaws that "any workman, bringing a certificate from any Carpenters' and Joiners' Society, may be admitted into the Society without the usual initiatory fee."[28]

This period of union activity came to an abrupt halt with the economic panic of 1837. From 1837 to 1843, there was a one-third decline in wholesale prices.[29] There are no comprehensive unemployment statistics for this period, but from partial evidence, it is clear that many jobs vanished. In New York, for example, 6,000 masons, carpenters, and other building workers were discharged in 1837. The ten-hour day survived by and large, but wages fell. Most trade unions simply went out of existence. This is everywhere the history of early American trade unionism; it would still be some time before trade unions had the means to enable them to weather severe economic recessions.

FROM 1843 TO THE CIVIL WAR

The recession bottomed out in 1843, and although the recovery was slow and unsteady, it did permit some revival of trade union activity. Much of the energy of those who sought to improve the position of the working man was dissipated in idealistic schemes for restructuring society. Among other ideas, producer cooperation emerged as a way to replace the capitalist organization of industry. As Commons observed, "during these years of unemployment [1837–1852], aggressive trade unionism almost disappeared, and the field was occupied by philanthropy and schemes of speculative reforms."[30]

In 1844 Pioneer Temple No. 1, House Carpenters Protective Association of New York, was organized. As with many unions of the time, it was a secret society, with grips, signs, and passwords. It was by no means antagonistic to employers, for its basic principles included the statement that "the interests of the employer and employee are one and the same when properly understood; therefore, the interests of the trade require that they should act together to overthrow those obstacles which depress Labor."[31] In March 1850 the association mounted a strike to force those employers who were paying less than $1.75 a day to reinstate that level of wages. However, the strike appears to have been lost, since it was reported in May that several employers had actually reduced wages to $1.62½ a day.

Several other carpenter groups were also formed in New York at about the same time. In 1849 the House Carpenters' Association of the City of New York came into being, and also expressed its willingness to cooperate with employers in the interest of mutual prosperity. It adopted a rule that was later to attain great importance in the labor movement; political and sectarian matters could not be introduced at union meetings.[32]

The Sash and Blind Makers' Protective Union was organized in 1850, its principal demand being $1.62½ for inside work and $1.75 for outside. The same year saw the birth of the Mutual Protective Society of Cabinet Makers of New York City, with an initial membership of 800. The Independent Society of Ship Sawyers of New York City, which appears to have been in existence for some time, resolved in 1850 that any sawyer who worked for less than $2.00 a day was subject to a $12.00 fine. Washington, D.C., Buffalo, New York, and Princeton, Indiana, also witnessed the formation of short-lived unions of carpenters.[33]

The economic upswing that began in 1851 ushered in a renewal of trade union activity. Again we see the impact of rising prices and a tighter labor market on worker militancy. Philadelphia was the scene of a carpenter strike in 1851. The demand was a wage increase from $1.50 to $1.75 a day, which would have yielded $10.50 a week. A newspaper account of the strike cited the following figures as the weekly cost of living for a workingman aged 30 to 40 years, with a wife and three children:

Flour	$ .62½	Fuel	$ .40
Sugar	.32	Sundries	.40
Butter	.62½	Household articles	.25
Milk	.14	Bedding	.20
Meat	1.40	Rent	3.00
Potatoes	.50	Clothing	2.00
Tea and coffee	.25	Newspaper	.12
Candles and oil	.14		
		Total	$10.37

And the query was raised: "Where is the money to pay for amusements, for ice creams, his puddings, trips on Sunday up or down the river, in order to get some fresh air; to pay the doctor or apothecary, to pay for pew rent in the church, to purchase books, musical instruments?"[34]

We do not know whether the Philadelphia carpenters achieved their objective, but fragmentary data from other cities suggest that there were substantial wage differentials among cities and crafts at the time. The Ship Joiners of New York and Vicinity adopted a motion in April 1854 establishing $2.50 a day as the standard for the area, an increase of 25 cents. The carpenters of Albany, New York, demanded an increase of 37½ cents a day in 1858, but the base rate is not known. In Scranton, Pennsylvania, carpenters were working ten hours for $1.50 a day for inside work and $1.62½ for outside work. Buffalo, New York carpenters were working in 1857 for wages ranging from $1.00 to $1.50 a day, and payment was often made in goods rather than in cash.[35]

There is available a fascinating account, dating from this period, of the history of the Ship Carpenters and Caulkers of St. Louis and Carondelet, Missouri, which throws a great deal of light on what working conditions were like for carpenters. In 1846 ship carpenters received $2.00 a day for work on the docks and marine railways; $1.75 a day for shipbuilding; and $2.50 a day on the levee for boat repairing. Working hours were from 7:00 A.M. to 6:00 P.M. from March 20 to September 20, and from daybreak to dark the rest of the year. Apprentices fared badly: they received $2.50 a week, out of which they had to pay for board and clothing. Most of them, moreover, were assigned to common labor work, and few stayed on for more than a year.

In 1847 the dockmen struck for $2.25 a day; they were joined by the ship carpenters and caulkers. After a brief stoppage the employers gave in, for labor was scarce. Two weeks later there was another successful strike for an additional 25 cents a day. The men resented the fact that the employers were cutting down on their meal time and working them overtime from five to ten minutes a day. To stop this, they purchased a bell of their own, which was rung at the proper times.

From 1848 to 1855 wages remained the same, but new work tended to move to the Ohio River, where both lumber and wages were lower. When the Ship Carpenters' and Caulkers' Association was formed in 1855, dock wages went up to $2.75 and repair work to $3.50 a day. The dock employers agreed with the union that if wages were not raised further, they would not seek to lower them, but in 1857 they nevertheless cut them back to $2.50 when an economic decline began. As for the association, "corruption found its way into its ranks; some of the journeymen became jobbers, other jobbers were allowed to become members. They ruled the association inside and broke it up from the outside.[36]

Another recession in 1854–1855 took a heavy toll on the trade unions, including the carpenters. "In 1853–1854 there was scarcely a trade in any of the eastern cities that did not have some sort of trade union. The depression of 1854–1855 marks the disappearance of nearly all of them."[37] Staying organized was a difficult proposition with many men out of work.

Just prior to the onset of the Civil War, unionism among carpenters took an upward turn, particularly among ship carpenters. Cincinnati, New Albany (Indiana), San Francisco, Chicago, and Baltimore were some of the cities in which ship locals were initiated. In fact, in 1860 a Ship Carpenters' and Caulkers' International Union was organized, with representatives from eight cities. But the East Coast unions did not affiliate with it, and it was not until 1864 that agreement was reached between the two groups, providing for mutually recognized traveling cards and a commitment to prevent the migration of workers from one area to another during strikes, as well as the provision of financial assistance. The eastern unions never formally affiliated with the International Union, which survived until 1866.

A number of national unions in other trades and industries were started before the Civil War, but apart from the National Typographical Union, they either went out of existence or merely acted as loose contact bodies among the locals. There apparently was an attempt to establish a national union of house carpenters in 1854, but no records have survived.[38] It was not until the great industrial advance that began after the Civil War that national unionism, based on more firmly established locals, became a viable proposition in the United States.

THE CIVIL WAR PERIOD

Wartime years have generally been favorable for trade union organization in the United States, and the Civil War was no exception. Rising prices and the shortage of manpower diminished employer opposition to trade union demands. The increased cost of living disposed workers to combine for higher wages. There are different estimates of how much the cost of living rose from 1861 to 1865, ranging from 62 percent to 75 percent. Real wages probably fell; average daily wages in industry increased by 50 percent to 55 percent, although construction workers did somewhat better, for their wages rose by about 60 percent.[39]

Commons records that between 1863 and 1864, the number of local unions of carpenters and joiners increased from four to seventeen.[40] However, there were certainly more than four locals in existence in 1863, particularly if the ship joiners and caulkers are included in the count. We know, for example, that in 1863 there were unions of building carpenters in Troy, Elmira, Buffalo, and Rochester, New York, and Saginaw, Michigan, quite apart from ship carpenters in Baltimore, New York City, and Philadelphia.[41]

There is no doubt that the year 1864 did mark a rapid advance in carpenter organization. Consumer prices took a big jump from 1863 to 1864, somewhere between 22 and 30 percent, and the tradesmen reacted by demanding compensating wage increases.

A typical product of the new union drive was Carpenters and Joiners Union No. 1 of Providence, Rhode Island, formed on January 29, 1864. This organization remained in existence until 1882, when it joined the Brotherhood, and eventually became part of Local No. 94. The initiation fee was set at $1.00. Members who were intoxicated or used profane language at meetings were subject to a $1.00 fine and possible expulsion. Although the preamble to the union constitution contained a reference to the "tyrannical oppressive power of Capital," strikes were deprecated as being injurious to the members and were to be avoided if at all possible. Wages were to be determined by a vote of the membership each February, to take effect at the beginning of the new building season in April.[42]

In November 1863 a more ambitious attempt at organization had been initiated. A convention was held in Rochester, New York, to form a New York State federation of carpenters. Delegates from five upstate New York cities came togetheer, "fired with the zeal that imitated our fathers of the revolution when they declared for freedom . . . We are willing to give our time, our money and our talent, to protect the rights of the mechanic, of not only the State of New York, but all through the length and breadth of the land."[43] By June 1864 eight unions were affiliated with the federation, which was instrumental the following year in attempting to bring about the formation of a national union. Among the other cities to witness the birth of a carpenters' union in 1864 were Lewiston, Maine; Louisville, Kentucky; Indianapolis, Fort Wayne, and Evansville, Indiana; St. Louis, Missouri; Detroit, Michigan; and Meadville, Pennsylvania. In Massachusetts a state federation was created, and by the end of the year eight locals were affiliated with it.

The new organizations were not welcomed by the employers. When a union of Ship Carpenters and Caulkers was formed in Buffalo, New York, a shipowners' association was established to resist it. In a statement issued on April 4, 1864, unions were denounced for taking positions "frightful in the extreme, and none can foretell the evil that will sooner or later grow out of them, if we sit still and deal out nourishment to them, and continue to submit to the ruinous and monstrously exorbitant demands they are constantly making upon our property and purses, totally regardless of our pecuniary ability to meet such demands. The day is not far distant when they will modestly ask an equal distribution of the property itself, if not arrested at this point. It is a fearful state of things when any Society (whatever may be its object) asserts that this man or that shall not be employed unless he first becomes a member of their union."[44] Employers were urged not to employ union members and to pay the maxium rate of $2.50 a day fixed by the shipowners.

In reply, the union denounced the situation in which employers "rioted in luxury and pleasure, rolling in carriages, and occupying seats in fashionable churches to enjoy discourses prepared for the privileged classes; leaving the workingman to labor and sweat, and receive a mere pittance to enable him to prosecute his weary journey through life." It argued that $3.00 a day was not an exorbitant demand, given the rise in the cost of living, particularly since two-thirds of the ship carpenters were working only eight months a year and had to provide their own tools, which had doubled in price. And they pointed out an attitude of employers that many workers were to experience in the future: "Every good citizen rejoices at the prosperity of an industrious, fair man, but it is too often the case that the man that once earned his bread by the sweat of his brow is the first to ape the aristocrat and play the petty tyrant."[45] A strike was called, but there is no record of the outcome.

THE CARPENTERS' AND JOINERS' NATIONAL UNION

Although there had been earlier ephemeral efforts to form a national union, the organization that was started in 1865 was the first to have any real chance of success. At about the same time several other trades, including the molders, the bricklayers, the plasterers, the printers, the locomotive engineers, and the cigar makers, were able to establish permanent national bodies.

On June 3, 1865, the New York State Carpenters' and Joiners' Union issued a call for a convention to be held on September 5. Some thirty –seven delegates, representing twenty-four cities, responded to the call. Officers were elected and a constitution adopted. Each state and local union was asked to contribute $5.00 toward building a national treasury.

The series of recommendations on strikes adopted by the convention indicates that the founders were cautious men. Strikes were to be avoided wherever possible, and when they were called, the demands were to be well-timed, prudent, and reasonable. No strikes should be called unless concurred in by two-thirds of the members in good standing. "Not more than one week should be allowed as reasonable time to employers to answer the demands, which ought to be courteously and firmly made."[46] One of the clues to the purposes behind the formation of a national union was the suggestion that before and during a strike, local unions ought to inform the National Recording Secretary of "the minutest facts," and keep him posted about the state of employment in their areas. The Recording Secretary was to notify all local affiliates about the strike, and these organizations would presumably keep their members away. The dissemination of employment information and strike notices was to become a function of the Brotherhood in later years, to offset the recruiting efforts of employers for strikebreakers. The Buffalo ship carpenters noted during their 1864 strike: "Unions are rapidly forming everywhere along the Lakes, and the capitalists will either have to consent to have their work poorly done, or employ good and true union men — for every workingman worthy of the name of *Man* is now enrolled or soon will be in a union."[47]

The Carpenters' National Union held annual conventions until 1871, when it dissolved. Part of the problem lay in its affiliation with the National Labor Union, which was formed in 1866. Under the guidance of William Sylvis, the head of the molders' union, the National Labor Union moved away from economic action and toward independent politics and producer cooperatives, alienating many of the trade unionists. The carpenters were split on the issue of affiliation, which weakened their own organization. Some locals withheld their per capita tax, and eventually there were no funds for even the limited operations that the National Union had been carrying on.

The carpenters were by no means unique in this initial failure to establish a national organization. The rise in wages that had characterized the Civil War period and its immediate aftermath gave way to a depression and falling prices in 1866. During the years 1866–1873 consumer prices fell by about 20 percent, and although wages tended to remain fairly stable, there was constant employer pressure for wage reductions. The national unions of the time were decentralized federations of autonomous locals, with no national benefit schemes. Low dues did not permit the accumulation of strike funds. Most of the strikes that took place were to counter wage cuts, which was never an issue likely to lead to union victory.[48] Once again, an adverse cyclical trend frustrated working men in their attempts to form protective associations.

It would not be accurate to say that the years 1866 to 1873 were completely devoid of union activity, although the positive results were meager. Carpenter locals were formed in such diverse places as Chicago, Milwaukee, Cincinnati, and New Haven. Peter W. Birk, a young man who joined Carpenters' Union No. 1 of Brooklyn, New York, in 1868, managed within a few years to gain for the organization an eight-hour day. But with the financial panic of 1873, "the union went to pieces. With no guiding force, no combination to hold the men together, no united action on any question, no understanding as to hours or wages, the men were at the mercy of employers and were forced to return to the ten-hour day."[49]

THE EIGHTEEN-SEVENTIES

The decade from 1870 to 1880 was not a good one for trade unionism. The financial panic of 1873 led to a rapid decline in prices, and this time in wages as well. Average daily wages in the building trades fell by between 25 and 30 percent from 1873 to 1879, when the depression bottomed out. There are no unemployment statistics going back that far, but contemporary accounts indicate that there was a great deal of idleness.

A year before the 1873 crash, another attempt was made to form a national union, this time with a more ambitious jurisdictional scope. The National Woodworkers' Union was organized at a meeting in Syracuse, New York, on January 18, 1872, its purpose being to unite all woodworkers into a single national body. The organizers attributed the demise of the Carpenters' National Union to its failure to develop a benefit system, and they pointed to the success of benefit-oriented unions in England as evidence of the need for this type of activity. The guiding spirit of the short-lived organization was Henry Wilkie, a coachbuilder for the New York Central Railroad. In March 1872 the new union reported that it was prepared to issue charters, and it set up an insurance and pension department. But the times were against it, and it apparently foundered in 1873.[50]

In New York, an organization known as the Union Carpenters of the City of New York mounted a strike in 1872 for an increase of 50 cents over the prevailing daily rate of $3.00, as well as for a nine-hour day. The strike was joined by other building tradesmen, and all construction work in the city was tied up. By June 1872 the employers had conceded the union demands and the men were back at work. But like so many others, this union also fell victim to the 1873 panic.

A more successful organizing effort, which survived the depression, was made on May 1, 1872, with the formation of the United Order of American Carpenters and Joiners. The United Order was the first union of more than local scope to surmount the economic vicissitudes that had proved fatal to previous efforts. Although its principal stronghold was New York City, where it had several lodges, it also had members in the neighboring states of New Jersey, Connecticut, and Rhode Island. Borrowing from the experience of successful unions in other trades, it instituted a benefit system that included payments for illness and burial and the insuring of tools. The latter was important because of the relatively high cost of tools and the general requirement that every man should supply his own. The Order was financed originally by a semiannual per capita fee of 50 cents, and its executive committee was authorized to levy assessments to pay strike benefits of up to $6.00 a week. The Order prospered, chartered a total of thirty-two lodges, and was eventually absorbed by the United Brotherhood.

Selig Perlman has written of the 1870s in the following terms:

> The business depression of 1873 to 1879 was a critical period in the American labor movement. The old national trade unions either went to pieces, or retained a merely nominal existence. Employers sought to free themselves from the restrictions that the trade unions had imposed upon them during the years preceding the crisis. They consequently added a systematic policy of lockouts, blacklists, and of legal prosecution to the already crushing weight of hard times and unemployment.[51]

This pretty well describes the situation of the carpenters during these years. Toward the end of the period, organizing efforts once more got under way, impelled by the adverse conditions of labor that employers had instituted. In Chicago, where men were receiving only $2.00 for a ten-hour day, well below the scale in other cities, the United Carpenters and Joiners Protective and Benevolent Association of Illinois was formed in 1877. Unfortunately, the new organization soon split on the basis of differing trade union philosophies. One group, the Protectives, believed that the main function of unions should be to seek higher wages and shorter hours, while the Benevolents believed that the emphasis should be on benefit schemes as a means of cementing worker loyalty to the organization. In 1879 the two groups went their own separate ways, and did not finally come together until 1882, after three years of bitter rivalry.[52]

A year after the Chicago carpenters began to organize, a similar movement began in St. Louis. Again, working conditions were very bad: ten to sixteen hours a day for $1.25 to $2.00. By 1880 the union had three branches, and on April 1 it mounted its first strike; its demands were a daily wage of $2.50, the abolition of piecework, and the right to quit work at 5:00 P.M. on Saturdays. It was successful in securing these demands, and in the spring of 1881 it struck again, this time for a $3.00 wage. This strike was of historic significance, because the secretary of the strike committee was Peter J. McGuire, who had moved to St. Louis in 1878. This was McGuire's first successful experience with union leadership; within two weeks of the beginning of the strike, most of the employers had agreed to the $3.00 rate.[53]

The ability of the St. Louis carpenters to raise their wages so rapidly was due in no small measure to an upturn in the business cycle that began in 1879. Wholesale prices rose by more than 10 percent between 1879 and 1881, and while consumer prices lagged behind, pressure for wage increases was growing. The building trades appear to have done particularly well; although the 50 percent increase in St. Louis was exceptional, perhaps because of the very low base in 1879, average construction wages for the country as a whole rose by 16 percent.[54] These favorable circumstances ushered in what was to be the beginning of permanent trade unionism in the United States.

The tempo of organization began to pick up in 1880. Carpenters in Charleston, South Carolina, who had been working up to twelve hours for $1.50, were able to combine despite intense employer opposition. Carpenters' Benevolent Union No. 1 came into existence in Cincinnati. The United Framers' Union of New York City, which came to be known as the German Speaking Framers' Union, was formed to protest against a $1.50 daily wage, and it soon succeeded in raising the wage to $2.50 and reducing hours to nine per day. This union remained independent until 1894, when it brought its membership of almost a thousand into the United Brotherhood. Shipwrights organized in Brooklyn and New York City, and carpenters in Buffalo, New York.[55]

THE UPSURGE OF 1881

The sentiment for trade union organization that had been stifled by almost a decade of economic depression finally burst forth in 1881. This

was a good year not only for the carpenters but for the other trades as
well. For the carpenters, it was the year in which organization overran
the boundaries of the major urban centers to which it had previously
been confined.

In city and town, local unions were formed with wage increases as their
principal goal. Table 1.1 contains data on going wage rates at the time of
organization in a number of cities, and for a few cases, the increases that
were obtained after union activity. The rate ranges reflect differences in
the quality of the journeymen; as the Cincinnati local reported, "The increase in wages carried with it $2.75 per day for first-class men and $2.50
per day for second-class men."

Some of the wage differentials are interesting. In New Haven, Connecticut, a strike for a 25-cent increase was lost even though the prevailing
rates were well below those in neighboring Bridgeport, where it was reported that "on account of scarcity of men and lots of work an advance
in wages was expected." The high Omaha and Denver wages reflected the
relative scarcity of labor in the West.

Gabriel Edmonston, who was to play an important role in the AFL as
well as in the United Brotherhood, described the conditions that enabled
him to organize a local in Washington, D.C., in the following terms:

> Prior to 1881, the condition of the journeymen carpenters was wretched in the
> extreme. The country was slowly recovering from one of our worst financial
> panics . . . Wages were so far below the cost of a decent living that the most
> skillful carpenters were often reduced to the point of beggary. The hours of
> labor were long. The introduction of the piece-work system, together with the
> constantly diminishing amount of yearly employment, owing to the multiplied
> use of machinery, was slowly but surely sapping the manhood of our craft.[56]

The reference to machinery exemplified a growing problem for carpenters—the rapid technological change in the building industry during the
1870s. New machinery facilitated subdivision of the craft through factory production of windows, doors, and other parts of buildings, where
the work could be done by unskilled labor. The prefabricated parts also
made it easier to introduce piecework, which led to the speedup. As Edmonston wrote on another occasion, "Piecework has steadily grown to
such proportions that in many cases the journeyman's day of hard, faithful work netted him just 90 cents. For instance, the bosses would offer
but 30 cents per square (100 square feet) for laying flooring, and three
squares were considered a fair day's work if properly laid."[57]

The years around 1881 witnessed the beginning of permanent organization in many crafts and industries.[58] Why these particular years should
have been so propitious for union organization and growth in general,
and carpenter organization in particular, is not an easy question to answer. As we have seen, there had been many organizing attempts earlier,

Table 1.1.   Carpenter wages in dollars, 1881 and 1891

City	1881: daily wage before organization	1881: daily wage after union activity	Daily wage in August 1891
Cincinnati, Ohio	1.50–2.50	2.50–2.75	2.50
Washington, D.C.	1.50–2.00	2.50	3.00
Cleveland, Ohio	2.00	2.25	2.50–2.70
Philadelphia, Pa.	2.00–2.25	2.50	2.75–2.97
Kansas City, Mo.	a	2.00–2.50	2.25
Denver, Colo.	a	3.00	3.00
New Haven, Conn.	1.75–2.25	b	a
Paterson, N.J.	1.75	2.50	a
Bridgeport, Conn.	2.75–3.00	a	a
Cambridge, Mass.	2.25	a	a
Indianapolis, Ind.	1.75	a	2.00–2.40
Pittsburgh, Pa.	2.25–2.50	a	2.75
Toledo, Ohio	2.00–2.50	a	a
Louisville, Ky.	1.75–2.25	a	2.50
Springfield, Mo.	2.00–2.50	a	a
Baltimore, Md.	2.00–2.50	a	2.50
New Orleans, La.	2.00–2.50	a	2.50
Omaha, Nebr.	2.50–2.75	a	a
Terre Haute, Ind.	2.00–2.50	a	a
Hamilton, Ontario	1.57–2.00	a	a

Sources: 1881: Frank Duffy, "History of the United Brotherhood of Carpenters and Joiners of America," pp. 108–111; 1891: U.S. Senate, 52nd Cong., 1st session, Report of Mr. Aldrich, July 19, 1892, pp. 1928–1937.

a. Data not available.

b. Failed to get increase following strike.

but apart from a few instances, all failed. Certainly the favorable employment circumstances that set in around the end of the 1870s played an important role, but there had been earlier periods of favorable economic activity, as, for example, during the years immediately following the Civil War.

The explanation may lie in the development of a critical minimum mass of craftsmen, sufficient to sustain permanent unions. For example, among the foreign language groups in the early history of the carpenters, the Germans were the most important. Between 1860 and 1880 the number of German-born in the U.S. population rose from 1.3 to 2 million, the largest rate of increase for any major group except perhaps for the

French Canadians. Many of the Germans brought with them not only their trade but a tradition of socialism as well, which gave them a higher propensity to organize than was true of other immigrant groups of the time. Gustav Luebkert, co-founder of the United Brotherhood and a socialist, was a leader of the St. Louis German carpenters.

John Dunlop has suggested that the period 1881–1886 followed a period of economic depression, which led to a basic dissatisfaction with the performance of the economic system.[59] There were undoubtedly other reasons as well—but whatever they were, the United Brotherhood came into being in the very year from which we can date the origin of the modern American labor movement. Not only its foundation but its subsequent rise as well mirror the history of trade unionism in the United States. To say that the United Brotherhood is the quintessential American labor organization is no exaggeration.

2

THE EARLY YEARS OF THE UNITED BROTHERHOOD

The initiative in bringing about the formation of a national union was taken by the St. Louis carpenters, under the leadership of McGuire and Luebkert. On April 24, 1881, a Provisional Committee was established to call a convention of the locals in various cities. As the first order of business, it started a newpaper, *The Carpenter*, which has since been published monthly almost without interruption for a hundred years. The first issue featured the call for a national union, written by Peter J. McGuire, the editor:

> In the present age there is no hope for workingmen outside of organization. Without a trades union, the workman meets the employer at a great disadvantage. The capitalist has the advantage of past accumulations; the laborer, unassisted by combination, has not. Knowing this, the capitalist can wait, while his men, without funds, have no alternative but to submit. But with organization, the case is altered . . . Then the workman is able to meet the employer on equal terms. No longer helpless and without resources, he has not only his union treasury, but the moneys of sister unions to support him in his demands . . .
>
> Carpenters, you have spent years to learn your trade; you have to furnish many tools; you lose a great deal of working time; you are continually subject to perils of life and limb, and to the exposures of climate. Is your severe labor worth no more than a bare existence? Should you have naught but a beggarly pittance? It is a shame to think that carpenters in some cities have to work for $1.75 or $2.00 a day . . .
>
> A national union will bring an understanding between the various cities, and will lead to uniform and higher wages generally. This spring, some cities with good organizations have the courage to demand higher wages than others. But there is danger that the high wages will tempt carpenters to come from the cheaper cities. Hence, every city should be organized, and the wages of all advanced to a uniform standard . . . Men will not then rush readily from one city to another to fill the places of their brothers on a strike. The state of trade in each city will be thoroughly known and the occurrence of a strike will be announced instantly. We can then maintain a monthly journal devoted to our organization; but, best of all, strikes will be less in number, for employers will then fear to oppose us.[1]

The man who wrote these words, setting forth so cogently the advantages of a national union, was one of the most remarkable figures in American labor history. Before going further, it is necessary to say something about his background. Peter J. McGuire was born in New York on July 6, 1852, the oldest of five surviving children of his mother's second marriage. He attended parochial school, where he must have received a very good grounding in English, to judge by his subsequent literary output. His father joined the army in 1862, obliging him to go to work to help support his family.

On the return of McGuire senior in 1864, Peter got a job at Lord and Taylor's department store, where his father was a porter. Peter kept a very sketchy diary, from which we can follow some of his movements during the next few years:

March 20, 1864	Went to Lord and Taylor's in Grand St.
1867	Worked in Fountain's India Store, Broadway near 14th St. for six weeks. Then back to Lord and Taylor's.
1868	Worked in Van Pelt's (private home in Clinton Place—two months taking care of horses).
1869	Left Lord and Taylor's finally on June 1, 1869.
1869	July 8, went as an apprentice in Haine's piano shop, 21st and 2 Ave.
1873	Out of time—Journeyman.
1873	Left Haine's, Oct., 1873, on act of reduction.
1872, May	Took part in 8 hours strike of piano makers.[2]

During part of this period McGuire attended evening classes at the Cooper Institute, an educational establishment for working men and women. It was at the Institute's debating club that he first met Samuel Gompers. McGuire also began to attend meetings of the International Workingman's Association, which was one of the centers of American socialism. This early introduction was an important factor in determining his future political philosophy, although his pragmatic attitude toward trade unions modified his socialist beliefs.

His first step toward becoming a professional agitator and organizer was in 1873, when the financial panic led to severe unemployment in New York, among other cities. Gompers later described the situation that resulted from the depression:

> The crash came in September when the Jay Cooke Company and Fiske and Hatch announced failure. The scenes downtown were wild on that rainy day . . . Thousands in New York City were walking the streets in search of a job. As winter came on the misery grew to appalling proportions. Public officials made gestures which might have had a value for political purposes, but did not give food to the hungry or solve the rent problem for those facing eviction. But the workingmen of New York had to find the necessaries of living and the labor movement took up the problem of rent and food for those out of work.[3]

On December 11, 1873, a meeting held at the Cooper Institute led to the formation of a Committee of Public Safety, with its task to secure government assistance for the unemployed. McGuire, who was only twenty-one years of age, was elected a member. "Night after night he spoke on corners and sandlots . . . After about a month of such activity, McGuire had become a public figure who drew frequent attention in the daily press."[4] By January 1874 he had become chairman of the committee.

On January 10 McGuire led a sit-in at the office of the police commissioner in an effort to secure a parade permit. His father had been in earlier to denounce Peter as a communist and a loafer. "McGuire, according to reports, wept at this and generally broke down in a combination of sorrow and rage . . . McGuire was shaken by the fact that his father had turned against him as indeed that gentleman had, for he denounced his son on the following Sunday, January 11, from the steps of the parish church."[5] Nevertheless, McGuire persisted, and called a mass meeting for January 13. This meeting, which was held in Tompkin's Square, had all the potential for disaster that led later to the Haymarket affair in Chicago. By 11:00 A.M. from 7,000 to 10,000 people were assembled, but before a formal program could begin the police marched in and dispersed the demonstrators, driving them into nearby streets. Thirty-five were arrested, many were injured, but there were no deaths.

This episode was crucial in shaping McGuire's career. In May 1874 McGuire joined with Adolph Strasser, who was to become head of the cigar makers' union, and a number of other radicals to form the Social Democratic Party of North America. This group was under the ideological influence of Ferdinand Lassalle rather than that of Karl Marx. Lassalle was a precursor of the philosophy of guild socialism as it was later developed in Great Britain by G. D. H. Cole. It was his view that industry would eventually be transformed into productive associations with ownership and control by the workers. However, unlike the syndicalists, he did not reject action through the existing state machinery; indeed, it would have to be through a workers' political party that the industrial transformation would take place. Trade unions were also crucial to the process, and it was Lassalle who was largely responsible for the creation of the General Union of German Workers, one of the earliest labor federations in Germany.

McGuire's diary for the years 1874 to 1881 presents a portrait of a young man who made his living as a carpenter in factories, but who devoted a great deal of time organizing on behalf of the Social Democrats. In 1874 he started a journal, *The Toiler*, together with Lucien Sanial, who was to become one of the most prominent socialists in the country. In October 1875 McGuire moved to New Haven, where he found work in a piano factory. By the following week he had organized a local branch of Social Democrats. At the end of the year he went "on tour of 6 weeks agitation" throughout the West, Southwest, and Middle States, and after returning to New Haven, began another five-week tour in August 1876.

His diary records that from August 1876 to June 1877, he was engaged in selling labor papers and tracts. During this period he made an extended tour of New England, and became involved in a strike at the Wamsutta mills in New Bedford. The enormous problem of trying to organize the unskilled mill operatives made him somewhat pessimistic about the outlook for the union that had managed to survive the depression: "I wish you to understand that I am in favor of trade unions upon a broad radical basis, but I do not claim that they are the most effective way of advocating our principles. Political action is our greatest means of agitation. It will force the labor question before the minds of the American people."[6] These years marked McGuire's furthest movement to the left. He was the most important English-speaking agitator in the Workingman's Party, which had succeeded the Social Democratic Party and was soon transformed into the Socialist Labor Party. But he never abandoned the Lassallean belief that a strong trade union base was essential to the success of economic progress for the workingman. And increasingly, as he became involved in the practical work of unionism, left-wing politics diminished in significance for him.

Despite his heavy schedule of work and lecturing, McGuire found some time for his personal life. His diary indicates that he met Nettie Shoemaker on May 10, 1877, but broke off with her in November. A few months later, he became acquainted with "Maggie," and while his diary is silent on his marriage, it records the fact that he arrived in St. Louis on Oct. 13, 1878, "with wife and little girl." But this marriage was not of long duration. In 1884 he wrote to a friend: "The death of my wife on Jan. 26 has released me from over 3 years of pain and trouble, for she was sick that long. But her death has been a severe blow to me, still I am getting over it."[7]

It is not clear why McGuire moved to St. Louis, but the fact that socialism was gaining strength in the Midwest, particularly among German-Americans, while it was losing ground in the East may have been an important factor. Whatever the reason, this move was opportune, for McGuire's arrival coincided with the end of the depression and the revival of trade unionism. He became active almost immediately in labor organization and in politics, and as a lobbyist for the labor movement, he was instrumental in securing the establishment of the Missouri Bureau of Labor. He believed that good statistics were essential to the improvement of labor conditions in that they made it easier to bring unfavorable conditions to public notice. For six months in 1879–1880 he was deputy commissioner of the Bureau.

McGuire also became involved in a movement that was to become an obsession with him—the fight for the eight-hour day. In May 1879 he organized an eight-hour-day parade in St. Louis, and on July 4 he addressed a crowd of 20,000 on the same subject. Four days later he spoke at a mass rally in Chicago. He continued to speak and write on this theme, which later became a pivotal issue in the collective bargaining demands of the Carpenters' Union.

On the political side, he joined the Greenback Labor Party, which was formed out of farmer and socialist groups in 1877 and achieved some success in the 1878 elections. McGuire was a delegate of the St. Louis socialists to the Chicago Greenback national convention in 1880, where he was a member of the platform committee. In July of the same year he was a delegate to the Missouri Greenback state convention; he was nominated to run for secretary of state, but declined. The Greenback movement collapsed soon thereafter, and McGuire finally turned his enormous energies to the trade union movement on a full-time basis.

As we have seen, McGuire became involved with a newly organized local of carpenters in St. Louis in 1881 and led a strike for an increase in wages. Since he was a carpenter by trade and had held jobs at a number of woodworking shops in the St. Louis area, this was a natural base for him. Soon after the successful conclusion of the strike he was drawn into the leadership of a strike by street railway workers, and after some violence had ensued when strikebreakers were employed, McGuire was arrested and charged with incitement to riot. He was released on bail supplied by the St. Louis Trades Assembly of the Knights of Labor, and the charges were dismissed a few weeks later when testimony of the principal witness against him was impeached.[8] On May 1, 1881, he was elected secretary of the St. Louis Trades Assembly, the most important labor post in the city.

By all accounts, Peter J. McGuire was indeed a remarkable man. He was charismatic, an eloquent speaker, a master tactician. He was imbued with a burning desire to help raise the standards of his fellow workers, and combined an idealistic philosophy with a down-to-earth knowledge of how to organize. At this stage of his career he was undoubtedly the outstanding American labor leader, and was generally acknowledged as such. Traveling about the country in response to requests for his presence, he began to show some impatience with administrative detail, a tendency that strengthened as he grew older. It was this failing, if it can be characterized as such, that was eventually to lead to his downfall.

Before turning to McGuire's role in organizing a national union of carpenters, a word should be said about another enterprise on which he embarked that was to have far-reaching consequences. In August 1881 a group of disaffected members of the Knights of Labor called for a meeting to be held on Terre Haute, Indiana, for the purpose of forming a rival labor body. McGuire attended as a delegate of the St. Louis Trades Assembly and, according to his diary, "framed call for Pittsburgh Convention to form Fed. of Labor." McGuire did not go to Pittsburgh, but he later played an important role in the organization that was created in November 1881, the Federation of Organized Trades and Labor Unions, which evolved into the American Federation of Labor five years later.

FOUNDING THE UNITED BROTHERHOOD

As a result of his successful leadership of the St. Louis carpenters' strike in the spring of 1881, McGuire achieved national prominence. In late April, together with Gustav Luebkert, he formed a provisional committee to organize a national union of carpenters. He brought out the first issue of *The Carpenter*, initially a small four-page paper with a circulation of a thousand copies. "Its editor, the writer of this sketch, was then engaged as a Journeyman at the carpenter trade, and after his day's work, devoted his time to his humble Journal and its work."[9]

McGuire felt that the time was ripe for national organization because of the "many evils" that had been developing in the trade:

> By the introduction of woodworking machinery operated in planing mills the old workshops and their handwork gave way very largely to machine-made doors and sash, machine-made mouldings, window frames, etc., so that consequently a larger amount of work could be done with less labor in a given time, resulting in protracted periods of idleness and unsteady work. And in addition, in many cities, the time-honored custom of day-work was rapidly giving way to piece-work, with the minutest sub-divisions of the trade into petty branches, lessening the demand for skilled mechanics, and making the introduction of unskilled labor not only a possibility, but more and more generally the rule. The entire absence of any apprentice system, or of any method of mechanical training, contributed to augment the evils.[10]

The provisional committee issued a call for an organizing convention to be held in Chicago in August 1881. During the intervening four months, McGuire was engaged in what he himself characterized as "incessant agitation." In the pages of *The Carpenter*, almost half of which was written in the German language, McGuire hammered away at the same themes. Piecework and subcontracting were denounced as leading to exploitation of the worker and botch jobs. The idea that wages were fixed by supply and demand and could not be affected by unions was ridiculed: "This terrible ogre of the 'dismal science' has become a scarecrow for the political economists to frighten workmen into submission." If this were true, McGuire argued, wages should move up and down with the business cycle, like the prices of pork or cotton. While capitalists reduced wages readily enough, they were much slower to raise them. "We may ask till doomsday for an advance in wages, and wait for a hundred years thereafter for the operation of supply and demand, and we will never get reasonable wages without organization."[11]

The Carpenter also began to carry the "stay away" notices that were to become a regular feature. Carpenters were advised to stay away from St. Louis, "as work is greatly delayed and the city is flooded with strangers."[12] In New York, trade was characterized as fair, but the labor market was overcrowded.

THE FIRST CONVENTION AND ITS AFTERMATH

The first convention of the new organization opened on August 8, 1881, at the Trades Assembly Hall in Chicago. "The spacious hall was artistically draped with flags and banners, and suitable emblems. A fine model of the lofty steeple of the Dresden Cathedral in Germany, with model of spiral staircase, ornamented the front of the stage, and attracted high encomiums for its excellent workmanship."[13] Thirty-six delegates from eleven cities were present, representing about 4,800 members, according to their own claims. The resolutions adopted by the convention indicate the issues that were uppermost in the minds of carpenters in 1881:

1. Shorter hours of work, "in order that labor-saving machinery may not be so extensively employed in reducing the compensation due to skilled labor."

2. No distinction should be made between winter and summer work. Physical danger increased in winter, so there was no warrant for reducing the daily wage because of the shorter working hours.

3. "That we abandon affiliation with either of the political parties, and pledge ourselves to support for public office only such men whom we are assured will best represent the laboring classes."

4. Eliminate laws that give monopolies the power to oppress the weak and defenseless.

5. Secure the passage of uniform lien laws to give first preference to wages.

6. "We recommend such action as may be necessary to consolidate all unions in the various building trades."

7. Establish employment offices in different cities to facilitate the finding of jobs.

8. Abolish the use of convict labor to produce competitive goods.

9. Stamp out subcontract and piecework.

10. Open correspondence with European unions in countries which were sending labor to the United States.

The constitution provided for a per capita tax of 2 cents a month to be paid to the national office, with a $5.00 charter fee. In the event of a labor dispute, the issues were to be first submitted to a local negotiating committee, and a strike could be called only by the national executive committee, consisting of the president, general secretary, and three vice-presidents, after a two-thirds vote in favor by the local membership. Locals were to set aside 10 percent of their income in a defense fund; strike benefits of $4.00 a week were payable from the general treasury, but only if the local had first received the approval of the general executive board. Gabriel Edmonston of Washington, D.C., was elected president, while McGuire became general secretary, the only full-time post, at a salary of $15.00 a week.

And thus the Brotherhood of Carpenters and Joiners came into existence. Its purposes and demands were not radical, even for the time. From the start, it was dedicated to collective bargaining and eschewed independent political action. McGuire wrote of the closing session: "And when the work was completed, a tremendous cheer rose up from the delegates . . . [The Brotherhood's] advent into the world means a bright future for our trade. No organization ever had such a hopeful beginning—so pregnant with cheer and so promising in results."[14]

It should not be imagined that all of this occurred without any controversy. There was a close race for the post of presiding officer of the convention between J. R. Smith of Cincinnati and R. W. Comfort of Chicago, the latter winning by a margin of two votes, and for the permanent presidential post between Edmonston and Comfort. McGuire's close associate, Gustav Luebkert, wrote a few weeks after the close of the convention: "I was for Comfort or any other Brother to beat Smith. The same [Smith] there to conquer the Presidency and was not able to fulfill it. Knownothing in the labor movement to put at the head of our organization would be digging the grave for it on its birthday."[15] Nor was McGuire exempt from criticism, even in his home base: "Our boys in St. Louis are all right now. They begin to see the wrong they done to me and McGuire. They were mad at us on account that we started the paper with the little money of the strike fund and not given it to them for division. Little men have little ideas."[16]

Before taking up the duties of his office, McGuire went to Switzerland as a delegate from the Socialist Labor Party to an international socialist congress, which some years later became the Second International. Even there, however, McGuire was thinking along trade union lines: "In America it is a fact that men are imported from Europe to over stock our trade, whenever our wages advance. This I want to prevent. And it can only be done by having an understanding between the organized workers of all countries."[17] The trip provided him with a brief respite from the arduous organizing schedule that he had been following. Defending his decision to leave for Europe so soon after the formation of the Brotherhood, he wrote:

> For eleven years I have been active in trade and labor circles, without a moment's rest or leisure. The time I should have for myself at night after my daily labor at the trade, has been spent to elevate my class — to organize and emancipate the toiling millions. I have not been alone in this work. I can point to the unmarked graves of Wm. H. Sylvis, John Siney, and others who have died from suffering, poverty, and overwork in the movement. They did not heed the warning of others when told to take rest and repose. Eleven years ago I was strong and sturdy; within three years past my system has been rapidly breaking down. My friends advised me to take this trip to not only go on with the good work, but to take a change of climate and scenery. This I have done and find myself another man. I have gained rapidly in flesh and strength. The trip has revived me and I expect to return to America fully recruited to make a vigorous effort for the organization and advancement of the Carpenters. Of course in my absence my salary is not to be paid to me.[18]

On his return, McGuire found that things were moving very slowly. There was not enough money to print *The Carpenter* in October and November, nor to pay his salary. He complained that it was odd for the Chicago locals to accuse him of neglecting his duties by going abroad when they had paid only $8.90 per capita for 1,200 members. The Cincinnati local split because "they have found Smith to be a dictator and had to leave the union because he packed the meeting and had things fixed so he was reelected President."[19] In Chicago, the merger between the Protectives and Benevolents that McGuire had attempted to bring about failed, and the two groups were at war. A member of the Protectives wrote Edmonston: "I am sorry to say that the boys have lost confidence in P. J. McGuire and don't care to send money to him until he makes his shortage good. We advanced the money for the Oct. number of the Carpenter but have not received it. I sent him twelve subscribers before the Convention and not one of them has received a paper as yet. So we can't blame them for feeling as they do."[20]

McGuire, for his part, attributed the Chicago struggle to conflicting leadership ambitions, with no real differences of principle involved. Nor was this confined to Chicago. He wrote Edmonston: "In regard to the troublesome fellows you have in Washington, we are afflicted with them everywhere, and in every organization . . . There is too much thirst for office on the part of some. That is a curse when men think of it at the sacrifice of all principle."[21]

The year 1882 started on a depressed note. McGuire wrote that his wife was very ill and that he had a bad cold. The money problem continued to plague him. "If I had a little more money I might be able to get along better. I need about $25 more. I could then get along and be able to start housekeeping. If I could get the loan of it for one month I could pay it back. But I do not know where to get it."[22] Six weeks later he complained again: "Of all the trials I ever had this position surpasses all. Hampered at every corner for want of money and embarrassed continually from the narrowness and meanness of a few, I feel sometimes almost in despair. But I rouse up and summon all my courage and make the best of it, for I know and am convinced that the cause is right and will prevail in spite of the vain and conceited and cantankerous ones in our own midst."[23]

McGuire had moved the national office from St. Louis to New York at the end of 1881, which helps explain his personal financial problems. To add to his difficulties, the New York locals withdrew from the Brotherhood, a move that McGuire attributed to the personal ambitions of John Ritter, the leading figure among the German carpenters. "He is a domineering bull dozing nature, and has lorded it over the German House Framers, who are a rough lot and require such discipline."[24] He felt that it might be possible to pry the German locals loose from Ritter, but it was clear that the differences between them were not ideological. "[Ritter] is now on the Executive Board of the Socialists and they are thoroughly disgusted with his cranky ways."[25]

Things improved a bit in March, when John D. Allen of Philadelphia, who was to become the second general president of the Brotherhood, advanced McGuire $50.00 out of his own pocket. McGuire had high praise for Allen. "I will tell you we ought to be proud of having such a Brother. And he is as true as steel and level headed and able."[26] Finances were still tight, however, and McGuire was particularly concerned about *The Carpenter*, which he regarded as crucial for cementing the national union. He wrote:

> Dear Edmonston: We must never think of giving up *The Carpenter*. Rather give up anything but that. I would sell my sewing machine and mortgage everything I have before that paper goes down. It is our life—our hope—our only power to hold the unions true to each other. I will work at the trade, give up my salary, kill myself at night to keep things going, if necessary to keep up our paper. There are over 500 men I know of who would give a dollar each rather than let it fail. But it won't fail, dear brother, I have got advertisments coming in April that will float it.[27]

All this time, McGuire was carrying on an active organizing campaign. He tried to get the United Order of American Carpenters and Joiners, which had nine locals in New York City and vicinity and 1,100 members, to come into the Brotherhood, but it was some years before this was achieved. New York was a difficult city for the Brotherhood to organize because of ethnic and ideological diversity. In addition to the United Order, there were locals of the Amalgamated Society of Carpenters and Joiners and the Knights of Labor. McGuire threatened to charter a Brotherhood local by May 1, 1882, if the United Order did not come in, but it was not until 1887 that the first Brotherhood local was formed in the city.[28] However, McGuire did not confine his efforts to New York; between March 16 and the end of April, he addressed meetings in Philadelphia, Camden, Buffalo, Toronto, Erie, Cleveland, Toledo, Detroit, Chicago, St. Louis, Kansas City, Indianapolis, Louisville, Cincinnati, Wheeling, Pittsburgh, Washington, and Baltimore. To say that he was indefatigable would be a gross understatement.

Amidst all this activity, he continued to edit *The Carpenter* and to write a good deal of it himself. The quest for an eight-hour day was a favorite theme. The argument was repeated that a worker could do as much in eight as in ten hours because of the physical exhaustion entailed in a long day. But more than productivity, a man's humanity was involved. After a ten-hour working day, men are "so tired and worn that as soon as the frugal meal is over they must lay their weary bodies down to rest to renew the strength of their weary limbs for the coming day's toil." But after an eight-hour day, a worker goes home "cheerful and not worn out, free from liquor, not too tired to enjoy the warm and glad welcome of wife and children, and not too tired to spend an hour enjoying their society."[29]

Immersion in practical union work did not weaken McGuire's belief in socialism. Pointing to evidence of declining business activity, he wrote:

> It is the natural outgrowth of the system of competition and cheapness perpetuated by the capitalists. And as long as that system lasts it will be the parent of haphazard production, badly regulated distribution and extravagant speculation. Under it the workman will always suffer, and the Kings of Industry will have it in their power to turn the screws and produce a panic . . . The only remedy for us is to organize industry on the basis of cooperative labor with State credit and social guarantees, with equal opportunities and steady employment for all, reduce hours of labor, and each worker to receive the full result of his toil. Think of it![30]

However, McGuire was no partisan of the small cooperative enterprises that the Knights of Labor were sponsoring at the time; what he had in mind was an eventual cooperative commonwealth, with large enterprises controlled by the trade unions. Any attempt of small cooperatives to compete with large capitalist firms he regarded as the height of folly. They lacked capital, and insubordination and dissension became inevitable because every member wanted to be the boss. In the end, a few individuals gained control and exploited the rest of the members. "The only hope is for workmen to secure government aid and credit to pur-

chase the means of labor and place them in the hands of national cooperative trade unions."[31]

In addition to philosophical advice, McGuire provided his readers with
guidance on more immediate and practical issues. A scab "is one who is a
moral pestilence to his fellows and should be suppressed at all
hazards . . . if there be any who will not go [on strike] voluntarily, we
must draft them."[32] Carpenters were urged not to fritter away their time
and money traveling from city to city in search of higher wages, but to
stay where they were and organize.[33] Locals were warned against giving
employers advance notice of intent to make demands, lest they import
labor from other cities.[34] Although McGuire felt that the union should
not stir up needless strikes, he set out a checklist of steps that were to be
taken in the event that a strike should become necessary:

1. Foremen should quit work with the carpenters, for their interests
were the same.
2. Make a strike list the first day, and enroll every man, whether or not
a union member.
3. Appoint an executive committee of about seven from among the
most active men.
4. Appoint pickets to watch jobs, railroad depots, and steamboat
landings. But do not allow men to picket their own former jobs.
5. Hold a daily strike meeting to keep workers informed of developments. Strike headquarters should be open daily in the event that employers wanted to sign agreements.
6. Men should keep away from the boss whose job they quit; let him
come to headquarters.
7. Never heed rumors; come to headquarters for information.
8. No one should return to a job until every man on that job gets what
he asked for. "When you come out in a body go back to work in a body."
9. When men return to work, they should contribute to strike benefits
for those still out.[35]

In general, strikes at best were like wars — undertaken only as a last resort, and optimally, short and decisive. To those who argued that strikes
do not pay, McGuire replied: "They forget that if men never struck
wages would be lower than they are. The very fear of strikes compels
many a boss to be fair." Moreover, strikes were educational. "Every
strike, every lockout, is a practical lesson to the worker. It teaches him
the value of collective power and breaks down the cobwebs of In-
dividualism. It is the drill ground of Labor's army; there we find who can
stand without flinching, who are the staunch and the true — the men we
can depend upon in the still greater conflict to come."[36] This is not bad
rhetoric for a carpenter with little formal education!

McGuire's literary efforts did not inhibit his organizational work. He went to Chicago in June, met with members of the competing Benevolents and Protectives, and secured agreement on a merger into a new Local 21.[37] When some dissidents tried to block the merger, McGuire recommended that they be suspended. "We have to make an example in some of these cases to restrain others."[38] Cincinnati was a tougher problem because of Smith, "a miniature Czar [who] rules by virtue of terrorism and brute force."[39] Smith soon lost out, undoubtedly because of McGuire's influence on the membership of the local. Back in New York, he helped out in a strike by Italian freight handlers by setting up a commissary. "It costs 17 cents a day per man for bread, cheese, meat and macaroni. Every morn at five o'clock we go down and keep the padrones from taking them off."[40]

THE ORIGIN OF LABOR DAY

The September 1882 issue of *The Carpenter* contained an unsigned article that McGuire subsequently revealed he had written.[41] Under the heading "The New York Demonstration," it told of a large parade sponsored by the New York Central Labor Union, and continued:

> It is now suggested that the first Tuesday in September shall become the labor holiday of New York and be celebrated every year by a parade and picnic. It is also proposed that this day should be likewise observed throughout the country, that labor by its own will should establish its own universal holiday . . . The ruling classes have their Decoration Days and Thanksgiving; why should not labor declare its own holiday?[42]

There has been some controversy over whether P. J. McGuire was in fact the father of Labor Day. William Green, when he was president of the AFL, accepted McGuire's claim. When a postage stamp dedicated to American workers was issued on Labor Day, 1956, McGuire's descendants joined President Eisenhower in the ceremonies, and a press release issued by the Labor Department stated explicitly that McGuire was the founder.[43]

In 1967 a retired machinist, George Pearlman, claimed, on the basis of some old records he had unearthed, that the true originator of Labor Day was one Matthew Maguire, who in 1882 was secretary of a Machinist and Blacksmith local in Brooklyn. This Maguire, who eventually ran for vice-president of the United States on a Socialist Labor Party ticket, had his own champions, including the Machinists' Union. Secretary of Labor Willard Wirtz was rash enough to tell the 1968 International Association of Machinists' convention: "My decision . . . is that there is no question as to who is the Father of Labor Day in this country. Officially, as of this moment, insofar as the Department of Labor is concerned, it is Matt Maguire, Machinist."[44] The historian of the Department of Labor took a more cautious position. After reviewing the events of 1882, he concluded: "Perhaps some new evidence may one day help single out a real 'father' of Labor Day."[45]

To the United Brotherhood, of course, denying McGuire's paternity was akin to blasphemy. In a lead article in 1974, *The Carpenter* pointed to a number of contemporary newspapers in which McGuire's claim was confirmed, and cited a letter from Samuel Gompers to McGuire which said in part: "I do not really know that the question is disputed that you were the author of Labor Day. But I have seen recently in some of the alleged labor publications even a dispute of that fact. Perhaps a brief history of the origin and growth of the day from you would undoubtedly be interesting and settle the question for all time."[46]

It is also interesting that a left-wing resolution introduced at the 1919 AFL convention advocating the switching of Labor Day to May 1 was supported by a Machinist delegate. Samuel Gompers, in his reply to this resolution, left no doubt where he stood: "The man in whose brain that thought for Labor Day was generated was one of those who helped to found the American Federation of Labor, the founder and organizer of the United Brotherhood of Carpenters and Joiners of America, a member of the Executive Council of the American Federation of Labor, the late P. J. McGuire."[47]

Many people, including Matthew Maguire, were involved in organizing the 1882 parade in New York that was the genesis of Labor Day. But that was quite another matter than the specific suggestion for the establishment of a permanent Labor Day at the beginning of September that P. J. McGuire made in the September 1882 *Carpenter*. No one has found a similar contemporary statement by Matthew Maguire or anyone else. If an arbitrator were called in to adjudicate the matter, he would have to find that Peter J. McGuire was the father of Labor Day.

THE SECOND ANNUAL CONVENTION

The second convention of the Brotherhood was held in Philadelphia on August 1–5, 1882. Twenty-four delegates were present, representing fifteen cities, four more than the year before. The executive board was reconstituted to include the president, the general secretary, and eight vice-presidents, each of whom was to be a district organizer. There was strong sentiment for setting up insurance schemes, and this was done. A death benefit of $250 and a permanent disability benefit of $100 were to be financed by a 10 cent per month per capita tax, thus beginning the long history of payments that was later to create financial problems for the Brotherhood. One must remember, however, that there were no government Social Security schemes in existence in 1882, and pressure for mutual protection induced most of the unions formed at the time to provide a minimum of assistance.

The convention raised the per capita tax to 5 cents a month and fixed the initiation fee at not less than $1.00 and the dues at not less than 25 cents a month. Locals were to be permitted to charge more if they wanted to, and could set up sick benefits if they could finance them. Any member who accepted piecework became liable to expulsion. Strike pay was limited to $5.00 a week for married men and $4.00 for single men, and the Brotherhood was to limit strike support to one strike at a time to protect its resources. Philadelphia was chosen as the new national headquarters, and McGuire's salary raised to $20.00 a week.[48]

McGuire, as a socialist, was somewhat skeptical of getting into the insurance business, but he recognized that it was a necessary step. When Washington Local No.1 refused to pay for benefits, he wrote to Edmonston: "There is this; that we want to place our Brotherhood on a permanent basis. Without a Death Benefit or something of the kind we can't hold our members. Hard times will come and scatter them; while a benefit of some general kind will interest the families and hold the members."[49] A few days later, urging that the Washington local support the national union, he wrote:

> I am fully aware that it would have been better for us had we fixed the Funeral Benefit at say $75 and the Disability Benefit at $250. As the law now stands it is very imperfect and shows very plainly that laws made in convention, on the whole are haphazard and incomplete. The only safe method is where the judgment of the member is called into action by a general vote.
>
> When you see the expressions of all the unions, you would see how much they think of the Endowment Fund. They have gained members on it everywhere and have held their members together where otherwise they would be lost.[50]

John D. Allen of Philadelphia was elected president of the Brotherhood when Edmonston declined to serve again, despite McGuire's urging that he stand for reelection.[51] Edmonston was elected as a delegate to the 1882 congress of the Federation of Organized Trades, and in 1884 became its secretary. When the AFL was established in 1886, Edmonston was elected treasurer, a post he held for three terms. He remained a close collaborator of McGuire's.

Allen was not a socialist, and in his first presidential message to the Brotherhood he emphasized the importance of staying out of politics. He pointed out that union funds could not be used for purposes other than those specified in the union constitution; nor could locals engage in any political demonstration or parade under the Brotherhood name. "In times gone by all trade unions have generally flourished until partisan politics were introduced. Keep politics out of our trade unions, and in a quiet manner individually, man for man, do your political work at the ballot box on election day."[52] Allen urged the members to be flexible on wages in the interest of securing a reduction of hours, and to establish libraries and lecture rooms in order to gain a better understanding of the economic conditions under which they worked.[53]

The second convention discussed but did not adopt a practice that McGuire favored—the equalization of funds. The idea was that the general secretary and the executive board would, on a semiannual basis, collect and redistribute local funds so that each local union would end up with the same amount per capita in its treasury. This would mean uniform dues and initiation fees, as well as uniform benefits. It would have meant, according to an editorial that appeared in *The Carpenter,* undoubtedly written by McGuire, "that we shall be a real Brotherhood, that the strong unions shall assist the weak whenever they need it. It is simply carrying out the true principles of unionism."[54] To the argument that this would encourage local unions to squander their funds, it was pointed out that the experience of the Cigar Makers' Union, which practiced equalization, had not borne this out. Not only was this scheme in line with the socialist proclivities of McGuire, but it would have provided the national union with a greater degree of control over the locals, which he also favored. The 1882 convention agreed that this issue should be discussed and resubmitted to a later convention.

After moving the Brotherhood office to Philadelphia, McGuire resumed his travels. In May 1883 he visited thirty-three cities. But trouble began to develop between him and Allen, which was almost inevitable because of conflicts in the ideologies and temperaments of the two men. Allen took the initiative in calling for a postponement of the 1883 convention in order to conserve union funds. This was agreed to in a referendum, and it was also stipulated that the general officers would be elected by a vote of the entire membership. Allen ran against McGuire for the post of general secretary but lost, provoking the following reaction from the latter in a letter to Edmonston:

> I only regret that you were not nominated as G.S., then I could rest secure in my faith that the B— —would not suffer. But for Allen to have the ambition he displayed is to me a surprise. That he had a case of *Big Head* I was sure of immediately after he became G.P. . . . Why more than once he sent me orders to stop the Journal. And he kept me busy trying to subdue the quarrels he needlessly provoked by his stupid letters.
>
> Now for Howard, Standford and others, who call me a *Socialist,* I forgive them for they deem it a term of disgrace while I regard it as a title of honor. To be a *Socialist* now is to be what an early Christian was among Heathens—re-

viled, hated, and persecuted. But to use the words of William Lloyd Garrison, "I will not equivocate, I will not deny, I will stand my ground and some day the world will hear me." *Socialism* is my faith, my religious ideal, and no man has a right to meddle with, no more than I have to reproach him for his. The only standard by which to judge me is Am I moral, am I upright and honest, am I better or worse than other men? [55]

McGuire complained that Allen had written him a twelve-page letter asking him to decline to be renominated, and threatening to split the Brotherhood if McGuire were reelected. Allen stayed on for a time as head of his local and eventually became one of the leading architects of Philadelphia, specializing in theatrical work and interior decoration.

Despite his professed devotion to socialism, McGuire was careful to keep partisan socialist propaganda out of *The Carpenter*. For example, the December 1882 issue of the paper offered, for 50 cents, a one-year subscription plus any one of the following books as a bonus: Macaulay's *Essays*; *On Self-Culture*; *The Imitator of Christ*; *The Light of Asia* (Buddhism); readings in Ruskin's *Modern Painters*; *Nutritive Care*. For one dollar, the subscriber could obtain Ruskin's *Letters to Workmen and Laborers*; Carlyle's essays; *Idylls of the King*; the orations of Demosthenes; or Oliver Goldsmith's *Letters from a Citizen of the World*. There was no mention of *Das Kapital* or any of the many socialist brochures that were circulating at the time. McGuire's selection shows him to have had high ambitions for raising the intellectual level of the carpenters.

John P. McGinley of Chicago replaced Allen as general president in 1883. McGinley apparently got along better with the general secretary; at least there were no complaints from the latter. The new president favored equalization of funds, but he advocated one reform that McGuire did not like: that the general president act as a full-time salaried officer of the Brotherhood. McGuire relished being the sole full-time officer of the national union.

Another economic recession set in during 1883, lasting until 1885, and although employment was not too badly affected, union progress was slowed. *The Carpenter* assessed the progress the Brotherhood had made in an editorial appearing in the May 1884 edition:

1. "Wages have maintained a more fixed standard, if they have not been really advanced, while the hours of labor have been reduced in most instances."

2. Piecework has been discouraged.

3. Fraudulent contractors have been brought to terms.

4. "The entire level of the trade has been elevated to a loftier, more dignified and respected position."

5. The Brotherhood has become a fixed factor in the American labor

movement. It helped create the Building Trades League and the Federation of Organized Trades. It was maintaining fraternal relations with unions in Germany, Britain, France, Switzerland, Spain, Italy, Denmark, Belgium.

6. "A few years ago we were a scattered lot of unions with only eight cities organized, and having no interest in common. Now we stretch as a unit from Halifax to San Francisco and from Canada to Texas, with over 60 cities organized and our membership increasing at the rate of 300 per month."[56]

This generally optimistic picture, which was more or less warranted, contrasted with the experience of most of the other national unions, which were losing members at the time.[57] But the Brotherhood was still struggling. The Victoria local struck for two months, and was defeated by an inflow of immigrants. Contemplated strikes in Seattle and Portland had to be abandoned "until a more favorable time, owing to the sudden influx of an immense immigration, either deluded or imported to those places by land speculators or the railroad companies."[58]

THE FOURTH ANNUAL CONVENTION

There was no third convention; that had been canceled. The fourth convention was held in Cincinnati in August 1884 and was attended by twenty-two delegates from sixteen cities — nothing like the sixty cities that were claimed to be organized, but a slight improvement over 1882. There were no delegates from New York or from any Canadian city, while some of the cities represented were small ones.

In his report to the convention, McGuire claimed that the Brotherhood represented 5,094 members, an increase of 1,866 over the previous year. Forty-seven locals in all had been chartered. He urged a number of constitutional changes, including a reduction in the number of men required to start a local from nine to seven, as well as an end to the prohibition of more than one local in each city. On the latter point, he noted that the Brotherhood had chartered a local of colored men in New Orleans, and some whites wanted a separate charter. Similarly, German carpenters in Chicago wanted a union of their own.[59]

Shortly after the convention had started, a black delegate was refused entrance to a nearby restaurant. This led to the introduction and adoption of the following resolution:

Whereas, — several of our delegates in company with Brother Rames of Charleston, S.C., last evening entered a coffee house in this city, at 77 West Smith St., known as the "People's Restaurant" and were told by the head waiter that he could not serve Brother Rames on account of color, therefore be it Resolved—that in this indignity offered to Brother Rames, one of our delegates, we recognize a gross and ignorant insult to our body worthy of the severest rebuke at our hands, and we call upon the citizens and working people of this city to do all in their power to withdraw all patronage from the "People's Restaurant" and hold it up to public execration.[60]

The Brotherhood never had any constitutional color bar, as many other unions of the time did. It followed the customary practice of chartering separate black locals in the South, but the national leadership was opposed to discrimination because of the pragmatic need to organize blacks in order to safeguard union wage and hour standards.

The 1884 convention undertook some major constitutional changes. A biennial convention replaced the annual one, and the terms of national officers were extended to two years. The general executive board (GEB) was to consist of five members elected by the locals within ten miles of the headquarters city. While this made it possible for the board to act more quickly, it set up tensions within the organization by bestowing too much authority on one locale. Following McGuire's suggestions, the number required to form a local was reduced to seven, and it was established that more than one local could be chartered in a city, provided the existing local approved. The death benefit was amended to provide for a $50 payment if a wife predeceased her husband, with the reservation that "in no case shall a member receive a benefit for more than one wife."

Perhaps the most important change was a new strike procedure that served to enhance greatly the authority of the GEB. In the event of a trade dispute, locals were required to appoint a negotiating committee to seek an adjustment. If this failed, a strike could only be called after a two-thirds affirmative vote of the members by secret ballot. In the event that the local contemplated a general strike within its jurisdiction, it was required to telegraph the GEB and seek the board's sanction. If this were granted, the local involved was entitled to strike support at the board's discretion, and if it exhausted its own defense fund it could call upon other locals for help, again with the board's permission. As a result of this provision, the GEB soon found itself swamped with appeals for sanction and financial assistance. But this procedure did serve to strengthen the national organization by overcoming local isolation.

The GEB was put into the business of upholding morals by the constitutional proviso that a widow "whose conduct is improper" was not entitled to a death benefit, although the union would see the deceased properly interred. There is no evidence that the GEB exercised this authority with any degree of frequency; denials of benefits were always couched in legal terms involving eligibility.

A long declaration of principles was adopted, including the usual ones with respect to wages and hours, but also a denunciation of tariff protection, on the straight economic ground that such protection was at the expense of all citizens. It was argued that protected industries should have no legal right to reduce wages or to employ foreign labor, since in effect they were being subsidized by consumers who should in return receive employment and wage benefits—a sophisticated position. Finally, on politics, "we hold that voting is better than striking, and that both are right and necessary, and that no wage earners should vote for any man or party that does not directly support the labor cause by thought, argument, and action."[61]

Joseph F. Billingsley of Washington, D.C., was elected president by a margin of one vote. He appears to have been a fairly conservative trade unionist in the Edmonston mold. McGuire was reelected by a margin of 15 to 5, while L. Z. Rames, the black delegate from Charleston, was elected fourth vice-president unanimously. Cleveland was selected as the next headquarters city, so McGuire had to pick up his household and move once again.

The year 1884 found Grover Cleveland running against James G. Blaine for the presidency of the United States, with Cleveland the winner by an exceedingly small margin of the popular vote, and 219 to 182 in electoral votes. McGuire, in an editorial, warned locals against getting involved; once unions became stronger, they could force the government to act in their behalf, "but in the present embryonic condition of the labor movement to drag in party politics, is to thrust in the wedge of dissension which will end in a grand smash up of our societies."[62] As for anarchists, who were regarded with general opprobrium, the membership were advised that "the only remedy is to take severe measures with them. They can't be reasoned with, nor are they subject to discipline. Kick them out wherever they are!"[63]

In commenting on the outcome of the presidential elections, McGuire noted that one consolation was that "wherever organized labor has asserted itself [by voting], it has helped its friends and defeated its enemies."[64] The latter phrase was to become famous as the political slogan of the AFL. It is not clear whether this precise formulation had been used earlier; McGuire may have been its originator.

The year 1885 was another lean one, marked by little improvement in working conditions. In his annual report for the twelve months ending August 1, 1885, McGuire claimed that nonetheless, forty-seven new unions had been organized with a gain of 1,425 members. There were few wage movements during the year, and many communications from locals appeared in *The Carpenter* urging carpenters to stay away from their areas because of lack of work. In an editorial that bears all the earmarks of McGuire's authorship, members were advised: "Stay where you are and don't tramp off expecting trade is better somewhere else than where you are. It is bad everywhere and tramping around only helps to reduce wages and makes things worse."[65] And a refrain began that was to be repeated many times in the future: locals were warned that unemployed members who were dropped for nonpayment of dues often became scabs, and they were advised to allow unemployed men up to four months before dropping them.[66] Members were also warned against entering into cooperative building associations; the example of Local 21 in Chicago, which had sponsored a cooperative venture with the loss of thousands of dollars, was cited. The success of such cooperatives in Britain was attributed to the homogeneity and geographic stability of the labor force, as well as many years of preparatory education.[67]

McGuire's frame of mind at the end of 1885 may be judged from what he wrote in *The Carpenter*. The General Secretary, he said, cannot accept all invitations to speak: "There is a limit to a man's exertions and to individual sacrifice. Railroad fare must be paid, hotel bills must be met and the General Secretary has paid about all that he can stand, and has not only overtaxed his physical strength, but has also taxed his pocket so far that he feels he cannot keep up the strain any longer, in justice to himself and family."[68] The GEB was keeping McGuire on a tight financial string. It would not pay his expenses for organizing trips, and indeed, had adopted a resolution that "only $1.00 at a time be placed in the hands of the G.S. to meet current bills, and he shall report to the E.B. as to its expenditure."[69] One can imagine how McGuire must have received this resolution.

The board did take one action that characterized its position on an issue of great importance — technological change. The general secretary of the Amalgamated Iron and Steel Workers asked Brotherhood aid in discouraging the use of steel nails as injurious to steelworker employment. The GEB replied that "while our Brotherhood sympathized with them, they did not deem it advisable to fight labor saving machinery, and recommended the Iron and Steel Workers to struggle for a reduction of the hours of labor to offset the evil."[70]

What was there to show at the end of 1885 for five years of backbreaking organizational work by McGuire in particular, but also by many other devoted national and local officials? Table 2.1 shows the revised membership data for the Brotherhood, given retrospectively by the general secretary. Membership had almost tripled, and the number of affiliated locals had grown substantially. But there was still a long way to go. The great majority of the nation's carpenters remained unorganized. There were a number of rival unions in the field: the United Order of Carpenters in the New York area, the British-based Amalgamated Society of Carpenters and Joiners, and above all, the Knights of Labor.

The main assets of the Brotherhood were the tremendous talents and indomitable spirit of its general secretary and a nucleus of devoted trade

Table 2.1. Membership in the United Brotherhood of Carpenters, 1881–1885

Year	Unions in good standing	Charters issued	Total membership
1881	12	—	2,042
1882	23	13	3,780
1883	26	11	3,293
1884	47	21	4,364
1885	80	50	5,789

Source: United Brotherhood of Carpenters, *Proceedings*, Fifth Annual Convention, 1886, p. 4.

unionists in the major industrial centers of the nation, prepared to spend their time and money on the enterprise of building a union. These may have not seemed much at the time; building contractors were not losing sleep out of concern for their freedom of action in setting wages and hours. But within a remarkably few years, the Brotherhood's stock of human capital was to prove sufficient to create a powerful organization, whose future viability would never more be in doubt.

3

THE FIRST GROWTH YEARS

The year 1886 was a landmark not only for the United Brotherhood, but for the entire American labor movement. A moribund Federation of Organized Trades was replaced by the American Federation of Labor. The Knights of Labor increased its membership from 104,000 in July 1885 to 703,000 a year later. There were many labor disputes, including a general strike for the eight-hour day on May 1, 1886.

Why such a burst of organization should have taken place at this particular time is by no means clear. Economic conditions had improved somewhat after a recession, but not spectacularly. McGuire reported in August 1886 that "the past year has not been as brisk for carpenter work as was anticipated in the early Spring . . . The railroad strikes in the South West had a very depressing effect on Speculative Capital, and it shrank from contemplated investments in Real Estate, causing quite a depression in the Building Trades." The shorter hours movement was difficult, and in a majority of cities, "the labor market is constantly overcrowded with idle Carpenters."[1] Nevertheless, by August 1886 the membership of the Brotherhood had increased fourfold, and its general secretary, described by Samuel Gompers as "a lovable, genial companion,"[2] was a principal mover in the formation of the AFL, with one of the largest national trade unions in the country behind him.

The close relationships that prevailed between the AFL and the Brotherhood in these early years is indicated by the fact that McGuire added the duties of AFL secretary to his Brotherhood obligations, while Gabriel Edmonston, who was the first president of the Brotherhood, was also the first treasurer of the AFL. McGuire relinquished his AFL position in 1889 because of the pressure of his Carpenter duties, but he became an AFL vice-president. Edmonston was reelected AFL treasurer for three terms; he then retired from the post at his own request, devoting himself thereafter to the Washington, D.C., Carpenters' organization.

THE EIGHT-HOUR MOVEMENT

Shorter hours of work had always been one of McGuire's main objectives, and it is no surprise that the Carpenters should have been the initial promoters of a widespread movement for the eight-hour day. At the 1884 convention of the Federation of Organized Trades and Labor Unions, Edmonston introduced a resolution to the effect that from and after May 1, 1886, eight hours should constitute a legal day's labor; it was adopted by a vote of 23 to 2. The idea was to conduct a general strike on May 1, 1886, if the eight-hour day had not been achieved earlier. Commons describes what happened then: "The eight-hour declaration was coolly received even at the hands of the trade unions affiliated with the Federation. So few unions acted upon the strike benefit proposal that the convention of 1885 did not venture to adopt it . . . By the time of the convention of 1885 only the Brotherhood of Carpenters and Joiners had voted upon the amendment."[3]

McGuire had sent out a notice to all Brotherhood locals asking for a membership vote on the following question: "Shall our Brotherhood favor the adoption of the Eight Hour rule of work in our trade, on and after May 1 (next) 1886?" There was a majority of 7 to 1 in favor of the proposition, but many locals failed to vote, suggesting a lack of enthusiasm for a strike. In his report to the members, McGuire merely stated that "it is more an expression of opinion than an emphatic decision in favor of a general strike next May." The vote was referred to the 1885 Federation convention, and Edmonston later reported that "the work of this session was directed mainly to strengthening national trade and labor organizations and preparing to put into actual operation the eight-hour work day. It was thought best not to order a general enforcement of this resolution on May 1, 1886, but to assist those who felt strong enough to carry their point. This alteration of the original plan became necessary for good and sufficient reasons."[4]

The Federation decided that before May 1, each affiliate would report to the Secretary on whether or not it would enforce the eight-hour day by that date, and on the steps it planned to take. It was agreed that the wage question would be dropped for the moment, and that only eight hours' pay for eight hours worked would be demanded. The Brotherhood decided that it would have to await the action of its next convention before deciding to call a general strike, but promised to provide financial support for those trades that went on strike. Although almost 200,00 workers did strike on May 1 in the country as a whole, and perhaps 20,000 succeeded in winning the eight-hour day, the strike was not a success. By the following year, most of those who had gained the concession had lost it.[5]

McGuire, although he had been advocating the eight-hour day for some years, was cautious about the eight-hour strike, as well as about strikes in general, reflecting the mood of his membership after several poor years. In an editorial entitled "Weigh Well These Words," he warned that reckless strikes constituted the greatest danger to new unions. To win, they needed money, discipline, and experience:

Avoid strikes, wait, have patience, organize more thoroughly, discipline your forces. When you move, let it not be too early in the season — May first is early enough — then if you go for more pay, go for 25 cents more at a time. Don't be too greedy or you may get beaten. Don't publish any notice of your demands in the daily papers, or it will flood your city with idle men — not alone that, it discourages building. The better plan is to send out a committee, canvass the sentiment of your employers, reason with them and by *moderate demands* and sensible action you can win them over. If you can't send out a committee then mail them a circular letter — but a committe is the most preferable.[6]

McGuire may have been a socialist in his ideology with respect to the eventual organization of society, but as a tactician he was a pure trade unionist.

The eight-hour movement did not yield the carpenters much in terms of reduced hours, but the leading role of the Brotherhood in the preparations for the movement gained it a measure of renown that was translated into rising membership. In February McGuire wrote to Edmonston: "Just think — seven new unions this month so far since Jan. 28."[7] Two months later he wrote again: "I can't leave here this week as I intended. Sickness in family will prevent. I am also terribly crowded with work. We are fairly booming. Fully 3,000 members more for April than we had for March. We will show over 14,000 members for April, and still growing. Let us congratulate!"[8] By August 1 he could claim a membership of 21,423 in 177 locals, compared with 5,789 members and 80 locals a year earlier.

During this hectic period McGuire was constantly on the road, and he claimed that he personally brought in 126 new locals. He hammered away at the theme that the carpenter was receiving grossly inadequate wages for the dangerous and difficult work he performed. "The carpenter works for little more than a common street laborer's pay, yet he exposes himself to the risk of falling from scaffolds and stagings, and works in perilous positions." To him, shorter hours and more pay were not only economic but moral questions: "As long as you find men with poorly furnished houses, with no carpets on the floor, no pictures on the walls, no books on the shelves, and working at starvation wages, after their day's work, in most cases, you will find them in saloons, where they find the light and gayety and imaginary comfort they found not at home."[9]

THE UNITED BROTHERHOOD VERSUS THE KNIGHTS OF LABOR

The United Brotherhood fought the first of its many jurisdictional struggles against an organization known as the Noble Order of the Knights of Labor, which had been formed in 1869. For fifteen years there was not a great deal of friction between the Knights and the various craft unions, since neither had a large membership. But as the Knights began to pick up members after 1884, both among the semiskilled and the skilled workers, they came into conflict with the craftsmen. The failure of the Knights to maintain clear craft demarcations, and their practice, on occasion, of organizing men of different levels of skill and drawn from different crafts into mixed units, exacerbated the difficulties.[10] In many ways, the reaction of the craft unions against the Knights in 1885–1886 resembled their attitude toward the CIO half a century later.

Even before 1885, while professing friendship with the Knights, McGuire was becoming concerned about their competition. In a letter written in 1882, he said: "I would not allow them [the Knights of Labor] to gobble up our unions and destroy all other organizations. I have taken solid ground against them and I have beaten all such tricks. The K of L cannot swallow us up unless our convention so decides. The very publicity of the K of L is its destruction. It is filling up now with politicians who think it some political machine for their use."[11]

By the beginning of 1886, in considerable measure because of friction between the Brotherhood and the Knights in Washington, D.C., McGuire published a summary of the relations between the two organizations. He pointed out that there was no official connection between the two, although some carpenters held membership in both and some Brotherhood locals were affiliated with Knights' local trade assemblies. The Brotherhood, he asserted, was not antagonistic to the Knights, and was prepared to work with it.

Our Brotherhood is not at war with the Knights of Labor, nor are we antagonistic to them. On the contrary, we recognize them as a factor in the labor movement, and we are ever ready to cooperate with them and work in harmony with them, as we are ever ready to lend a helping hand to all branches of honorable toilers. This we have proven again and again in the time of any trade trouble, or when necessary to "enjoin an oppressor" by a boycott. Our local unions have always demonstrated their practical sympathy with the cause of United Labor. Our local unions make it a rule to affiliate with the Trades Assembly, Central Labor Union, or whatever else it may be called, of their respective localities . . .

The Trades Unions and Knights of Labor should work together in parallel lines without any collision or quarrels. And if any clash occurs, it is due more to the lack of understanding each others' legitimate functions, or it is caused by the over-zealousness of some K. of L. organizer with an unquenchable desire, as one said "to capture every Trades' Union or bust them up" . . . We fully recognize the good work being done by the K. of L. and we extend to it the hand of fraternity. Let each organization pursue its own chosen course and work together in its own lines, without any desire to raid each other or to clash.[12]

Behind this conciliatory statement was a history of growing antagonism between the two organizations that can be illustrated by the events in Washington, D.C.[13] Brotherhood Local No. 1, which had been organized by Edmonston, was unwilling to go along with the benefit system that the national union had initiated in 1882. Capitalizing on this discontent, the Knights formed Local Assembly 1748 in opposition to Local 1. At first the Brotherhood local remained dominant, but in 1885, in protest against what it regarded as an illegal assessment instituted by McGuire, Local 1 stopped paying per capita dues and elected a strong opponent of the benefit system as president. In January 1886 some of its members decided to use its funds as an admission fee to the Knights, whereupon McGuire authorized Edmonston to safeguard local property pending reorganization and to institute court proceedings to conserve the funds, if that proved necessary. McGuire wrote to Edmonston: "The mere proposition of a motion to appropriate money to join the K of L is manifest *intent* to misuse the funds and divert them from the original purpose of their collection. That ought to be good grounds for securing an injunction or restraining order."[14]

When Edmonston was expelled from Local 1 as a result of this action, McGuire wrote that the local was suspended and had no authority to expel anyone from the Brotherhood, and that, moreover, any disciplinary action could be appealed to the GEB.[15] The dissident local members were given the option of reinstatement if the expulsion of Edmonston were rescinded and their back dues paid up, but many of them refused and joined the Knights' Assembly 1748. Thereupon a new charter, numbered No. 196, was issued to Edmonston in June. A month later, McGuire urged him to attend the forthcoming Brotherhood convention: "In the next convention we will be between two fires — one, the K of L will send a committee to try and capture us, as they have done in every Trade Union convention this year . . . The other is a strong clique formed by San Francisco, Washington, and a dozen other cities to pull together for common devilment in our B."[16]

On April 26, 1886, while these events were taking place, a group of trade union leaders, with McGuire's name at the head of the list, issued a call to the craft unions of all trades for a conference to be held in Philadelphia. In an interview, McGuire admitted to being the author of this call, but disclaimed any intention of starting a "Labor War." The object of the meeting was to bring about a close alliance of the trade unions in order to work more harmoniously with the Knights. "If our object were hostility, we would not have called the Conference to Philadelphia, the headquarters of the Knights of Labor."[17]

Invitations had been sent to forty-three unions. Twenty-two delegates appeared, representing twenty national unions, while twelve additional unions sent letters of support. A committe of five, including McGuire, was appointed to draft a treaty of peace for presentation to the Knights on behalf of the trade unions. This treaty was presented to the General

Assembly of the Knights in May; its first clause specified that "in any branch of labor having a national or international organization, the Knights of Labor shall not initiate any person or form any assembly of persons following said organized craft or calling without the consent of the nearest national or international union affected." The Knights would also have been required to revoke the charters "of any trade having a National or International Union," and to urge the craftsmen to join the national craft union that had jurisdiction. In sum, the trade unions were asking the Knights to turn over to them all skilled craftsmen, and to stop any intervention in labor disputes conducted by the trade unions.

The Knights rejected the treaty, but issued a conciliatory statement and proposed further negotiations. However, when the Knights' General Assembly met in October—at a time when the organization was in the full flush of success, reaching a high point of 700,000 members—it not only refused to accept the treaty but issued what amounted to a virtual declaration of war against the craft unions.[18] This led to a call for a national labor congress to be held at Columbus, Ohio, on December 8, 1886, the meeting that resulted in the formation of the American Federation of Labor. McGuire was secretary of the trade union committee that issued the call.[19]

THE FIFTH CONVENTION, 1886

In the midst of these events that were to prove so crucial for the future history of the American labor movement, the Brotherhood assembled in convention in Buffalo. Some eighty delegates representing sixty-three locals were in attendance. In his report to the convention, the general secretary, after detailing improvements in wages and hours that had been achieved during the past two years, made the following complaint: "For several years, the General Secretary has furnished the office and rent free, saved the cost of fuel and light, and avoided the expense of office assistance, until the work lately became so great as to overtax his health by long hours."[20] As a result, McGuire's salary was raised to $125 a month. The Brotherhood was coming of age.

The most important work of the convention was to rewrite the constitution. McGuire favored uniform monthly dues to finance benefits so that traveling members did not lose their benefit rights, as well as the equalization of funds among locals.[21] The convention authorized the GEB to draw from the funds of each local union a sum not to exceed 10 cents per member to replenish the general fund of the national union when deficiencies in the latter were created by increases in death benefits. Strike benefits were set at $5.00 per week for married men and $4.00 for single men, payable after the second week of a strike, provided the strike was sanctioned by the GEB. Each local was required to set aside a monthly per capita amount of 5 cents for a protective fund to finance strikes. Although the funds were to remain with the individual locals, they could be drawn upon by the GEB to support strikes of other locals, if need be. These financial steps clearly marked a significant centralization of the Brotherhood. Thenceforth, the GEB was to spend a good deal of its time answering appeals for funds.

To prevent locals from becoming too large and difficult to manage, a constitutional change provided that when a local reached a membership of 400, another local should be chartered. The new locals were to be independent; no branches of a local were to be permitted. The benefit structure was rearranged. The GEB was authorized to draft new laws where necessary and submit them to a referendum, with a two-thirds majority required for approval. The general officers were to consist of a president, general secretary, and two vice-presidents, while a treasurer and five GEB members were to be elected by the local unions in the headquarters city, continuing the previous practice. Philadelphia was designated as that city for a ten-year period, which gave the locals that were located there great potential power. The purpose of this provision, at a time when travel was slow and difficult, was to permit the board to act quickly in emergencies and to be able to meet frequently to handle death benefit claims, requests for strike assistance, and various appeals permitted under the constitution.

William J. Shields of Boston was elected general president. After he served his term, he had a long and distinguished career in Boston and represented the Brotherhood at many AFL conventions. McGuire was reelected unanimously. The problems that he had anticipated failed to materialize. The dissident leaders of Local 1 were permitted to address the convention, which then upheld McGuire's actions by an overwhelming majority.

THE AMERICAN FEDERATION OF LABOR

When the founding convention of the AFL met in December 1886, McGuire was at the peak of his career. He was the undisputed leader of one of the largest trade unions in the United States. It was in large measure through his initiative that the constitutional convention of the AFL had been called. It was he who called the first session of the convention to order, and when he declined the temporary chairmanship, the convention unanimously insisted that he take it. He could have been elected to the presidency of the AFL if he had wanted the job, but given the poor performance of the Federation of Organized Trades and the uncertain outlook for the new organization, his permanent post as head of the Carpenters' Union undoubtedly seemed preferable. Samuel Gompers came to the AFL job from a nonsalaried post in the Cigar Makers' Union. McGuire was elected the first AFL secretary, which was not initially a salaried position, and Gabriel Edmonston was elected treasurer. One of the outstanding historians of the American labor movement had the following to say of McGuire's role:

> He was Gompers' co-worker in the most difficult time in the A.F. of L.'s history. Not only was he shrewd and knowledgeable in the affairs of the labor movement, but he was a vigorous and astute organizer who gave his services freely to many organizations besides his own. McGuire was ready, during the time he was an executive of the A.F. of L., to help both in organizing workers in other crafts and callings, and in their negotiations with employers. Many labor organizations owe their beginnings to him, and many others profited from his tireless efforts on behalf of the cause. Shrewd, eloquent, and well-educated, he possessed a wide knowledge of foreign trade union movements as well as reform philosophies. More than any others, Gompers and McGuire can be regarded as the architects of the labor movement.[22]

McGuire continued as secretary of the AFL until 1889, when he resigned because he could not give sufficient attention to the job. He was then elected a vice-president of the AFL, a post that he occupied for the remainder of his active life as a trade unionist. He worked closely with Gompers, though occasionally displaying some exasperation with Gompers' leadership. Writing to Edmonston, ensconced as AFL treasurer, he opposed the publication of an AFL newspaper until sufficient finances were assured. "Let us be sure of 50,000 to 55,000 men for whom tax will be paid before we rush into expenses regardless of the 'wherewith.' I have written Gompers this again and again, but he is over-sanguine or else lacks business tact."[23] He had urged that the AFL constitution be issued as an inexpensive leaflet in both English and German, but nothing was being done. "This thing galls me, to think the Federation is not pushed."[24] He opposed hiring an assistant to Gompers: "I shall vote No until I see some work done and some results."[25] When Gompers complained that the member unions were not paying their dues to the AFL, and that he might have to quit the presidency and go back to his trade, McGuire wrote Edmonston: "Sam is not energetic enough in correspondence. I would suggest you write him and urge him to work. I have done so verbally to him today. I would hate to see the Federation drop. But I must say it looks very discouraging for the Federation unless more push is shown."[26]

McGuire was measuring Gompers against a very demanding standard of activity—his own. Nevertheless, relationships between the two men continued to be cordial, and many years later Gompers wrote of this early period of struggle: "Sometimes I was fighting single-handed to hold the line of federated organization. The only other officer of the Federation who felt a real responsibility for the work was P. J. McGuire. Though McGuire's office was in Philadelphia and he was the mainstay of his own struggling organization, he found time for the problems and the work of the Federation."[27]

AMALGAMATION AND GROWTH

With the AFL on its way, McGuire turned his attention once more to the affairs of the Carpenters' Union and resumed his arduous schedule — traveling, writing for the journal, corresponding with locals, organizing. It must have pleased him greatly when the *New York Sun*, a leading newspaper, wrote:

> The model organ among all the official and other papers published in the interests of trade unions is in all probability the *Carpenter* . . . As these organs are both numerous and powerful, and the *Carpenter* is the most nearly perfect of them all, it will be interesting to see what its reading matter consists of . . . It is a peculiarity of this little journal that every line that is written for it is distinguished for the simplicity, directness, and good taste of the language used.[28]

It was not the best of times, but there was work. Some of the terms used to characterize the state of the trade in various cities in 1887 were "fair," "middling," "very dull," "improving," "too many here," "stay away," "not brisk," "quiet," "very good," "busy," "slack," "flat," "crowded."[29] When the Coronado Beach Company announced its intention of building a large hotel in San Diego, the local warned that there were already too many men looking for work, that the cost of living was at least 40 percent higher than in the East, and that rents were exorbitant.[30] This showplace was eventually built by Brotherhood members.

Faced with rising prices, local unions sought higher wages and a reduction in hours. By April the Carpenters found themselves embroiled in major strikes for the first time in several years. Two thousand men struck in Cincinnati for a nine-hour day and an increase of 28 cents an hour. After six weeks the GEB urged the locals to permit the mill men to return to work for ten hours, but to maintain the nine-hour day for outside men, "until such favorable time as we can call out the mill men and sanction them in their demands." This advice was followed.[31] A few months later, however, some outside contractors reinstated the ten-hour day, and the GEB agreed to donate $150 to support a renewed strike, provided the locals raised $250 a week on their own.[32]

Employer resistance in Chicago proved even stronger. After 7,000 men were on strike for three weeks, it looked as though the Carpenters had won their demand for an eight-hour day and a 35 cent an hour increase. But the Master Builders' National Association precipitated a lockout in order to reinstate the nine-hour day, using Chicago as the focal point in a nationwide struggle. In addition to the United Brotherhood, locals of the Knights of Labor were involved. Substantial financial help, amounting to over $3,000, was authorized by the GEB for the Brotherhood locals, and since there was not enough money in the national protective fund, other locals were asked to send in part of their protective funds.[33] The eight-hour day was maintained in the end.

A Toronto strike did not turn out as well. Again the issue was an attempt by the employers to raise hours, here from nine to ten. McGuire provided the local $1,000 out of general office funds and instructed other locals to forward their protective funds. Depletion of the national fund forced the GEB to shut down assistance, and the ten-hour day was reinstated.[34]

The board then adopted a policy of sanctioning only defensive strikes in view of the limited funds at its disposal, refusing to approve strikes in Quincy, Illinois, and Galveston, Texas, among others. This provided McGuire with an argument for a larger central strike fund; he reported to the 1888 convention that the board would have been willing to do more if the protective fund were in better shape. The Chicago and Toronto strikers received only $4,335 from local protective funds when in fact there should have been $7,400 available.[35]

One of the consequences of the Chicago lockout, which had affected several trades in addition to the carpenters, was an initiative toward the formation of a national building trades labor federation to counter the Master Builders. The GEB agreed to send delegates, but not to be bound by any actions of the founding convention unless its local unions had first ratified them.[36] McGuire was cautious about this enterprise, objecting to the additional expense and urging instead that isolated construction locals affiliate directly with the AFL.[37] However, Brotherhood locals were urged to join AFL central labor unions and local building trades bodies.

Friction with the Knights of Labor continued, particularly in Washington. The secessionist group there was working together with the Knights local, and joined in attempting to have loyal Brotherhood members dismissed from their jobs. Terence V. Powderley, the national president of the Knights, agreed to look into the matter, but later reported to the GEB that no victimization had occurred.[38] Gabriel Edmonston at that time held the position of official carpenter of the U.S. House of Representatives, and the Knights tried to have him removed. McGuire urged that he be retained in that position, arguing that "Mr. G. Edmonston is a sober, industrious first-class mechanic, a thorough trades union man, a member of our Brotherhood for the past six years and a half, he was our general president for one year."[39] All of this prompted McGuire to lash out against the Knights:

> One thing has been done by the American Federation of Labor. It has at least well defined the essential and radical difference between the glittering promises of the Knights of Labor and the practical, every-day work of the trades unions. And so well defined is the difference, that the Knights of Labor are now taking lessons from the trades unions, and are forming themselves on National Trade District lines, which are simply skeletons of trades unions without either their flesh or blood.[40]

The attempt by the Knights to organize what were essentially national craft bodies did not prove successful. Many of the carpenter locals that did join the Knights soon switched their allegiance to the Brotherhood. By the beginning of 1888 McGuire felt that the Brotherhood did not need to be concerned any longer: "Out of sixteen unions admitted to our Brotherhood last month, five of them were K. of L. Assemblies, and four more are this very month underway to join us."[41] In what was in fact a postmortem analysis, McGuire attributed the decline of the Knights to top-heavy organization, the placing of unlimited power in the hands of a few general officers, and the admission of political adventurers. "When the Trade Unions were attacked, we took up the gauntlet for the maintenance of their autonomy, for we then knew that their absorption in the Knights of Labor would mean their injury whenever the crash came in the Order."[42]

The 1886 convention had reached a favorable resolution on the idea of forming Brotherhood districts to coordinate local activities, and the GEB urged that such councils be established wherever possible. However, McGuire opposed the formation of a state organization of carpenters in Massachusetts: "When our national organization numbers 100,000 members and it becomes too unwieldy to do business—then state organizations will be in order—and not until then!"[43] His goal was to build up a strong national body, and he did not look kindly on the formation of competing power centers.

The years 1887 and 1888 marked the beginning of the long march toward the establishment of the Brotherhood as the exclusive representative of carpenters in the United States. Up to then, the Brotherhood had had its hands full trying to set up local unions throughout the country. As more cities and towns came within its orbit, competition not only with the Knights of Labor but with other organizations as well began to develop. McGuire had never been an imperialist; indeed, his socialist convictions led him to take a relatively favorable view toward all other organizations of workingmen. But by the time of his report to the 1888 convention, his attitude had begun to change:

I maintain most firmly that while we should be ever ready to help all other sister labor organizations, and do practically recognize the common fraternity of interests that exist between all branches of honorable toil, yet, in the management of our own trade affairs, we should never make ourselves subordinates to any other organization, nor should we ever allow a dual form of organization to exist in our trade, for if we do, sooner or later, one will be bound to come into conflict with the other to the disadvantage of the workmen's best interests.[44]

The Amalgamated Society of Carpenters and Joiners, which had originally been established in 1868 as a branch of a British trade union of the same name, had several thousand members in 1887, mainly in New York. McGuire wrote the general secretary of the English union proposing a mutual exchange of working cards so that British carpenters coming to the United States to work where the Amalgamated had no locals could work under the Brotherhood's jurisdiction; this procedure would have curbed the growth of the Amalgamated.[45] The reply was cordial, but it rejected the suggestion on the ground that the Amalgamated constitution barred the admission of members of other unions, and that the constitution could not be changed until the next convention, not to be held until 1892. The reply also contained the pointed remark that "we have several societies of carpenters and joiners in this country, but we have never experienced any difficulty in working amicably together."[46] But McGuire persisted, and by October 1887 an exchange of cards had been negotiated.[47]

This did not mean that the Brotherhood was prepared to recognize the Amalgamated as a legitimate competitor on the national scene. When the Amalgamated applied for an AFL charter in 1888, the Brotherhood objected successfully "on the ground that to do so would sanction a dual organization with divided authority in trade affairs."[48] What it did mean, however, was that the Brotherhood was provided with a means of entering the New York labor market, from which it had been barred by the independent United Order of Carpenters. Its adherents could now work in New York with Amalgamated cards.

McGuire began to apply pressure on the United Order by proposing a meeting in February 1888 for the purpose of bringing about an exchange of working cards. This meeting was not held because of the opposition of several United Order lodges, but when one was finally arranged for April, at which the Brotherhood proposed a merger, negotiations were broken off. A month later, however, an agreement was reached after persistent approaches by McGuire. The United Order was reluctant to change its name, so the compromise adopted was to add the word "United" to the Brotherhood's title. In effect, the United Order accepted the Brotherhood's constitution and benefit system. In a referendum vote, eighteen United Order lodges voted for the merger, while four were opposed. Dissenting members attempted to block the merger by securing an injunction, and they did manage to get a temporary one. The injunction was dissolved, however, and the Order sent seven delegates to the 1888 convention.

When these delegates entered the convention hall, they were greeted with enthusiastic applause. But a controversy over the agreement developed when the convention refused to accept the new name for the organization. The United Order insisted that the agreement be carried out in full, and after a two-hour debate, the convention reversed its previous position and, by a vote of 69 to 20, approved the name "United Brotherhood of Carpenters and Joiners of America." Needless to say, McGuire argued strongly in favor of the change.[49] And thus was consummated the first important merger in the Brotherhood's history, bringing in about 3,500 new members organized in twenty-five lodges.

The city of Washington continued to be a thorn in the side of the Brotherhood. The newly established Local 190 was unable to make headway against the combined resistance of the now independent Local 1 and the Knights of Labor local assembly 1748. Assistance was sought from the GEB, but there was little it could do.[50] By March 1889 the members of Local 190 were considering disbanding, particularly after Billingsley, the former general president of the Brotherhood, switched his membership to Local 1.[51] Although this was averted, it was to be some years before the controversy could be resolved.

During this period the GEB was meeting several days each month and passing on many detailed matters, including all death benefit claims. It dispensed some money and a great deal of free advice. For example, it approved a request from S. H. Whiteside of Kansas, who was prepared to organize two locals each week for $15.00 and expenses; the board considered this a bargain.[52] Members of a Pittsburgh local complained that the council there was forcing them to stay out for the full wage advance demanded, which was only 2 cents an hour more than their employer offered: "The Council are doing an injustice both to the employer and the members in his employ by insisting they should stand out for such a slight increase. They should be permitted to go to work."[53] The sum of $200 was sent to help the members of the Jacksonville, Florida, local who could not find work at home and could not leave town because they were quarantined as a result of an epidemic.[54] The GEB was in the process of becoming a kind of general staff for the Brotherhood, judging local issues from a national point of view.

The constitution originally provided that no more than three strikes in different cities were to be permitted at any one time, the purpose being to conserve the protective fund and make certain that strikes could be financed. In 1886 this was reduced to two simultaneous strikes, and the board turned down a number of requests for sanction because it could not exceed this limit. However, locals that were prepared to finance their own strikes could do so without board permission.

Even at this early date, the board was insistent that constitutional procedures be followed precisely in disciplinary cases. When Local 157 of Elizabeth, New Jersey, inquired whether members who were working ten hours during a strike for nine hours could be expelled without trial, the Board ruled that charges in writing would have to be preferred and a fair trial held. In this case, the Board recommended that "the action of the members in question be overlooked, close down the strike, hold what members they have and take a better opportunity of pressing their demands for shorter hours."[55] The emphasis was on preservation of organization, even at the expense of yielding on immediate demands.

THE FIFTH GENERAL CONVENTION, 1888

A British carpenter who was visiting the United States at the time of the fifth convention recorded the fact that trade was not brisk, and that thousands of Italian and German immigrants, prepared to work for very little, were entering the country. He offered the following advice to his compatriots:

> If anyone wishes to succeed here he must make up his mind to rough it, and to have the means in his pocket to go far West or South, and there forego all comforts of Old England, and prepare to build up a home, and with about 20 years sojourn, with strength to stand the climate, probably may gain a little standing. The American Free Country is not all Gold, as it is often painted, give me Old England with all its faults.[56]

Despite this rather bleak picture of a carpenter's life in the United States, the convention opened on an optimistic note. Total membership in good standing had risen to 31,916, a gain of more than 10,000 since the 1886 convention. The number of locals in good standing had risen from 177 to 464. Some idea of the heterogeneity of the Brotherhood's membership can be gained from the fact that thirty locals transacted their business in German, six in French, four in Scandinavian, three in Czech, and one in Polish. There were eighteen locals in Canada, and of the sixty-eight in the South, eleven were composed exclusively of blacks.[57]

McGuire attempted to estimate the financial benefits of unionism as an argument for raising dues. He claimed that over the past two years, wages had advanced from 25 to 75 cents a day in 268 cities, affecting 26,000 men. At an average of 50 cents a day, this would make a gain of $2.8 million annually on the basis of a nine-month work year. If the nonunion men who benefited from the increases were added, the total would come to $4.5 million. Regardless of what economists have said about the trade union impact on wages, McGuire was in no doubt that the Brotherhood had substantially raised the economic status of carpenters.

The convention took several steps to shore up the finances of the Brotherhood. The charter fee was raised from $5.00 to $10.00, and the minimum amount of dues that could be charged by a local was increased from 25 to 35 cents. The structure of the national union was changed by the establishment of seven geographical districts, each of which would be headed by a vice-president who would also act as an organizer. While each vice-president was required to be a member in good standing of a local in his district, he was to be elected by the entire membership of the Brotherhood, a provision that was to cause a great deal of future controversy. At this time, the move was essentially a concession by McGuire to those who wanted less centralization of authority in the national office. However, the new vice-presidents were vested with little authority. The central protective fund was to be augmented by a new requirement that each local send half its protective fund contribution (2½ cents a month) to the General Office.

McGuire was reelected by acclamation, and his annual salary raised to $2,000. Shields declined reelection; his place was taken by D. P. Rowland of Cincinnati, who won by a fairly close vote. Rowland had been active in union affairs since 1880, and served several terms as president of his local. After his service as general president, he became business agent of the Carpenters' District Council in Cincinnati and was eventually elected to the Ohio General Assembly. He has been described as "a plain 'matter of fact' man, careful, conservative and cautious. He was not given much to writing but was a good practical speaker."[58]

CONSOLIDATION OF THE ORGANIZATION

The growing competition from immigrants was becoming acute. The 1888 convention resolved that although those who planned to become citizens were welcome, "we urge organized labor everywhere to endeavor to secure the enactment of more stringent immigration laws." During the year ending June 30, 1888, some 5,500 carpenters entered the country out of a total of 52,000 skilled migrants—the largest number of any skilled trade.[59] New York, the main port of entry, was particularly hard hit:

> In this city . . . it will probably be a tough struggle with Castle Garden [the immigration center] to contend against, unless Congress passes some law to protect us . . . No man of family with living expenses at fifty to sixty dollars a

month for rent and food can compete with the "birds of passage" from Great Britain, who hire a room for $1.50 a week and cook their own meals. Then when the dull season comes they take flight to Europe with the earnings of several months to live comfortably, while American citizens must walk the streets idle in search of work to keep their families.[60]

The need to restrict immigration became a constant theme in the publications of the Brotherhood, and with virtually all other trade unions, as the flood of immigrants rose.

The tempo of strike activity picked up in 1889, and a queue had to be formed for national assistance. The GEB consistently counseled caution. For example, Local 480 of Washington, Pennsylvania, was advised: "We think you would be justified in accepting the proposition of the Bosses as that is of more importance to you than the advice of strangers. We have already found that it is better to have the Employers your friend than your enemy and congratulate you on your good judgment."[61] When Boston Local 33 informed the GEB that it was asking for an increase of 35 cents an hour, a nine-hour day, and eight hours on Saturday, it was told: "It is the opinion of the GEB that you have too many non-union men to fight and you had better agree to the concessions of the Bosses if you can get the price per hour."[62] When Vice-President Wood asked financial support for organizing in the South, he was turned down on the ground that "the E.B. does not think it advisable to proceed in the South during the hot weather."[63] The board also turned down a number or requests by locals that it finance visits by McGuire, saying that it was inadvisable for him to leave the office for any length of time. The board was economy-minded as well as conservative.

Although McGuire consistently favored centralization of Brotherhood funds in the national office, he did not apply the same logic toward attempts by Gompers to build up a central strike fund in the AFL. Despite his leading role in bringing about the formation of the AFL, he shared the view of most of the other craft unions that such a development might compromise their independence. He felt that "the plan in itself is crude, and we fear will not work satisfactorily. If adopted it would encourage undisciplined and weak unions, without funds of their own, to enter fruitless strikes, in the prospect of being supported from the funds contributed by other societies."[64] The GEB went so far as to withhold part of its per capita dues to the AFL that were intended for organizing purposes. Gompers agreed to desist but asked in return that the AFL be given some control of organizers in the employ of the Carpenters. He was informed in no uncertain terms that "we do not think it proper for the officers of the AF of L to have any control over our finances or organizers. While wishing to work in harmony with other labor organizations we do not feel justified in paying the AF of L full tax as we think it excessive and desire to retain 50 percent of the same and use it as in our judgment it may seem best for organizing."[65]

The Brotherhood was becoming too large and diverse for all local policy to be determined at the central office. Several events that occurred early in 1890 presaged the future controversy over the allocation of administrative authority, and constituted the first challenge to McGuire's leadership. In April, General President Rowland was sent to Chicago to assist striking local unions. He was followed by instructions to put the men back to work if they could get a portion of their demands, but Rowland refused obey: "I telegraphed [the GEB] that to comply with their order and set a small portion of men to work before the negotiations were completed would prove disastrous, lose the strike, and disrupt the locals in Chicago."[66] The response of the GEB was to threaten to withhold any further strike assistance until its orders were obeyed. But Rowland stuck to his guns, and the GEB backed down and sent additional funds.[67]

The second confrontation came from an unexpected source—the vice-presidents. A board of vice-presidents had been established in 1888, with no specified authority, and meeting only once a year. This group challenged the authority of the GEB to withhold strike assistance from their districts. In a sharply worded reply, the GEB informed them: "The G.E.B. maintain the ground taken that the board of Vice Presidents are not a superior body to the G.E.B. . . . Feeling that we have met the situation in proper spirit while you have refused to accept, therefore the G.E.B. must courteously decline the invitation of the board of Vice Presidents to meet with them."[68] The board asserted that the vice-presidents had never brought their requirements to its attention, and invited them to furnish the names of the localities which they claimed had been neglected. Replies came back promptly, whereupon the board appropriated amounts ranging from $50 to $200 for the individual districts.[69] There was no doubt that the board had the constitutional authority to grant or withhold funds; what was being challenged was their parsimony in doling out the funds.

It took some years for the relative authority of the general officers and the GEB to be clarified. Every national union faced the same problem, and each eventually resolved the conflicts on the basis of its own requirements.

THE EIGHT-HOUR STRIKE

Gompers reported to the 1890 AFL convention that the AFL executive council had chosen the United Brotherhood to spearhead another drive for the eight-hour day. "The organization in question responded with alacrity to the Executive Council's decision, and of its own accord put a number of lecturers in the field."[70] It would probably have been more accurate to say that McGuire had volunteered the services of his organization to achieve what he had always regarded as his most important single goal. The 1888 Brotherhood convention had resolved that the eight-hour day should be put into effect not later than June 1, 1890, and the AFL had contributed to the movement by sponsoring meetings throughout the country during the preceding year, many of them addressed by McGuire or Gompers.

When Gompers announced the selection of the Brotherhood to lead the struggle, he wrote McGuire: "There is no doubt in my mind that few of the historians of the great events in the history of the development of our people will accord a higher place of honor and distinction than to the United Brotherhood of Carpenters and Joiners."[71] Beginning on March 1, carpenter locals began to go on strike. The GEB was flooded with appeals for assistance, most of which it granted. Additional funds were raised by an emergency assessment of 50 cents per member. By the end of May there were 141 strikes in all, involving 208 locals and 54,852 men, both union and nonunion. By August McGuire claimed that the hours of labor had been reduced to eight in 36 cities, and from ten to nine in 234 cities, many of the latter involving mill hands rather than outside carpenters.[72]

Employer resistance was strong, and success was by no means universal. When the Boston district council, after several months on strike, complained that the GEB was not being sufficiently supportive, the latter replied with some asperity that they had sent $9,600 and continued: "The G.E.B. would respectfully ask the Boston unions to send the communications to the General office in which the G.E.B. ordered them out on strike."[73] Repeated requests for additional funds were turned down, and the New York locals, which requested permission to help Boston by striking several jobs, were told that this would be unfair to their employers, who had conceded their demands. Strikes had to be called off in a number of cities; by the end of July the GEB had dispensed the very substantial sum of $68,150, and had little left.[74]

Despite some defeats, McGuire was able to tell the August convention that "the results after all are very gratifying. For through this we have gained a large number of new unions and thousands of new members, and advanced the wages and reduced the hours of labor in numerous cases . . . In the front rank of industrial battle this year our members have stood with undaunted courage as pioneers in the world-wide movement for shorter hours of labor. And though defeated in a few instances, the grand array of victories won will be an inspiration for the future, and the few defeats of this season we are satisfied next year will be turned into cheering successes."[75]

McGuire attributed the losses to the organized resistance offered by the local Builders' Exchanges, led by the Builders' National Association, which he accused of precipitating many of the strikes by refusing to negotiate. These local Exchanges were threatening to blacklist and boycott dealers in building materials who sold lumber or nails to Brotherhood members. He also deplored the fact that some locals had struck against the express wishes of the GEB, notably in Buffalo, Dayton, and Denver, leading to their eventual defeat. The failure of the AFL to help was ascribed to the fact that many of its affiliates would not or could not pay an assessment levied on behalf of the Carpenters.

THE 1890 CONVENTION

The sixth general convention found membership in good standing at 53,769, an increase of almost 22,000 over the previous convention level. The number of locals in good standing rose from 464 to 697. McGuire boasted that the Brotherhood "is now the largest and most powerful organization, numerically, of any specialized trade in the whole civilized world . . . All that now remains is to perfect its financial resources to be equal to any occasion, and the United Brotherhood will be as invulnerable as it is massive."[76]

The convention spent most of its time revising and simplifying the constitution. There were several important changes. McGuire had recommended an increase in the minimum dues and the per capita tax, and the convention agreed. On the other hand, the pressure for a broader dispersion of authority, which McGuire did not favor, led to the creation of a new general executive board consisting of five members representing the various districts, plus the president and only two vice-presidents; this new board was to meet quarterly. The purpose was to consolidate the existing GEB and the troublesome board of vice-presidents.

These changes required ratification by a two-thirds vote of the entire membership, and this process led to the first open controversy between McGuire and other Brotherhood officials. The constitutional changes did not receive the required majority, and the GEB declared the previous constitution still in force. In a signed editorial, McGuire complained that the many good innovations had been defeated because of a reluctance to raise the dues. "To declare the Chicago Constitution completely defeated is to maintain the old system with all its imperfections, inconsistencies, and drawbacks. It is to look backward instead of forward."[77]

McGuire accused the GEB of failing to make an accurate count of the ballots, which led the board to respond angrily:

Any member of the Brotherhood who reads the November issue of The Carpenter can understand the smears and insults contained therein. If the G.S.

has an honest difference of opinion with the G.E.B. he would meet with them instead of writing over the country to irresponsible persons thereby trying to sow discord and contention and bring the board located in Philadelphia in contempt with the Brotherhood located outside of Philadelphia . . . It may have been said in the past that the G.S. ran the G.E.B. but anyone that knows of the workings of the organization at present would not accuse him of running the Board at present located in Philadelphia.[78]

One of the specific issues involved was the refusal of the GEB to count the votes of locals that had failed to affix their seals to the tallies. When challenged again by McGuire, who questioned the authority of "four men from one city" to thwart the entire organization, the board agreed to permit a new referendum. Now that he was at odds with the GEB, McGuire argued for the new system, saying that GEB meetings would cost less since they would be infrequent, and that it would prove "far more satisfactory than the old system of a G.E.B. all from one locality."[79]

The second referendum resulted in approval of the new GEB scheme, but apart from two other minor amendments, all the other changes adopted by the 1890 convention were defeated. McGuire thanked the locals for supporting him in his controversy with the old GEB,[80] but the fact remained that he was not able to get his way on the financial questions, which were among his principal concerns. However, the Philadelphia board was replaced by one that was prepared to work more harmoniously with the general secretary, and internal dissension was allayed for the time being.

Rowland had declined reelection as general president in 1890, and he was replaced by William H. Kliver of Chicago. Kliver had joined the Brotherhood in 1884 and served as vice-president. "He was a man of Herculean build, with a powerful voice. His whole soul was wrapped up in the Labor Movement."[81] He later moved to Indiana, where he served two terms in the state legislature, and from 1909 to 1914 was building commissioner of Gary.

One further feature of the 1890 convention was a bitter attack on the Knights of Labor by McGuire, who accused them of slurs and innuendos against the Brotherhood. The situation in Washington, D.C., had still not been resolved. On August 15, 1889, the Brotherhood had chartered a new local, No. 531, but several of its members were discharged from jobs at the behest of the dissident Local 1 and the Knights of Labor assembly. Conferences between AFL and Knights leaders failed to resolve this and other controversies, and it was not until 1893 that the Brotherhood was able to make any real progress. In February of that year, when there was a good deal of work available, Edmonston brought in some out-of-town Brotherhood members and merged the two existing Brotherhood locals. Nevertheless, Local 1 and the Knights of Labor group continued in existence until the turn of the century, in spite of pressure by the AFL central labor unions in the city to bring about their demise.[82]

Despite his inability to convince the Brotherhood membership to provide the organization with greater financial means, McGuire could feel satisfied when he looked over the first decade of the Brotherhood's existence. Starting from scratch, he had built up a solid organization that spanned the entire country. The Brotherhood was firmly established as one of the nation's largest and most affluent labor organizations.

4

THE END OF AN ERA

The final decade of the nineteenth century was marked initially by retrogression in the fortunes of the United Brotherhood. A severe economic downturn, beginning in 1893 and lasting five years, cut membership sharply. The reversal of the growth pattern that had marked the first decade of the Brotherhood's existence led to internal discord. McGuire, the undisputed leader of the organization, found his policies challenged, although he still retained his popularity with the rank and file. Above all, changing technology was transforming the work of the carpenter and made the Brotherhood more concerned with appropriate trade boundaries. Several unions of woodworkers had grown up alongside the Brotherhood, and in seeking to protect the job opportunities open to its members, the Brotherhood began to question the rationale of the independent existence of these other unions.

The amount of strike support that the GEB authorized was limited by the availability of funds. The amount fell from $75,000 in 1890 to $44,000 in 1891, and still further to $26,000 in 1892. Locals appealing for help were advised to accept any fair employer offers, even though they were well below the local's demands. McGuire reiterated what was to be a constant theme for the next decade: "Until we have a system of higher dues this condition of affairs is apt to occur. And it really appears as if the law of 'the survival of the fittest' would truly apply to trade organizations, for where unions do not make ample financial provision, by having dues high enough to cover more than their running expenses, they are not likely to survive any great length of time."[1]

It was becoming apparent that an organization as large and diverse as the Brotherhood needed more than one full-time official. As Robert Christie pointed out, "In 1890 the national office had no regional representatives of any kind, no paid and full-time president or executive board, no organizers. There existed no lines of communication between McGuire and the various district councils except that which McGuire provided personally."[2] By 1892 district councils, which had been made mandatory, existed in 32 cities. Local unions had been established in 724 cities; New York led the nation with 93, followed by Pennsylvania with 80 and Ohio with 74. There were 16 locals in Canada and 127 in the South, 10 of which consisted exclusively of black carpenters.[3]

The need for a more formal structure and clearer lines of administration was increasing, but McGuire resisted. "McGuire was jealous of his powers and was a despot of sorts, albeit a benevolent despot. If anyone tried to usurp his powers, he had but to appeal to the rank and file, many of whom were his personal friends, to eliminate the pretender."[4] The GEB, now a regional body rather than confined to Philadelphia, began to act more independently, and McGuire found himself fighting a rearguard action.

NATIONAL MOBILITY

The tendency of local unions to restrict entrance into their geographical jurisdictions had been a problem from the start, even though one of the major reasons for the formation of the national union had been to facilitate mobility.[5] Time and again, the GEB had to remind locals that they were obliged by the national constitution to admit members of other locals holding valid traveling cards. The movement of carpenters, as well as that of other building craftsmen, was a function of fluctuating demand for labor in an industry that was subject to both seasonal and cyclical swings. Data on the precise volume of carpenter mobility are lacking, but it appears to have been substantial. Men were drawn to cities where they believed jobs were available at good rates of pay, although information was often inaccurate. Indeed, local unions often claimed that employers put out false information in order to attract migrants and create an oversupply of labor in an effort to keep wage rates depressed.

The traveling card was the method devised by the Brotherhood and other unions to maintain control over migratory workers. A carpenter leaving his home could secure such a card provided he was in good standing. This enabled him to secure work on union jobs in other cities without payment of an initiation fee. Initially, travelers were provided with confidential national passwords, signs, and hand grips which enabled other locals to confirm the genuineness of their membership. These were replaced in 1886 by passwords issued quarterly by the national union to the locals, of which the traveler was informed upon his departure. After experiments with various types of temporary cards, the Brotherhood installed a clearance card system in 1888; this card was valid for three months, and if it was not renewed or deposited with a new local within this period, the holder was suspended from membership. The migrant was obliged to deposit his card within thirty days of his arrival in a new locale.

Differences in skill standards among locals, as well as variation in dues and benefits, complicated the administration of the clearance system. There was a growing reluctance on the part of locals to accept itinerant carpenters, and procedural changes were made in an effort to discourage them. In 1891 clearance cards were made valid for thirty days instead of three months, and an applicant for a card who had been a Brotherhood member for less than three months was required to pay a $5.00 fee. Travelers were required to deposit their cards immediately upon obtaining a job, and their admission to the new local was made contingent on an examination of their qualifications by a local committee. There was considerable controversy over the right of locals with high initiation fees to levy charges against migrants coming from locals with lower fees, and eventually the constitution was amended to permit such a tax where the difference was greater than $5.00.

Over the years, the system was modified in several respects. At the present time it is compulsory for local unions to issue transfer cards, valid for one month, if a member is in good standing. If a strike or lockout is in effect, however, it is optional with the local involved to issue cards; in such cases, a member may seek work in another jurisdiction by securing a temporary working permit there. The presumption is that the transferee will return to his home base after the termination of the dispute. A working permit may also be secured in the absence of a strike for a member who does not wish to transfer his membership. He may be charged an amount equal to the dues of the recipient local, unless the permit is for five days or less, in which case there is no charge. Detailed constitutional provisions regulate the conversion of working permits into full membership as well as other membership rights of transferees.[6]

The present-day carpenter is less likely to move from one city to another than his nineteenth-century brother, although commuting among different local jurisdictions has been facilitated by the greater ease of travel. Moreover, the rise of contracting firms operating on a national basis has created a new group of traveling workers, and has required national union action in order to preserve the jobs involved for the union. Particularly in places where work is scarce, local unions may still seek to preserve the work that is available for their present membership. One of the most important factors behind the formation of the Brotherhood in 1881 remains a valid force a century later.

THE SEVENTH GENERAL CONVENTION

By 1892 membership in the Brotherhood had fallen to 51,313, a decline of 5,600 from the high point reached a year earlier. On the other hand, the number of locals in good standing rose to 802, just about the same as 1891 and more than a hundred greater than the 1890 figure. McGuire attributed this indifferent performance to the fact that "in the past year, there was not the same general widespread agitation and consequent public awakening on the eight-hour question, that prevailed in

1890."[7] The reason for this was the lack of financial resources, in his view.

By this time McGuire was completely converted to the importance of a benevolent system. He produced the following figures to show the growth of benefit expenditures over the previous decade:

Year	Number of benefits paid	Amount paid (dollars)
1883	6	1,500
1884	9	2,250
1885	36	5,700
1886	45	9,200
1887	139	16,275
1888	172	18,750
1889	224	25,575
1890	254	32,267
1891	374	44,732
1892	620	72,613

"These facts," he concluded, "should speak in thunder tones to every man, and prove an all-convincing argument as to the good, the value and benefits of trade unions."[8]

An important issue of structure arose at this convention out of the desire of McGuire's opponents to reduce his authority. In some unions, regional representatives who sit on the national executive board are elected by the delegates from each district separately. Thus, a popular regional official can win election to the board even if the national officers oppose him. Whether this is a more democratic system than election of regional representatives by the convention at large is a debatable question; there is no doubt, however, that it opens the way for more internal conflict. Regional officials elected separately have an independent power base from which to challenge national officers, and they can do so with impunity as long as their district support remains firm. Opposition slates are often drawn from this source.

A motion introduced at the 1892 convention would have installed such a method of electing executive board members. The fact that it lost by a margin of only two votes indicates the strength of the opposition that was building up. It is likely that even if the convention had approved this motion, McGuire could have opposed it successfully in the required referendum, but it was becoming clear that there would have to be a wider dispersion of authority within the organization.

Another interesting issue involved concern with the appropriate form of political action. The committee on organization recommended the establishment of a legislative committee to work for an eight-hour day law and other desirable legislation. The report was concurred in, but when it came to establishing a committee, those who feared that this might involve the organization in politics prevailed: "The general feeling of the delegates was that while a Legislative Committee was vitally essential, still it should be guarded by evey possible safeguard, and that particularly now in a Presidential Year such a committee might be subject to improper influence. Hence a motion to lay the entire subject on the table was made and adopted by a large vote."[9]

The convention did raise the minimum dues to 50 cents, after defeating a resolution for a 60-cent minimum. It was also decided that each local union should set aside 5 cents per month per member for a strike fund, to be forwarded monthly to the general secretary until $12,000 had been accumulated, after which the funds were to be retained in the local treasuries at the call of the general secretary. This did meet McGuire's request for a national strike fund that could be used quickly when authorized by the GEB.

In line with the established practice of rotating the presidency, Henry H. Trenor, who had come into the Brotherhood as a member of the New York–based United Order of Carpenters and had been a leading advocate of merger with the Brotherhood, was elected the eighth general president. After serving a two-year term, he returned to his local union and eventually became a successful contractor.

THE PANIC OF 1893

The financial panic that set in during the summer of 1893 and the ensuing depression left their mark on the Brotherhood, as they did on the entire labor movement. A great many strikes took place in 1894, most of them defensive, including the famous Pullman strike, which was defeated by a combination of federal court action and the dispatch of United States troops by President Cleveland.

Before the recession set in, the GEB devoted itself to routine matters, including the appropriation of organizing funds for different parts of the country, the actual expenditure to be supervised by McGuire. When a local union wrote in for permission to use opening and closing songs as part of its ritual, the GEB decided "that it is in order for the members of the Union to sing all they please in connection with the work of said Union."[10] A circular was sent out to all locals urging that they bring pressure to bear on the congressmen in their districts to secure the enactment of more stringent immigration laws, since "the wholesale importation of the pauper criminal and servile classes is a serious menace to our liberties. None can deny lowering as it does the moral tone of our people."[11]

As the depression deepened, the board became cautious in dispensing funds. A request of the Massachusetts district council for $500 for organizing expenses was refused because it was inadvisable "in view of the present depression to expend money in this direction just at the present time."[12] In an attempt to secure more members, the general secretary was empowered to allow locals to admit members at reduced fees, and to permit suspended members to rejoin on "reasonable terms." Locals could be given an extension of time for payment of their per capita, or excused entirely from such payment.[13] However, this did not mean that locals could do as they pleased; Local 10 of Detroit was suspended for refusing to abide by a GEB decision that one of its members had to receive treatment consistent with the Brotherhood constitution in a discipline case.[14]

With economic conditions continuing to deteriorate, the board, at McGuire's urging, authorized locals to use their strike funds to sustain their memberships in any manner they deemed best, "viz., by holding mass meetings, social entertainments, assistance of members or any way that will protect their members by upholding the rules of the Organization."[15] However, when a Cincinnati local asked permission to pay strike benefits to unemployed members, the board held that only those who were working at the inception of a strike were eligible for benefits.[16]

THE EIGHTH GENERAL CONVENTION

It was against a rather grim economic background that the eighth general convention convened in Indianapolis in September 1894. Two-thirds of the Brotherhood members had been idle for months. Membership had fallen to 34,000, a decline of more than 17,000 since the previous convention, erasing all the membership gains that had been achieved since 1889. The number of local unions had fallen from 798 to 716; a large number had simply surrendered their charters. This was a low point in the history of the United Brotherhood.

The convention was marked by a renewal of the opposition to McGuire's hold on the administration of the union, and by an effort on the part of proponents of independent political action to move the Brotherhood in that direction. On the former issue, the outgoing president had recommended that his post be made full time. "Elect a man to the office, have him work all day in a shop or building and travel fifteen or twenty miles of an evening after a day's work, and he will be of but little use."[17] This suggestion was turned back by a decisive margin. A renewal of the attempt to make the members of the executive board full-time salaried officials also failed. Finally, an effort to prevent national officers from succeeding themselves went down to defeat. McGuire was still in control of the convention.

On the political side, the unemployment and wage cuts resulting from the depression emboldened those who favored a radical transformation of the economic system. McGuire returned to his Lasallean belief in trade union cooperatives. When he addressed the opening session of the convention, he asserted:

> With enlightened effort and united forces the workingmen will eventually eliminate all middlemen standing between the worker and the full result of his toil. With the growth of intelligent labor sentiment would come united power at the ballot box. There never could be any complete solution of the labor question until labor became possessed of the tools of production and controlled the avenues of exchange and distribution, free from the control and exaction of money and power.[18]

Two political resolutions of an advisory character were adopted by the convention and later ratified in a referendum vote.[19] The first recommended endorsement of the political program that had been adopted by the 1893 AFL convention, which the Carpenters' delegates had supported. This resolution pointed out that the trade unions of Great Britain had opted for independent political action, and as a result had won such things as the eight-hour day, employers' liability for injuries, municipal ownership of streetcars and gas and electric plants, and nationalization of some utilities. It was resolved that the American labor movement should move in the same direction, including "the collective ownership by the people of all means of production and distribution."

The second resolution stated that the strike and the boycott were not sufficient or powerful enough weapons to improve existing economic conditions. It endorsed a long list of demands, including free use of all new inventions, nationalization of public transportation and communication, abolition of capital punishment, scientific management of national resources, making public lands inalienable, progressive income and inheritance taxes, equalization of women's wages, adoption of occupational health and safety laws, and the institution of the device of recall for all public officials. "We call upon all American citizens who are rightfully alarmed at the gigantic encroachment of corporate power and monopolistic greed upon the rights of the people to unite with the Socialist Labor Party in a majority effort to dethrone the representatives of monopolies and trusts now holding office."[20]

This resolution put McGuire in a quandary. He was still a socialist of sorts, but he feared the intrusion of partisan politics into the trade unions as a divisive force that was bound to weaken them. He had voted for the 1893 AFL program and indeed had supported the candidacy of John

McBride, an officer of the United Mine Workers who had the endorsement of the socialists, for the AFL presidency against Gompers.[21] A year later, however, he must have had second thoughts. Many trade unions had supported Populist candidates in the 1894 elections, and while the AFL convention that year rejected a statement directly endorsing socialism, the socialists were able to join with the Miners and other supporters of McBride to elect him to the AFL presidency, the first and only time in his long career that Gompers was defeated for the post. This time, however, the vote of the Carpenters' delegation was split. McGuire and one other delegate voted for Gompers, while the other two delegates supported McBride.[22]

McGuire's fear of the dangers of political fission was expressed in several *Carpenter* editorials that appeared after the 1894 Brotherhood convention. In the first, headed "Politics of Trade Unions," he pointed out that only 8 percent of workers were organized. To hurl them into politics would be suicidal, and would deprive minorities of their rights. Those who wanted a labor or socialist party were perfectly free to form them, but not with union funds.[23]

He also attempted to minimize the significance of the political resolutions that had been adopted at the 1894 convention:

> The votes on the . . . sets of political resolutions submitted as shown above are merely expressions of opinion and have no binding force on the members nor on the policy of the United Brotherhood as an organization. Each candidate for membership in our Order was most positively assured before joining that, in becoming a member, we would not in any way conflict with his political opinions . . . of course it is very desirable and proper that trade unionists should vote for labor measures and labor men and act as an independent factor in politics. But there are political organizations for that purpose outside of the trade unions, and our members as citizens are at liberty to join them, if they choose. But as wage workers, as carpenters and joiners, we must be united in the United Brotherhood, regardless of party politics, creed, or nationality.[24]

McGuire was undoubtedly shaken by the depression following the 1893 panic, but Robert Christie's interpretation of his actions on this and subsequent occasions misses the central core of McGuire's philosophy. To say that he was "neither socialist fish nor trade union fowl" and that had he "completely surrendered either his ideology or his pragmatic trade unionism he might have avoided a tragic fate"[25] is to misunderstand his fundamental commitment to trade unionism, first and foremost. The socialist commonwealth might emerge at some distant future date when trade unions embraced a substantial majority of workingmen, but until then, the trade unions had to be preserved against the intrusion of partisan politics, socialist or otherwise. Any other course might be fatal.

This was never expressed more clearly than in a powerful speech he made at the 1898 AFL convention, at the twilight of his career, when he said:

> We have never made it a test of membership in the American Federation of Labor, or in any trade union, that a member should belong to any political party or endorse any economic creed. To do so now at this Convention would be to break faith with our members . . . Because some of us stand firmly by the historical uniting and cohesive character of the trade union movement, we are called "pure and simple." Better any time a pure and simple trade unionist than an impure and complex confusionist. Let us put an end to this continual political spirit-rapping, guided, as it is, by theoretical, speculative extremists. It is time we notified the men of isms and schisms in the labor movement that the trade union can never be side-tracked or befogged by economic theories or debatable small-potato politics. In this country let us use all political parties . . . Elect trade unionists to every office and advance them on every possible occasion. Divest ourselves of petty jealousies. Arouse cohesive class feeling among the workers by industrial conflicts when forced upon them by unfair employers. Make the industrial struggle more intense so that the ties of unionism will grow stronger, and out of it will eventually come the Labor party—a party clear-headed and conscious of its status and its rights—not bleary-eyed and befogged by the seduction of any isms.[26]

There has never been a better statement of the basic rationale for the kind of trade unionism that won out in nineteenth-century America. McGuire remained an idealist until the end, but he rejected any sweeping social changes except those that came about through the efforts of trade unions and the democratic use of the ballot box. To write, as Christie does, of "his inability to choose between practical trade unionism and socialism"[27] is to denigrate McGuire's role, perhaps even more important than that of Samuel Gompers, as an architect of the American labor movement. From the moment he threw himself into the difficult struggle to organize the carpenters in 1881, and to keep them organized against employer resistance, McGuire had no difficulty in choosing between practical trade unionism and socialism.

The 1894 Brotherhood convention agreed to a reduction in the benefit scale and a registration fee of 50 cents for each new member to be paid to the national office, but these changes were rejected in a subsequent referendum vote. The result was that it became necessary to levy a 30 cent per member assessment in February 1895 to meet benefit claims. McGuire did not press for any dues increase; in fact, he argued against high dues and initiation fees, and urged locals to reinstate members on easy terms in order to build up the membership level. He warned against hastily called strikes, and suggested that no stoppage be called "without first having a conference committee of level-headed men wait on employers." He favored the suspension of locals that called general strikes in their locality without first consulting the general officers.[28]

These counsels of caution were reinforced by the GEB, which notified all locals in January 1895 that no application for strike sanction would be approved for any purpose other than the eight-hour day.[29]

The shortage of jobs raised more sharply than ever before the issue of jurisdiction, and particularly the effects of technological change. In his report to the convention McGuire called attention to the potential loss of carpenter jobs due to the introduction of iron and steel frames and staircases, and tile floors. The convention declared that the Brotherhood had "the supreme right of governing all workers in the craft in the United States and Canada, and that hereafter no other organization of carpenters be recognized." However, in an uncharacteristic move that was later to cause problems, the Machine Wood Workers International Union was ceded jurisdiction over all mill hands except carpenters who occasionally were employed at mill work, millwrights, or stair builders.[30]

Charles E. Owens, who came into the Brotherhood through the New York United Order, was elected general president. As the result of an amendment to the constitution, the offices of general secretary and treasurer were combined, and McGuire was elected to the new post by acclamation. However, three of the five men elected to the GEB had opposed various policies of McGuire's in the past. Even though McGuire clearly controlled the convention, opposition continued to mount.

CONTINUING EFFECTS OF THE DEPRESSION

The years 1895 and 1896 were difficult ones for the Brotherhood. Unemployment was high; strikes were discouraged. Nevertheless, McGuire continued to urge organizational efforts. "Within the past few months," he wrote in June 1895, "the United Brotherhood has shown a decidedly encouraging increase in membership in a number of localities. Fully 35 percent of our unions have had gains . . . In a number of cases it has been caused by placing a good, wide-awake man in the field as business agent or walking delegate to drum up the men."[31] The next month, he reported that "where we maintained our locals intact, wages have been upheld, the men have been better treated, and the cutting of prices and wages has not been as ruinous to contractors and men." And he added: "Now is the time to organize . . . We will pay any reasonable expenses you may incur in case you organize."[32]

Declining membership meant a reduction in financial resources for the organization, and a campaign was undertaken by both the GEB and McGuire to pave the way for remedial action at the next convention. The board instructed McGuire to prepare a circular showing the union's financial condition, emphasizing the fact that almost the entire income of the national office was going for benefits. "We are fast drifting into a mere beneficial order . . . Under our present income we cannot support our present benefit system and carry out the work outlined by us. If forced to do so, we must soon be bankrupt. Either we must raise our dues or reduce our benefits. Which shall it be?"[33]

McGuire himself began to promote an idea that he had not been able to sell to the membership in the past: equalization of funds. The following statement appeared in large type in successive issues of the *The Carpenter*: "To build up the United Brotherhood to be a power, we must have uniform dues, uniform initiation fees, uniform sick benefits, and an out of work benefit for our members when unemployed. We must have higher dues and an equalization of funds annually. Then we will prosper." He explained that this scheme did *not* mean that every local would have the same absolute amount of money in its treasury, but rather the same amount per member. At the end of each year the general secretary would determine the average per capita amount of funds held by all locals, and would reallocate funds from the surplus to the deficit locals. For example, if the Brotherhood had 50,000 members and $500,000 in all local treasuries, the average per capita would be $10. If Chicago had 5,000 members and $55,000, while New York had 50,000 members and only $45,000, Chicago would remit $5,000 to New York. Each local would be allowed specific amounts for its expenses, determined and monitored by the general secretary. "Under this system we would be a United Brotherhood indeed in every sense of the word. Now we are simply a Federation of local carpenter unions."[34]

To McGuire, this was a completely logical system. He pointed out that the Cigar Makers Union and the Amalgamated Carpenters had employed it successfully. Strong locals would help the weak, and honesty would be ensured by headquarters' auditors (he called them "financiers"), who would be constantly on the road. And, of course, it would give the national office much greater control over the locals.

Equalization of funds has never been customarily practiced by American trade unions. It implied a greater degree of self-denial on the part of the wealthier and better-organized districts than was normally to be found. Locals were prepared to turn over funds that were specifically earmarked for strike relief to fellow locals in financial need, but they were not willing to share their funds on a more general basis. McGuire was usually in touch with rank and file opinion, but in this case his idealism overshadowed his perception of the strength of local separatism.

Apart from finances, the GEB was called upon to interpret various constitutional provisions. One clause in effect at the time barred from membership anyone who sold intoxicating liquor, a fairly common trade union rule at the time. The board ruled in one case that a carpenter temporarily tending bar could not remain a member.[35] In a second case, the rule was extended so that if any member of a carpenter's household was selling liquor, he was similarly disqualified.[36]

THE NINTH GENERAL CONVENTION, 1896

The poor economic climate that had characterized the 1894 convention had not yet dissipated two years later. Between the two conventions 115 new locals had been chartered, but 236 had surrendered their charters, for a net loss of 121.[37] Membership had fallen by more than 4,000. McGuire offered the following explanation for the drop: "The trying hard times, the industrial and financial stagnation, the lack of commercial confidence and shrinkage of enterprise and credit which fell on us with crushing force over three years ago, have since only become more marked and intensified."[38]

A strong effort was mounted once again at the convention to reduce the authority of the general secretary, and this time it succeeded. The Committee on the Consititution recommended that the general president be made a full-time officer, as well as the chairman of the GEB *ex officio*, and that he be vested with the power to appoint district organizers in consultation with the GEB. The power to approve new local charters, bylaws, and trade rules was also given to the president by the convention. The proceedings do not record the discussion on these amendments, but that the division was a close one is indicated by the vote on the question of the presidential salary. A motion to set it at $1,800 a year was defeated; the sum of $1,500 a year lost by a vote of 46 to 52; and a $1,200 salary carried "by a close vote."[39] McGuire's salary at the time was $2,000, so he would have continued as the highest paid national officer in any event.

As it turned out, these amendments were defeated in the referendum vote. According to Christie, this resulted from "the strong grass roots support which McGuire had built up over the years . . . McGuire had vaulted over the heads of the local leadership and found behind them a rank and file unquestioningly loyal to him."[40] In fact, all other amendments providing for any increase in dues and expenditures, including those supported by McGuire, were also defeated. The failure to establish a salaried presidential post was probably due mainly to rank and file financial conservatism in the midst of a recession. Indeed, a few months later, McGuire, noting that the Cigar Makers had raised their dues to $1.20 a month, commented: "Just fancy carpenters paying $1.20 per month dues, when they recently voted down a proposal to make their dues only 75 cents per month all round. The carpenters will have to learn the value of higher dues, as did the cigar makers and other unions. It may take time, but we will keep at it."[41]

The issue of whether local unions should have the right to charge for traveling cards arose at this convention. The New York (Manhattan) district council had imposed a $1.00 charge for working cards on Brooklyn district members. The GEB had ruled that while this was lawful under the constitution, it was likely to lead to "endless bickering and strife,"[42] a prediction that turned out to be true. Although the convention sustained the decision of the GEB, it recommended that the two districts formulate joint working rules and drop the imposition of working card fees. The New York district responded to this recommendation by setting a $5.00 charge for a quarterly working card a few months later, and was warned by McGuire that building a wall of exclusion around New York or any other city was contrary to the spirit of trade unionism.[43]

The incoming president was Harry Lloyd of Boston, described as "an able orator, in fact one of the best in his day. He was eloquent, convincing and powerful in his advocacy of Trade Unionism."[44] Lloyd represented locals in Toronto and Boston, and had served as first vice-president from 1888 to 1890.

THE SHARPENING OF JURISDICTIONAL CONFLICT

As long as the Brotherhood was a small organization struggling to organize the building carpenters, and jobs were relatively plentiful, there was no real effort to stop the peripheral wood trades from forming separate unions. However, as the construction industry became better organized and new technology began to transfer site work to the factory, the Brotherhood became more concerned with the total pool of jobs that might be made available to its members.

The impact of new technology was vividly described by McGuire in testimony before the United States Industrial Commission in 1899 (see Appendix A). When asked about the effect of new and improved machinery upon his trade, he replied:

It has been very injurious. There was a time we cut out the doors and the sash and the wooden trim that goes around the wash-board and base in a room — all the wood-work in the construction of a building was prepared by hand or motor power machinery. With the use of steam-power machinery the doors and the sash and the wood-trim come from the northwest sections, where children are employed at tender years . . . There are scores of places in [Wisconsin, Minnesota, Michigan, and Indiana] where the material is prepared so cheaply that they can send it over to England . . . On some machines the product, for instance on a planing machine, is fourteen times greater than by hand . . . In cutting out moldings the capacity of a molding machine is about twelve times greater than by hand. So it ranges, tiling, grooving, and flooring, all through the manifold branches of woodwork.

When further queried about the resultant displacement of carpenters, he added:

> In New York City the only work left for a carpenter on the good buildings is to go in there and hang the doors, first case around the doors, hang the doors, put in the sash and the inside blinds. All the other work in the larger buildings that was formerly of wood is now either of stone, terra cotta or tiling or metal of some kind . . . the board and sash and trim work, which used to be prepared or gotten out in the shop twenty-five to thirty years ago when the men could get that material ready in the winter and in the dull seasons when they could not work out of doors, and on rainy days, and then go in the buildings and put them up, that is being furnished them now by the well-equipped machinery and latest improved machinery.[45]

With the emergence of skilled and semiskilled operatives in the increasingly specialized wood product factories, the Brotherhood began to take an interest in organizing them, particularly because many of their building construction members moved back and forth between site and factory work, as McGuire pointed out. By 1890 the Brotherhood had twenty one locals composed of planing mill hands, three of sash, blind, and door makers, and five of stair builders.[46] The constitution was amended to define eligible members to include stair builders, planing mill bench hands, or any carpenter running woodworking machinery, thus setting up a broad jurisdictional claim.

In 1890, the same year the Brotherhood began manifesting its interest in the mill men, a group of independent locals met in St. Louis and formed the Machine Woodworkers' International Union. The word "machine" was included in the title when some of the delegates feared that offering membership to all mill men might antagonize the Brotherhood and another woodworking union, the Furniture Workers. Thomas I. Kidd, who was to play a major role in the organization throughout its existence, was elected general secretary-treasurer. The new union applied to the AFL for a charter, which in later years would have evoked an immediate and sharp reaction from the Brotherhood. McGuire objected at first but then changed his mind, much to the regret of later generations of Brotherhood officials.[47] For disputes soon arose between the two organizations, and did not end until the Wood Workers were absorbed by the Brotherhood in 1912.

In his report to the Woodworkers' 1891 convention, Secretary Kidd reported that Carpenter locals were placing obstacles in the way of new mill-hand organizations, but said that he had "borne their opposition patiently, believing that the United Brotherhood will in time accord us the right to control our own trade and refrain from encroaching upon our jurisdiction."[48] The following year a Woodworker delegation appeared before the Brotherhood convention to present its claim of jurisdiction, only to be told that the Carpenters had jurisdiction over the entire trade. Kidd warned that the Woodworkers would not sit by idly while their rights were being impaired, and threatened that if the Brotherhood insisted on admitting machine men, "it might be advisable for us, at some future time, to consider the question of opening our doors to mill carpenters."[49]

Kidd appeared once again at the 1894 Brotherhood convention to repeat his request for jurisdiction over both machine and bench hands. The Committee on Organization came in with a recommendation that "it would be in the best interests of the U.S. and organized labor in general that the M.W.W.I.U. be granted entire jurisdiction over all mill hands, except carpenters who may at times be engaged in mill-work, or mill-wrights or stair-builders," and further recommended that the Brotherhood support the Woodworkers in their organizational efforts.[50] This report was adopted by the convention, and came to be known as the Indianapolis Agreement. The Carpenters later claimed that this agreement was never ratified by the Woodworkers,[51] but there is little doubt about the intentions of both parties.

This did not end the controversy. The Machine Woodworkers began negotiations with the International Furniture Workers' Union, looking toward amalgamation. The latter had been organized in 1873, the impetus provided by the German Cabinet Makers' Union of New York City. After early vicissitudes, it attained a membership of 10,000 by 1886. Several years later the Furniture Workers proposed that its monthly journal be merged with *The Carpenter* as an economy measure for both organizations, but McGuire firmly rejected this idea: "If we were to join interests with your society in having one Journal together, there would be disputes . . . that now under our present system can be avoided. We would very much like to help your organization in any way possible or practicable, but we cannot consent to giving your organization two or three pages of space in our paper under no circumstances whatever."[52] The Furniture Workers made several attempts to amalgamate with the Brotherhood, and when they were rebuffed, turned to the Machine Woodworkers. While the two were negotiating, several New York locals of both organizations moved their charters to the Carpenters, which did not serve to improve relationships. In June 1895 a consolidation was achieved, and the Amalgamated Wood Workers International Union emerged, with Thomas Kidd as general secretary.

Resolutions were introduced at the 1896 Brotherhood convention that would have annulled all previous woodworker agreements, but they were defeated. Friction mounted when the Amalgamated Wood Workers claimed exclusive jurisdiction over all inside work, something the Brotherhood was not prepared to concede. In July 1897 Kidd appeared before the GEB to protest the granting of Brotherhood charters to two locals of mill men in New Orleans, and to discuss the possibility of a new agreement.[53] The Amalgamated was willing to cede jurisdiction over those employed in planing mills or woodworking establishments where steam, electric, or other power was used. The GEB rejected this proposal and formulated one of its own, whereby the Amalgamated would have jurisdiction over all mill hands except carpenters who might at times be engaged in mill work and all mill hands who were members of the Brotherhood. In return, the Amalgamated was to recognize the Brotherhood's right to outside carpentry work, as well as the fitting up of offices and stores.[54]

This proposal, which was accepted by the Wood Workers, was clearly doomed to fail; it left too many ambiguous demarcation lines. The Wood Workers felt that they had made a major concession in renouncing jurisdiction over the fitting of offices and stores, work that had been done in the past by cabinet makers.[55] On the Carpenter side, complaints began to come in that the Amalgamated was undercutting the Brotherhood scale. For example, the New York district council charged that the Newark Amalgamated local was working at from 20 to 28 cents an hour below the New York shop scale, which had the effect of diverting work from New York to Newark. The GEB refused the New York request to set up a new local in Newark, but it told the Amalgamated that its Newark scale had to be equal to that of the Brotherhood.[56] This is how matters stood between the Brotherhood and the Amalgamated when the 1898 Brotherhood convention opened.

Another union that cut right across the Brotherhood's jurisdiction was the Amalgamated Society of Carpenters and Joiners. This union was actually a branch of a powerful English organization, a situation that has few parallels in American labor history. It was formed in 1867 and received an AFL charter in 1890. Although its total membership never exceeded 10,000, its strength was concentrated in the larger cities of New York, Philadelphia, and Chicago. When the Brotherhood was first organized, cooperative arrangements were often made at the local level, but conflict inevitably arose as time passed. The Chicago district council complained to the 1894 convention that the Amalgamated was undercutting its wage scale and supplying strikebreakers. There had been a United Carpenters Council, consisting of locals of the Brotherhood, the Knights of Labor, and the Amalgamated, but this dissolved in 1888. By 1892 the Knights were down to 171 members and the Amalgamated to 157 in Chicago, while the Brotherhood claimed 4,000. The 1894 Brotherhood convention resolved that no other organization of carpenters would be recognized in the future, and urged all locals and district councils to secure exclusive jurisdiction for the carpenters.[57]

This did not necessarily imply all-out warfare, however. In January 1895 the GEB, together with the general president and secretary, decided to confer with representatives of the Amalgamated in an effort to bring about an adjustment of their differences, and by July it proved possible to reach an agreement. This settlement provided that neither organization would admit to membership anyone who had been fined, suspended, or expelled from the other. The Brotherhood agreed to recognize Amalgamated cards, except that where no branch of the Amalgamated existed, Amalgamated members would be required to pay the same fee to support a Brotherhood business agent that Brotherhood members were charged. "It is mutually agreed that the two aforesaid bodies have common interests as trade unions, and shall in the future as a matter of policy, and for purposes of common defense, oppose as far as possible the formation or advancement of any other organization of carpenters in the United States or Canada."[58]

This agreement was largely the work of McGuire, who had always been an admirer of British trade unionism and favored the maintenance of good relations with the British parent of the Amalgamated. But many locals on both sides were not prepared to accept this settlement, and relationships deteriorated. In April 1897 the GEB directed McGuire and the board chairman to meet with Amalgamated officials for a discussion of the many problems that had arisen. Nothing important appears to have resulted from this meeting, and McGuire reported to the 1898 convention:

> Our agreements with the Amalgamated Society of Carpenters and Joiners and the International Union of Wood Workers have been kept inviolate, but in many localities our Locals find considerable fault that on the other hand these two organizations in a number of localities have not always been so rigid in the observance of their part of the agreement. Still, whenever complaint has been filed by me the general officers of these societies have ever shown a willingness to make investigation and to secure proper respect for our agreement with them.[59]

Despite the unwillingness of many district councils and locals to go along with the various agreements that had been reached, McGuire continued to be conciliatory, always trying to avert direct confrontation. In the end, he had to give way to those who took a stronger view of the importance of strict jurisdictional boundaries.

THE TENTH GENERAL CONVENTION

The tenth convention was held in a year of continuing economic stagnation, although there were signs of revival. In his report to the convention, McGuire characterized the current situation of the carpenters in the following terms:

For five years and more, the bulk of our members have experienced hard and trying times and have made immeasurable sacrifices to uphold Union principles. Some have dropped by the wayside unable to pay dues or worn out in the intense struggle. While others, weak and irresolute, have yielded to the most debasing conditions, work at any price and on any terms, sinking manhood and honor in a brutal struggle for bare bread.[60]

Nevertheless, membership rose by more than 3,000, although it still remained far below the 1891 peak.

In preparation for the convention, the GEB had submitted a number of propositions to a referendum vote in order to gauge the views of the membership better. These were as follows:

1. Shall we establish a uniform initiation fee for all locals? It had been pointed out that initiation fees ranged from $5.00 to $25.00.

2. Shall we have a uniform reinitiation fee for all members?

3. Shall we inaugurate a uniform system of monthly dues in all locals? Dues ranged from 50 cents to $1.00 per month.

4. Shall the dues be increased to 75 cents a month for all unions paying less?

5. Shall we create a uniform system of sick benefits for all locals? These ranged from $3.00 to $6.00 a week, for 10 to 52 weeks.

6. Shall we have a special annual tax to pay the transportation expenses of convention delegates? The purpose of this was to enable smaller and poorer locals to be represented.

7. Shall the general president be a full-time paid official?

8. Shall we have two paid general agents in the field the year round to organize and examine local accounts?

All of these propositions were defeated by a margin of 2 or 3 to 1. The constitution committee drew the following conclusions from this poll: "The Local Unions are not in favor of a strongly centralized system of uniform initiation fees, uniform dues and uniform benefits. They favor the present system of local regulation on these subjects, above a certain minimum."[61] Despite his popularity with the members, McGuire was not able to achieve his goal of uniform dues and benefits, let alone equalization of funds.

The outgoing general president, Harry Lloyd, remarked on the fact that virtually all the work of the previous convention, including the proposal to elect a full-time president, had been nullified in the subsequent referendum. His solution was to submit to a general vote only matters of broad principle, leaving to the convention the authority to make the consequent detailed changes. Needless to say, this recommendation was not approved by the convention; it was an article of faith held by the membership that the right to vote on convention decisions should be left intact.

Lloyd also made a number of other recommendations that McGuire must have approved:

1. An out of work benefit: "I firmly believe that if a small out-of-work benefit, such as the cigarmakers pay, was introduced in our organization, and that would enable our members who are out of work to pay their dues, this dropping of our old and best members would cease."

2. Equalization of funds: "The small isolated Unions in the small towns or villages are continually coming into and going out of existence for want of the support that they very properly look for to the large Unions."

3. Higher dues: "You cannot run a good and profitable organization on wind or bluff; unless you faithfully fulfill your contract by giving all that you promise you will certainly meet with defeat."

4. Continuous employment of organizers on the road.[62]

The convention failed even to consider these recommendations. The referenda had shown that the membership was in no mood for substantial changes where money was involved.

However, the convention did begin the process of reducing the power of the general secretary. John Williams, who was elected general president, recounted the significance of what had happened in a review written several years later:

Everybody who is familiar with the affairs of the Brotherhood up to 1898 knows that it was dominated in every vital respect by one forceful personality. All administrative functions were united in one official. That many members realized and resented such a condition is evidenced by the action taken at several preceding conventions, when proposals to make the General President a salaried officer were adopted by the delegates without much opposition. The purpose of such a change was to divide administrative authority and responsibility and thereby eliminate the one-man control that to many had become quite irksome. Every attempt to effect the change by a direct amendment of the constitution failed, however, for the simple reason that when submitted to the vote of the membership the cost of such a change was always so featured in explanatory notes accompanying the proposition to invite its defeat. The serious minded and earnest men who in various sections of the country were leading the forces of the Brotherhood knew that growth such as they had a right to look for was out of the question until the organization was unshackled and freed from the bonds of a form of paternalism that repressed its spirit and stunted its growth. I know that this was their feeling, but they withheld expression thereof because of their loyalty and affection for a great leader and a masterful mind. But relief had to come and what could not be attempted by direct action had to be accomplished by strategy.[63]

A few words about this statement, made by a leader of the opposition to McGuire, are in order. To say that the time was ripe for a dispersion of authority within the Brotherhood was a fair statement; from a geographical point of view, the organization had become too complicated for even a first-rate administrator to run single-handedly — and McGuire's strength lay in organizing rather than administration. He preferred being on the road, speaking at meetings and encouraging local initiative, to shuffling papers at headquarters.

But to put the blame for the Brotherhood's vicissitudes during the 1890s on McGuire's administration of its affairs is neither fair nor accurate. Membership fell simply because of the severe economic depression and the ensuing unemployment. The Brotherhood was not the only organization to suffer. Membership in the Bricklayers' Union fell from 27,000 to 19,000 between 1892 and 1894; in the Brotherhood of Railway Trainmen, from 28,000 in 1893 to 19,000 in 1897. Similar losses were experienced by the Locomotive Firemen and the Railway Carmen, all of these being relatively well-established unions.[64] Indeed, had it not been for McGuire's perennial optimism and unquenchable energy, the Brotherhood might well have suffered even more serious losses.

The first and most important step taken at the 1898 convention was "conversion of the office of general president from a mere figurehead into an active and essential figure in administration." The general president was given the power to examine and approve all local rules and bylaws; to decide all grievances and appeals; and to sit with the GEB, although not to vote.[65] The vote in favor of these amendments was not close — 64 to 22. The strategy that was followed to improve the chances for confirmation in the referendum was to refrain from making the general president a salaried official, so that the extra cost would not be a negative factor in the vote. "The Committee also requested and received authority to prepare the official circular submitting the amendments adopted by the convention to a referendum. By this means any scheme that might have been contemplated to defeat the plan was frustrated."[66] The strategy worked; the constitutional amendments were approved by a vote of the membership.

McGuire suffered another defeat when his conciliatory jurisdictional policy was rejected by the convention. New York Local 476 had filed an appeal in opposition to the agreement that the GEB had negotiated with the Amalgamated Wood Workers. Although the Committee on Grievances and Appeals reported in favor of sustaining the action of the GEB, the convention, after a lengthy discussion, defeated the motion by a margin of 69 to 40. Instead, the convention adopted the following resolution:

> Resolved, That it is the sense of this convention that no other carpenters' or woodworkers' organizations of any kind be recognized by the Brotherhood

and that no agreement be entered into with other carpenters', woodworkers' or machine-hands' organization by our General Officers and further, that all agreements now existing be annulled.

Resolved, That this convention cancel all previous agreements with the Amalgamated Society or any other society of Carpenters' and Joiners', and that no agreements be entered into with any society except by a majority vote of the Union, or District Councils in that locality.[67]

The majority for the resolution was overwhelming — 72 to 32. Among those who voted against the resolution was Frank Duffy, who was to play an important role in future jurisdictional strife. And on another note of things to come, the electricians were condemned for claiming "to do everything in sight where their wires are laid."[68] The convention rejected an appeal to join the recently established National Building Trades Council, the stated purpose of which was to bring about closer cooperation of the building trades, although it had promised not to interfere with the internal affairs of its constituent national unions.

Some interesting statistics were presented to the convention. Since 1881 the Brotherhood had issued 1,055 charters, about one-quarter of which were still in existence. Some of the cancellations were a result of consolidation, but the figures do suggest a high mortality rate for this early stage of the union's existence. Second, the following data on strike benefits paid out by the national organization show clearly the impact of the business cycle on strike activity;[69] the Carpenters were well aware that a period of high unemployment is no time to press for higher wages:

1886–1888	$10,311
1888–1890	75,497
1890–1892	71,336
1892–1894	53,437
1894–1896	15,015
1896–1898	8,697

Despite his defeat on various issues, McGuire was reelected general secretary by acclamation, but John W. Williams of Utica, New York, who had been in the opposition, was elected president by a close vote. Williams held the post for only one year; in 1899 he was appointed head of the New York factory inspectorate, and later became the state's commissioner of labor. William D. Huber of Yonkers, New York, who had been elected first vice-president, automatically succeeded Williams.

RISING FORTUNES

Beginning in 1898 and lasting until 1903, a period of very rapid member-ship growth set in for the Carpenters. By 1903 membership had risen to 167,200, from a low of 28,200 in 1897, a phenomenal sixfold increase. There was a general expansion of all trade union membership during these years — membership for unions in the aggregate grew by a factor of four — but the Carpenters substantially exceeded the average. The causes were primarily economic. Unemployment estimates are available only since 1900; they show that the rate of unemployment fell from 5 percent in 1900 to 2.6 percent in 1903. Average hourly wages in construction rose by 15 percent from 1897 to 1903. The depression of the 1890s was forgot-ten in the euphoria of the upswing.

For the national office of the Brotherhood, this meant a much greater volume of business. McGuire reported in August 1900 that "at no time in our history has the United Brotherhood had a larger number of trade movements than in the early spring and summer of this year; 214 cities under our jurisdiction, involving over 36,000 members, were on strike at various periods up until this very month."[70] Requests for strike sanction flooded into Philadelphia. At first the GEB was cautious; locals demand-ing higher wages were advised to concentrate instead on getting an eight-hour day. When the Pittsburgh district council asked for strike sanction to secure a half-holiday on Saturday while working a nine-hour day the rest of the week, it was advised to make a stand for eight hours, "which is one of the fundamental principles of our organization."[71] But as the months went on and the number of requests rose, the board began to ap-propriate substantial sums of money for strike assistance: $3,000 to Chicago; $800 to Brooklyn, New York; $500 to Galveston; $1,000 to Atlanta; $500 to Utica, New York. In order to finance these grants, an assessment of 25 cents per capita was levied in April 1900.[72]

What it was like to be an organizer for the Brotherhood in those years is illustrated by a report by Albert Cattermull, a general organizer who was also a member of the GEB.[73] On June 8, 1899, he went to Marion, Indiana, where there was a strike. A few days later he managed to put about half the striking members back to work at an increase of 22½ cents a day, but the rest of the contractors held out until June 15, when they agreed to the higher wages. He then proceeded to Indianapolis, where he spent a day advising Local 281 on its trade rules, and then moved on to Belleville, Illinois, for a conference with the officers of Local 433.

From June 20 to 27 Cattermull remained in Milwaukee, assisting the district council and business agent on the enforcement of trade rules. After a week in Chicago, he moved on to Scranton, Pennsylvania, where a strike was in progress involving all the building trades. "I was actively engaged night and day, speaking at the various meetings and assisting to strengthen our position all along the line." After a visit to Philadelphia to attend a GEB meeting, he returned to Scranton on July 21 to find two-thirds of the carpenters at work, having won an eight-hour day. From July 23 to 26 he visited Buffalo, New York, where he discussed the possibility of a charter for the car builders. He then went to Detroit on July 27 to investigate local conditions, to Chicago on July 28, "and until August my time was fully occupied in visiting adjacent towns endeavoring to organize new local unions."

From August 3 to 9 he was in Indianapolis attending meetings of Locals 281 and 60, which were considering new demands. In Cincinnati (August 10–15) he found conditions disappointing, and conferred with some contractors, who expressed a desire to cooperate with the district council "in preventing a further continuance of the present ruinous competition." He reached Louisville, Kentucky, on August 16, where he secured an agreement for a shorter working day. He proceeded next to Evansville, Vincennes, and Terre Haute, all in Indiana, where he addressed meetings. Catermull expressed himself as much encouraged for the future; he felt that a substantial membership increase was in the offing. Organizing for the Brotherhood was not a sedentary occupation, but it was a satisfying one, at least in those good years.

An internal conflict of some magnitude arose with the New York City locals, which appeared to have some difficulty adjusting to their status as subordinate bodies of a national union rather than as units of the formerly independent United Order of Carpenters. Trouble began when the New York district council declared as unfair the products of an upstate mill, despite the fact that the latter employed only Brotherhood members, in an effort to prevent the import of out-of-city fabricated wood. Complaints also began to come in that New York was refusing to honor clearance cards. The New York locals had raised their initiation fees to $20.00, and they attempted to charge the difference between that figure and the customary $5.00 for any man coming into the city.

The New York locals were ordered to stop this practice, aimed at restricting the labor market to local residents, but they refused to comply. As a result, the general president suspended five New York locals on June 25, 1900, after they had withdrawn their delegates from the New York district council. The seceding group formed themselves into an organization called the Associated Carpenters of New York City, and proceeded to negotiate with the Brotherhood. The GEB provided $2,000 in financial assistance to the district council to enable it to get back on its feet, and restored several of the dissident locals to good standing at the request of members who remained loyal to the Brotherhood.[74] The rebellion eventually simmered down and the recalcitrant locals were rechartered, but the conflict flared up once again in more virulent form fifteen years later.

Problems also developed with the ship joiners, who at the time were independently organized. Many of these workers were carpenters who went to work on ships when building jobs were unavailable, and were obliged to withdraw from the Brotherhood when they did so, since its constitution prohibited any Brotherhood member from becoming a member of any other labor organization. Requests for charters from ship joiner locals were turned down on the ground that the trade should not be broken down into specialized compartments.[75]

The ship joiners persisted in their desire for craft identity, and in 1902 they organized the International Union of Shipwrights, Joiners, and Caulkers, at the suggestion of the AFL executive council. They were granted a charter in October 1902 on the understanding that their members would be required to join the Brotherhood when working on buildings.[76] But as we shall see, this arrangement was not able to stand up against the expansionist proclivities of the Brotherhood during the next decade.

THE ELEVENTH GENERAL CONVENTION

The convention that assembled in Scranton on September 17, 1900, was marked by still another attempt to revise the governing structure of the Brotherhood. In his opening report General President Huber urged that the position he occupied be made a salaried one, in view of the great volume of work involved. He also recommended that the general secretary, "who has been in failing health for some time past," and who found onerous "the task of answering the vast amount of correspondence and the many calls on him to go to different places," be provided with an assistant.

The convention adopted these recommendations, in the latter case by providing for the post of general treasurer in addition to that of general secretary, and by transferring a good many of McGuire's functions to the treasurer. A minimum initiation fee of $5.00 was stipulated, as well as minimum monthly dues. It was resolved that in the future the general officers would be elected by referendum rather than by the convention. The convention also voted to move the general office to Indianapolis, a shift that was motivated by the desire to establish the office nearer the center of gravity of the membership, as well as being designed to move the organization out from under McGuire, whose personal situation would have made it difficult for him to change his domicile.

In submitting these changes to a referendum vote, McGuire included a personal statement opposing the move to Indianapolis as being without good cause, and objecting to the creation of the post of treasurer as an unnecessary expense. Albert Cattermull, by this time chairman of the GEB, sent out a similar statement, also opposing the various changes on the ground of expense. Almost inevitably, the changes were defeated by the vote of the membership.

However, objection to McGuire's rule was rising. An almost unprecedented event occurred — someone was nominated to run against him. McGuire was reelected by a vote of 78 to 55 and once again proved that he had the support of the membership. In the past the opposition had resorted to attempts to amend the constitution to whittle down his power, but this overt challenge was a portent of things to come.[77]

There was an interesting debate at the convention on the union label, illustrating an important economic issue. Attempts by the locals in New York and other big cities to keep out mill work produced in smaller towns gave rise to a resolution recommending that the union label be given to all mills employing union labor, on the ground that a tariff wall could be built around the cities. A minority of the resolutions committee proposed that the GEB be given authority to determine what a fair minimum rate should be, and to withhold a label if granting it would have the effect of reducing prevailing wages in the large cities. The majority view prevailed, but this conflict between the higher-wage cities and the smaller towns — often, in fact, city suburbs — continued.

The convention did adopt a resolution calling for an annual pension, the amount to be fixed by the GEB, and this was approved in the subsequent referendum. The carpenters were strongly in favor of an extension of the benevolent system, although they were not always prepared to pay for it.

A Brotherhood committee had been appointed to negotiate with the Amalgamated Wood Workers in the mounting struggle between the two organizations, after the 1898 convention had repudiated all previous agreements. Amalgamated was willing to cede to the Carpenters jurisdiction over the fitting of offices and stores, except for saloon, bank, and drugstore fixtures manufactured in shops under contract with them. In its report to the convention, the Brotherhood committee recommended against any such agreement: "The time has come for the U. B. of C. & J. to assert its right of jurisdiction over all Carpenter work as specified in our Constitution, believing that division of control by two organizations of one trade can not be tolerated, particularly where the standard of wages of one is lower than that maintained by the other."[78] The report was adopted, and it was agreed that charges would be brought against the Amalgamated at the next AFL convention. The restraining hand of the general secretary had been lifted.

Finally, there was an interesting political debate, representing one of the last efforts to move the Brotherhood in a more radical direction. A majority of the committee on resolutions, noting that appearances by labor representatives before congressional committees had failed to win favorable legislation, proposed instructing Brotherhood delegates to AFL conventions to vote against any further appearances, "and to impress the Convention and the wage workers at large of the necessity of keeping aloof from all capitalistic political parties, and to enter into independent political action on their own behalf."[79] The resolution was defeated on the floor of the convention, and there was never again any question of the Carpenters straying into the pitfalls of independent political action.

THE PALACE REVOLUTION

Frustrated in their efforts to reduce McGuire's authority by constitutional change, his opponents, who now controlled the GEB, resorted to direct action against him. By this time McGuire, although not yet fifty years of age, was in ill health. Christie records that "McGuire returned to liquor to ease his troubled mind. He remained away from his duties for days at a time, brooding over his fate in Philadelphia saloons."[80] McGuire himself told the 1902 convention that "a man wears out like a piece of machinery . . . I am not lost entirely in this world, but I have had enough to wreck me physically, destroy me mentally."[81]

Early in 1901 Huber, who had been elected to a full term as president at the 1900 convention, relayed to the GEB compaints by local unions of failure by McGuire to transmit moneys due them, alleging a breach of duty. The board looked into the charges and found them groundless.[82] However, friction developed when McGuire began ignoring GEB requests for documents that were needed for the conduct of business, and the board resolved to consult a lawyer on its right to secure the material.[83] A few days later the general secretary failed to appear for a scheduled meeting with a board committee, pleading that he was ill. He was given until the next morning to bring the documents, and when he failed to comply the union's bonding company was notified, a legal step that was required in the event of a shortage in McGuire's accounts.[84]

With this step, the die was cast. A full-scale revolt was in progress. The general president was asked to suspend McGuire pending investigation, the board noting that "he has refused to produce cancelled cheques for monies he claims to have paid."[85] McGuire was suspended, and Frank Duffy, a GEB member from New York, was appointed temporary general secretary. All locals were notified of the change, with the following explanation: "Charges will be preferred against [the general secretary] and submitted to the local unions in due form and time. At present Bro. McGuire is at home sick in bed and has been for some time past."[86]

An audit by a board-appointed accountant revealed a shortage of $10,074. McGuire and his attorney met with the board on July 25, 1901, but no agreement was reached. "Bro. McGuire was taken ill shortly thereafter and confined to his home for a long period."[87] When Duffy assumed office, he "found the General Office in a deplorable condition; that is, as far as the work of the General Secretary-Treasurer was concerned." Death and disability claims were unpaid, letters were lying around unanswered.[88]

The next round in the controversy came in September. Duffy prepared an issue of *The Carpenter* and then left Philadelphia for a meeting. In his absence, McGuire asked the printer to visit him at his home in nearby Camden, and induced the printer to insert an article in which McGuire stated that he was ill but would resume his office in a few months. When Duffy returned he stopped the mailing of the revised issue, and substituted the one that had originally been approved.[89]

On October 19 formal charges were sent to McGuire, to which he was asked to reply by November 4. When he failed to do so, the charges were circulated to the locals, and they were asked to vote on whether McGuire should be suspended permanently. McGuire responded by sending out his own circular, denouncing the suspension as illegal. His bitterness is reflected in the wording of the circular:

> Not alone as Founder of the United Brotherhood of Carpenters and its Chief Executive Officer for twenty years, during which time I have handled several million dollars and given rigid accounting for the same, but as an humble worker in labor's ranks for over thirty years, through all the ups and downs of the movement, I am certainly entitled to more than passing consideration.
>
> I have known persecution in many forms for my humble part in taking up the cause of the persecuted and wronged. Still it has been left for the later years of my life, now to suffer at the hands of some men a persecution all the more galling because it is calculated to injure my character and to harm the labor movement. To have my name heralded far and wide in the newspapers as a defaulter, to be dragged into court as a criminal, to be classed as a fugitive from justice, is a persecution undeserved.[90]

The board decided to attempt to recover the shortage fom the bonding company, which insisted that McGuire would have to be formally charged and arrested before it could make restitution. McGuire wanted the matter settled in court, and it was agreed that there should be no publicity. McGuire promised on three different occasions to appear at a magistrate's court but failed to do so. The day a warrant for his arrest was sworn out, he appeared at the office of the Brotherhood and asked for more time to allow his accountant to go over the books; this was agreed to, with the understanding that he would return the next day. But once more he failed to come, and on December 9, 1901, he was indicted by a grand jury for embezzlement of union funds.[91]

A referendum on McGuire's permanent suspension was tabulated in January, with 14,347 members voting for suspension and 12,702 against.[92] Since a majority of two-thirds was required to carry the resolution, McGuire's status remained the same; he was on temporary suspension. McGuire demanded reinstatement but was refused on the ground that his bond had been canceled. This placed the board in a dilemma; it had been unable to remove him from office, but it had no desire to drag the case through the courts, if only because of the harm it would cause to the union.

Finally, at the end of April, McGuire's attorney indicated that he was prepared to settle out of court. McGuire agreed to renounce all claims against the Brotherhood for salary, to resign his post, and to pay the union $2,000 in compromise of its claim against him. The board accepted with alacrity; as Huber stated later:

> Was it not better to accept what restitution Bro. McGuire was able to make? Was it not better than to have this organization, as well as the labor movement in general, stigmatized for years to come on account of the downfall of one whom we have held in the highest esteem . . . The further prosecution of the case of Bro. McGuire would be a strong weapon in the hands of those antagonistic to Union Labor. It would hold Union Labor up to ridicule, and show that even the greatest leader of them all was not to be relied upon.[93]

While negotiations for the settlement were going on, Samuel Gompers sent Frank Morrison, who was secretary-treasurer of the AFL, to Philadelphia to see what could be done to help his old comrade in arms. (McGuire had refused renomination as first vice-president of the AFL in 1900 because of ill health.) Morrison reported back to Gompers:

> Looking at the matter myself, and from what I know, I was convinced that a trial of P.J. at this time meant a conviction. This can be best explained when Lambert, P.J.'s closest friend, and one of his counsels, states that if there were a thousand lawyers in Philadelphia and the opinion of the thousand was asked, and 999 said that they believed that they could get an acquittal upon the evidence that would be adduced, he would say in the face of all that he believed that there would be no possibility of an acquittal. While he would believe that P.J. was in no way intentionally dishonest, and maybe morally innocent, nevertheless the books showed a shortage.

Morrison reported that the lawyer who was handling the case for the Brotherhood was also the counsel for the Contractor's Association of Philadelphia, and that he had been instrumental in helping elect the district attorney. "It means that the whole city government, and the influence of the contractors of the city of Philadelphia, is used against McGuire, and you know that in the city of Philadelphia the politicians, if they want to convict a man will do it, whether he is guilty or not."[94] Morrison concurred in the view that the matter should be settled out of court.

The settlement had to be approved by the next Brotherhood convention, and McGuire indicated in a letter to Gompers that he might fight for reinstatement so that his name could be cleared. But the situation he found himself in was made clear in the following request to Gompers: "Meanwhile I would like to have something to do to pay expenses and keep family. I am 'broke', but do the best I can. The $50 you sent me came very timely and it was too much to expect of you and don't wish you to send me any more . . . Could you arrange to send me out for the A.F. of L. to a few big cities, can I be of any service in the anthracite coal fields?"[95]

The convention was an anticlimax. For three days it acted as a trial court to determine whether the GEB had acted properly and whether McGuire's resignation should be accepted. During the debate McGuire was absent, sick at home. The debate, which was printed verbatim as an appendix to the convention proceedings, was extremely bitter. On behalf of the GEB, A. C. Cattermull accused Fred C. Walz, who was also a GEB member, of having started the whole affair by refusing to sign the 1900 audit, resulting in the invalidation of McGuire's bond. When Walz charged that Cattermull had threatened him with retribution if he refused to sign, Cattermull said Walz was lying. He continued:

From the day the [1900] New York Convention met I made an enemy and so did another poor unfortunate, and that man has been an implacable, bitter enemy from that time to this . . . If you will take my advice, you will not stop at the receipt of the Finance Committee's report. You will not let it rest there. There is a great deal more yet to be told of the most contemptible methods ever resorted to by mankind to gall another, to put him in his grave.[96]

Cattermull claimed he had been opposed to McGuire's indictment, and urged instead that he be given a six-month vacation to recover his health. "I wanted to save his name from being dragged through the mud and mire of newspaper discussion; but no, they willed it otherwise. They wanted publicity, the whole world to know it . . . I shall always have in mind the feeling that he was not given the opportunity to show—that should have been given him for clearance of his name of these charges."[97]

McGuire appeared before the convention on the last day of the debate. According to Christie, "He was afflicted with rheumatism, dropsy, and gastric catarrh, he could no longer stand erect."[98] He denied the charges, but made no attempt to explain the discrepancies. The Finance Committee then proposed that the $1,000 payment already paid by McGuire as part settlement be accepted in full payment of the entire debt, and that the preceding course of events, including McGuire's resignation, be approved. During the vote, "there was a great deal of confusion, and order was maintained with difficulty." The motion was adopted by a margin of 198 to 137. Even after an exhaustive discussion of all the details, a substantial minority of the delegates either refused to believe the allegations that had been raised against McGuire, or voted with him out of sheer loyalty.

Were the basic charges against McGuire true? Christie, a strong defender of McGuire, concedes that "it is likely, but not proven, that McGuire used some of the union's funds to defray the cost of his illness." He then adds: "The circumstances under which he did so, however, could constitute a crime, or even a reprehensible act, only to the most caviling of critics."[99] That would seem to be a fair judgment. The idea that McGuire deliberately appropriated union funds for his own personal living expenses is so at variance with his whole career, with his complete dedication to the Brotherhood, and with the great personal sacrifices that he made on behalf of the organization, that it is simply not credible.

McGuire suffered the ignominy of rejection by the union he had founded and nurtured because he had attempted to hold on to power too long. The men who brought about his downfall may well have been personally ambitious, but they were correct in their judgment that the Brotherhood had become too large and complex an organization to be run by one man. The transfer of power is never an easy matter in any organization, and it can only be managed gracefully when by custom or constitution there is not much room for maneuver. This was unfortunately not the case with the United Brotherhood in 1900. Although McGuire's ability to function had been seriously impaired by alcoholism and its attendant ailments, there appeared to be no way of removing him from office in a manner that everyone concerned would have vastly preferred. He could not be voted out of office as long as the carpenters in the cities and towns of the country continued to revere him as the founder.

To a considerable extent, the conflict between McGuire and the GEB arose out of the McGuire's unwillingness to share authority with the group of professional officials that was formed as the Brotherhood membership expanded. There was some feeling that he was not sufficiently tenacious in upholding the Brotherhood's jurisdictional claims. Basically, however, McGuire stood in the way of younger men who wanted a larger role in the formulation and administration of the union's policies, rather than having any fundamental disagreements with the directions in which the organization was going.

Christie argued that "for all his trade union success, McGuire was not at heart a trade unionist. The coming day he saw dawning was not a trade union, but a Socialist day."[100] This view is contradicted by the success of the Carpenters under McGuire's leadership. McGuire was one of the great trade unionists of nineteenth-century America, when some degree of utopianism was necessary to sustain efforts to organize. He was always careful not to inject his personal political views into union affairs, and although more conservative officials may have been irritated by his occasional philosophical forays, they could not point to any deficiencies when it came to organizing a "pure and simple" trade union. He was as fully aware as was Gompers of the limitations imposed by the contemporary economic and political environment of the United States.

Peter J. McGuire died on February 18, 1906, at the age of 54, in want and poverty. The executive council of the American Federation of Labor marked his passing with a laudatory resolution that constitutes a fitting epitaph for this remarkable man:

> His career in the trade union began at a time when it was almost considered a disgrace to be a participant in the labor movement. He was one of the Pioneers who blazed the way on this North American Continent for the great army of organized labor that is now marching towards the accomplishment of the object for which the trade union was founded. No body of men in the trade union were closer to P. J. McGuire or knew him better than the members of this Council, and we desire to give expression to the fact that we found him at all times an earnest, efficient and valiant co-worker in the cause of labor. His ability was very great; certainly none in the movement were more able, and few were his equals.
>
> The particular monuments that he has left behind him are the Brotherhood of Carpenters and Joiners, and the American Federation of Labor, for it can justly be said that his great force of mind and character were particularly devoted to the upbuilding of these two organizations, and that he was a very great factor in promoting their success, and when, in the future, the history of the trade union in America is written, we are sure the name of P. J. McGuire will be found very close to the head of the list of those who made the movement a success.[101]

We do not know who drafted this resolution, but it was probably Samuel Gompers.

The Brotherhood took little official notice of McGuire's death, merely noting it briefly in *The Carpenter*. The scars that remained were very deep. Duffy, who had replaced McGuire as general secretary, wrote to Gompers about the consequences of the controversy:

> When poor, old P. J. McGuire was suspended from office the labor world and the labor press went wild and said many very unpleasant, unnecessary and uncalled for things about General President Huber and myself. For a time we

were looked upon as something very bad and unfit to associate with. In fact, we got the cold shoulder on more than one occasion from men who are leaders of their respective organizations.

At the Convention of the American Federation of Labor in New Orleans and Boston, I was looked upon with holy horror, and it got to such a pitch that certain individuals would not take a drink with me, nor even a smoke. But irrespective of this I performed my duty as I saw it.[102]

On the request of several locals, the GEB sent out a circular asking for donations to help support McGuire's widow and children. Almost $4,000 was collected, mostly in small amounts.[103]

Gompers wrote to Duffy commending the Brotherhood on raising funds for McGuire's widow, and added:

Let me say that I never attached any blame to either yourself or President Huber for anything in the procedure resulting from Brother P. J. McGuire's conduct. I knew him in his boyhood, young and mature manhood, his rare gifts and quality of nature and character. I did not believe, and do not now believe that he was intentionally dishonest and any shortage in his account was brought about by the awful mental strain and responsibility devolving upon him during the years of struggle in behalf of the Brotherhood, and of the general labor movement. His mental and physical condition were such that he could not give the organization the benefit of his ability for a considerable time. I believe it to be justifiable that he should be supplanted by one competent and willing to perform the duties of the office of Secretary-Treasurer.[104]

The Brotherhood later appropriated funds for the repair of the McGuire residence, and in 1914 provided Mrs. McGuire with a small weekly allowance.[105] But it was many years before the union gave appropriate recognition to the achievements of its founder. In 1952, to mark the centennial of his birth, the GEB recommended, and the convention adopted, a resolution calling upon the Brotherhood and all its subordinate bodies to celebrate the event and authorizing the erection of an appropriate monument "so that his memory and the good he accomplished for the American worker may be perpetuated for all time."[106] A stone monument was built at Arlington Cemetery in Camden, New Jersey, where McGuire is buried.

5

ROUNDING OUT THE JURISDICTION

Up to 1900 the Brotherhood had followed the British pattern of union leadership, in which a long-tenured general secretary is the chief executive officer while the presidency is an honorific position with frequent turnover. Beginning in 1900 the general president of the Brotherhood was the top operating official, with the general secretary playing a subordinate though important role, which is the common American practice. This change was in part, at least, a reaction to the dominant position that McGuire had occupied, combining in one office all executive and ministerial authority.

William Huber had attained the general presidency in 1899 when his predecessor resigned, and was first elected in his own right in 1900. Huber was born in a small New York town in 1852. As a young man he became a carpenter, and after several moves settled in Yonkers, New York, where he was a charter member of Local 726. After a brief tenure as first vice-president of the Brotherhood he became president, a post he held until 1913. He presided over the Brotherhood during a period of rapid expansion, and had a firm belief in the destiny of the organization. Never a socialist, he was what might be called a pure and simple trade unionist. Huber expressed an attitude toward the left from which the Brotherhood never deviated thereafter. Commenting in 1906 on the Industrial Workers of the World, which had been formed a year earlier and had gained some adherents in a few western locals of the Brotherhood, Huber wrote:

> Where cooperation and unity should prevail, we find division and enmity. And by whom? By a few fool dreamers who, hiding their nefarious designs under the guise of organized principles, now ordain and declare that trades unions, as presently formed, are ages behind the times, and boast that our unions must succumb to the inevitable and be disrupted, divided and torn apart . . . I certainly, as General President of this U.B., have at all times fought these mushroom growths, here today and gone tomorrow, labor heresies, and will continue to do so as long as I am connected with the labor movement . . . we decline to be dominated by any fanatical heretics and refuse to assume the burdens of political intrigues . . . There is no room in this country for such an organization to exist, and our local unions are only being contaminated by even permitting one of their disciples to pollute the floor of their hall and the air of their meeting room with his foul harangue of sedition.[1]

McGuire's successor as general secretary was Frank Duffy, who was destined to play a major role in the organization for many years. Born in Ireland in 1861, he moved to New York with his parents when he was two years old. He came into the Brotherhood via membership in the United Order of American Carpenters and Joiners. He worked his way up the union ladder, becoming president of the Bronx district council, business agent, and general organizer for the Brotherhood. In 1900 he was elected to the GEB. He was not overly modest. He assessed his own personal attributes as follows: "In his youth, he received a good public school education, and afterwards took a course in the higher branches. He is a student of economics, an able speaker and a fluent writer, and, like his predecessor, he addressed thousands of meetings in all parts of the country, and is always in great demand on public occasions."[2]

He quoted a close friend as saying of him that "his ability as a defensive speaker is above the average, and his attacking power is masterly and superb. He has the happy faculty of correlating his facts in such a form as to denote care and preparation, carrying with it conviction when delivered from the platform."[3] Duffy was very proud of his intellectual abilities. When the *American Federationist* carried a speech on industrial education, one of his main interests, that he had given at Columbia University, they made a few editorial changes; Duffy, in a letter to Gompers, objected vigorously to the blue penciling:

> I may not be much of a scholar but I assure you I utilized my spare time in years gone by and do so now in studying history, poetry, mathematics, and at least some of the sciences. I may not be a Shakespearean scholar, but when I quote from Shakespeare I make it a point the quotations used are correct.
>
> In my address at Teachers College I spoke of the make-up and character of men. I said that Mark Anthony, the great Roman orator, standing over the dead body of Brutus, said, "He was the greatest Roman of them all, he was a man," but to my surprise and chagrin, I find the word Caeser substituted for the word Brutus . . . Now Brother Gompers, Mark Anthony gave a great oration over the dead body of Caeser to which I did not refer or allude to at all. I admire every word of that oration. I think it was great, but I think too that the few words he used over the dead body of Brutus still greater . . . In the future before changing or allowing anyone to change matter for publication sent you, be sure that you are absolutely correct before said changes are made. Mind you Brother Gompers don't think for a moment I am scolding or fault finding for I am not. But I cannot help telling you that I am greatly disappointed with the change made.[4]

We do not have Gompers' reply, but he undoubtedly exercised great care thereafter before altering any of Duffy's writings.

Of his appointment to the post of general secretary, Duffy wrote that when he heard that he was under consideration after the suspension of McGuire, he visited Huber and asked that his name be withdrawn for several reasons: because he did not want to move from New York, where he had a good job at the trade above the established rate of wages, and because he and McGuire were old friends.[5] But when the GEB approved his nomination the next day, he accepted with alacrity the position that he was to occupy for forty-seven years.

CONSOLIDATION OF THE CARPENTRY TRADE

During the first decade of the twentieth century, a great deal of effort was devoted on behalf of the Brotherhood to the task of absorbing independent unions that were operating in areas considered part of the craft of the carpenter. This kind of activity is to be distinguished from the more common jurisdictional dispute in which the issue is the marginal one of whether a particular type of work belongs to one well-established craft or another — for example, the carpenter or the iron worker. But in cases where the entire work jurisdiction of another union was deemed to be within the carpenters' orbit, that union was branded as dual and its very right to continue in existence challenged.

This section will be concerned mainly with the dual union controversies, almost all of which were settled to the satisfaction of the Brotherhood and were in considerable measure responsible for its continued membership growth. Looking backward, it was almost inevitable that clashes would occur once the Brotherhood had become established as a major and permanent fixture on the American labor scene. When it was a young and struggling organization, the Brotherhood had its hands full attracting members in its traditional stronghold — building construction. But as time went on, technological change tended to blur the line between work at the site and work at the factory, and there was deepening concern about protecting the jobs of the traditional craftsmen, particularly when they themselves moved from one industry to another because of seasonal and cyclical factors.

The Carpenters' Union has often been described as aggressive in pursuit of its jurisdictional claims. This is not an unfair characterization, but it applies equally to virtually all successful trade unions, in Europe as well as in the United States. The absorption of small unions by larger and stronger ones has been a universal feature of union growth, although the circumstances behind union mergers have differed from case to case. In Great Britain, for example, the great general unions of semiskilled workers took over local bodies in industry after industry when the skilled crafts insisted on preserving their exclusivity. If the American carpenters and other craft unions had refrained from exercising jurisdiction over semiskilled woodworkers, and other skilled crafts had followed suit, it is not inconceivable that American trade union structure would have

moved in the same direction as that of Great Britain, which is characterized by a multiplicity of small craft organizations.

But they did not; beginning about 1900, the Brotherhood, for one,
began moving toward a structure that was an amalgam of craft and industrial unionism, although this was not clearly recognized until the
great upheaval of the 1930s. Up to 1900, although the Brotherhood had
admitted some factory workers into membership, it continued to tolerate
the coexistence of independent unions of semiskilled factory hands.
Thereafter, unification of all the woodworking trades, regardless of level
of skill, became its unvarying goal.

The Amalgamated Wood Workers' International Union

As we have already seen, the Wood Workers were chartered by the AFL
in 1890,[6] after the Brotherhood had first objected and then withdrawn its
objection. The agreements of 1894 and 1897 served temporarily to
stabilize relationships between the two organizations, though there were
many skirmishes at the local level. The abrogation by the Brotherhood of
all existing agreements with other woodworking unions led to serious
conflict.

This controversy first came before an AFL convention in 1901, when
the Wood Workers asked that they be given full jurisdiction over all factory woodworkers, pursuant to their agreements with the Brotherhood.
The convention found that the Brotherhood had violated the agreements, but refused to take any further action until the two organizations
made further efforts to settle their differences.[7] This was the convention,
incidentally, at which the famous Scranton Declaration was adopted,
affirming that organization along "trade lines" was optimal for American
labor, and making an exception only where, "owing to the isolation of
some few industries from thickly populated centers where the overwhelming number follow one branch thereof, and owing to the fact that
in some industries comparatively few workers are engaged over whom
separate organizations claim jurisdiction, we believe that jurisdiction in
such industry by the paramount organization would yield the best results
to the workers therein." The purpose of this exception was to accom-
modate the claims of the United Mine Workers, who had already staked
out an industrial jurisdiction.

Negotiations produced no agreement, and in April 1902 the AFL executive council stated flatly that the Wood Workers had exclusive
jurisdiction over cabinet makers and machine and factory workers. This
served only to exacerbate the situation; at the next AFL convention, the
Brotherhood asked that the Wood Workers' charter be revoked for
"strikebreaking" activities, but it was finally agreed that a committee of

eleven would be appointed to adjudicate the matter, five from each side and the eleventh to be selected jointly.[8] Some idea of the attitude of the delegates toward the competing organizations may be garnered from the fact that when Duffy ran for vice-president against Thomas Kidd, who was the incumbent fifth vice-president of the AFL and president of the Wood Workers, Duffy was defeated by a margin of almost 5 to 1.

The first meeting of the joint committee was held in February 1903, and P. J. Downey, president of what was then the Amalgamated Sheet Metal Workers' Association, was chosen as umpire. He awarded the Wood Workers jurisdiction over employees of planing mills and furniture and interior finish factories, but gave the Brotherhood control over all work on new and old buildings and installation of store and office fixtures. Downey found that the Brotherhood had violated its earlier agreements, and had not claimed jurisdiction over mill men until 1898. The Brotherhood members of the joint committee refused to accept the award, arguing that since Downey had been involved in the dispute as a member of the AFL convention grievance committee that had dealt with the matter earlier, he was not in fact a neutral.[9] They claimed that he had misread the evidence in not appreciating that the agreements with the Wood Workers had never been ratified by the Brotherhood convention.[10] Charges of "scabbing tactics" against the Wood Workers were renewed.

The Brotherhood's concern was undoubtedly heightened by the fact that the membership of the Amalgamated was growing rapidly at the time, from 12,300 on January 1, 1902, to 31,230 two years later. The number of its locals had risen from 68 in 1899 to 242 in 1903, distributed among 30 states and 171 cities. As long as the Amalgamated was concentrated in a few industrial centers—over a third of its membership in 1898 was located in Illinois—the Brotherhood could ignore it. When the Amalgamated began to expand and threatened to take over all factory woodworking, the specter of a potentially powerful rival evoked a more aggressive policy on the part of the Carpenters.[11]

The Wood Workers appealed to the AFL executive council, which decided that unless the Brotherhood lived up to the award, Gompers was to notify all AFL unions that the Brotherhood was in violation of its agreement, and that the Wood Workers should be given their support.[12] When this was communicated to Duffy, he replied that "the [AFL] convention was informed in very forcible and plain language . . . that the laws of the United Brotherhood of Carpenters and Joiners of America were made by the highest authority in our organization—the referendum vote of our members—and that the delegates representing the United Brotherhood could not alter or amend them in any shape or form."[13]

The Wood Workers brought the matter once again to the AFL convention in 1903, and demanded that the Brotherhood charter be revoked.

The convention grievance committee upheld the Wood Workers, and after a long debate, its report was confirmed by a vote of 8,243 to 3,315. In reporting back to their constituents, the Brotherhood delegates listed the names of the unions that had voted against the Carpenters, and warned that they might not receive any future support.[14] AFL or no AFL, the Brotherhood was not going to yield. Its delegates to the AFL convention recommended that organizers be put in the field immediately to charter new locals of mill men, and recommended "that it would be advisable not to voluntarily withdraw from the AF of L but rather to accept suspension if inflicted, thus forcing the opposition to become aggressive against the UB."[15]

At the behest of the AFL executive council, Gompers wrote asking that a conference between the two organizations be held during the 1904 Brotherhood convention, but this suggestion was rejected. Instead, a convention committee report stated: "We hope and trust the day is not far distant when [the Wood Workers] may see the error of their ways and realize the necessity of transferring their membership and unfurling their flag under the protective arm and the banner of the U.B."[16] The Wood Workers refused to accept this advice. Instead, they appeared once more before the AFL convention, which resolved that "the Executive Council require the organizations to comply with the decision of the Boston Convention or stand suspended until decision is complied with."[17] When a Brotherhood delegate was asked why the Carpenters had not requested an appearance before the convention grievance committee, to which the complaint had been referred, the reply was that this would have been useless.

At the next meeting of the AFL executive council, it was decided that breweries in Milwaukee should be notified that the Wood Workers had jurisdiction over mill men and cabinet makers, and the Brotherhood was warned that unless it complied, suspension would be considered. When this produced no results, the executive council resolved "that this question is of such importance to the labor movement that the whole matter be referred to the next convention of the American Federation of Labor."[18] But the next convention merely recommended that further meetings should be held between the two organizations under the chairmanship of Gompers. When the Carpenters made it clear that they would accept no solution short of merger of the two unions, although they were willing to allow a period of grace, a tripartite continuation committee was established to carry on further negotiations.

A plan of amalgamation was agreed upon at the 1906 AFL convention, and it looked as though the controversy was finally over. Under the terms of this agreement, which was subject to ratification by both parties, there was to be a complete merger of the two unions by November 1, 1908. Hostilities were to cease immediately, and there was to be mutual assistance in trade disputes. The executive council of the AFL would supervise the execution of the agreement. Duffy told his membership: "By this agreement we lose nothing, but after a given time gain everything we have been looking for. Instead of continuing a war of extermination on the Woodworkers, detrimental to the employers of labor, and involving other trades, we bring about a final settlement in a quiet way."[19]

The celebration proved premature. By a margin of 2 to 1, the Wood Workers rejected the amalgamation plan after their officers had failed to recommend its adoption. Hostilities broke out once more. The general secretary of the Wood Workers complained to Gompers that the Carpenters were refusing to handle wood made in Wood Worker shops, with the result that nonunion material was replacing it. He demanded that the Downey award be enforced.[20] Building contractors wrote in to complain that they were caught in the middle, and also urged Gompers to act. The Brotherhood contended that the rejected agreement had nullified the Downey award, and Huber informed Gompers that:

> no one regrets, more than your humble servant, that we were forced to take measures against the A.W.W. to protect our members as well as those who voluntarily desired to become such; but as they were the aggressors it was certainly up to us to defend ourselves . . . I regret this controversy for the bad effect it has on the labor movement in general. *But we will protect our own.*[21]

Duffy wrote Gompers that the Carpenters' membership had ratified the merger plan, and accused the Wood Worker leadership of creating "ill feeling, venom, bad blood, and misunderstandings between the two organizations, resorted to those despicable tactics in order to have the agreement voted down."[22] The tables were now turned, and the Wood Workers were placed on the defensive. There was still some sympathy for them within the AFL, which, at its next convention, defeated a resolution that would have required both organizations to recommend amalgamation to their members.[23] It is clear in retrospect, however, that the Wood Workers made a fatal mistake in agreeing to the merger and then repudiating it. All the Brotherhood had to do was stand firm for the merger plan, and its superior numbers and strength were bound to prevail. The 1909 AFL convention recommended that Gompers and Brotherhood officials attend the next Wood Worker Convention to convince them of the desirability of consummating a merger.[24]

During these years the membership of the Wood Workers was being slowly whittled away by the Brotherhood, which used its control of the building site effectively. In Chicago a Carpenter boycott of Wood Worker trim led to the virtual takeover of the latter's membership by Brotherhood locals. Huber reported in 1908 that apart from one or two mills, all the mill men in Chicago were under Carpenter contract. The Brunswicke-Balke-Collender Company, the largest trim manufacturer in New York City, switched from the Wood Workers to the Brotherhood after being put on the unfair list by the latter, and by 1910 the Wood Workers had been virtually eliminated as a factor in New York.[25] As a result of the warfare between the two organizations, total membership in the Wood Workers fell from 30,000 in 1903 to 4,000 in 1909.[26]

Although Gompers attended the Wood Worker convention that was held in March 1909, the Brotherhood failed to send representatives. Several months earlier the AFL executive council had recognized the Brotherhood label for factory-made building trim, in direct derogation of the Wood Worker label. The response of the Wood Workers' Union was to redefine its jurisdiction in such broad terms as to bring it into conflict not only with the Carpenters but with the Piano and Organ Workers, the Wood Carvers, and the Carriage and Wagon Workers, all independent unions at the time.[27] This only added to their difficulties; at the next AFL convention, the adjustment committee brought in another suggested amalgamation agreement, and in the debate the following exchange occurred:

> Delegate McKee: I want to have information before I vote. Does the report provide for compulsory assimilation?
> Delegate Lewis, secretary of the committee: The committee has come to the conclusion that where certain men in this labor movement do not see the benefit of concentration, we believe it is time to compel them to get into line for their own good.[28]

On a roll call vote, the resolution carried, 11,203 to 1,707. Support for the Wood Workers had all but vanished. They continued to fight a rearguard action, only to have the AFL executive council recommend that the Wood Workers' charter be revoked unless they agreed to merge. The 1910 AFL convention was not yet willing to take the final step, and sent the matter back to the executive council for further study.[29] But the next convention directed Gompers to revoke the Wood Workers' charter by July 1, 1912, if it failed to amalgamate. The president of the Wood Workers, D. D. Mulcahy, tried the next year to convince the delegates to delay matters:

> You cannot amalgamate successfully two sets of men whose feelings are so far apart. I have had enough experience in the factories to know as much about this as anyone else. They have already driven over twenty thousand men out of the labor movement. They have got some men, it is true, but in the vast number of industries today there is very little effort toward organizing those men, and there is less success attained than ever before.[30]

The long struggle came to an end on April 1, 1912, when a merger between the two organizations was consummated. All beneficial members of the Wood Workers were given immediate beneficial standing in the Brotherhood. Wood Worker locals were given the option of maintaining a separate status, subject to the approval of the district councils concerned. The records of the Wood Workers were shipped from Chicago to the Brotherhood headquarters in Indianapolis, and, as Duffy recorded, "thus ended the famous Carpenters–Wood Workers' jurisdictional fight that caused so much annoyance and trouble in the labor world for many years."[31]

Deibler, in his history of the Wood Workers, asserted that the Carpenters had expended large sums of money in this controversy and wasted an enormous amount of energy that might better have gone into improving conditions for its members. "The carpenters must be condemned for arbitrarily taking a stand, and fighting for this until the bitter end. There is no hard and fast line of jurisdiction in the industry to justify the tenacity with which they have insisted on controlling the entire field."[32] He rejected the argument that the building carpenter wage scale was endangered by the lower factory scale, and concluded that the conflict would "always remain an indication of mistaken policy on the part of those who have insisted on waging the fight."[33] Writing many years later, Christie echoed these views, and maintained that as a result of the fight 13,000 workers dropped out of the labor movement. "There can be little doubt that the Brotherhood took in only those Wood Workers it wanted, allowing the rest to drift."[34]

This was not at all the perception of the men who were leading the Brotherhood. They were convinced that complete organization of the mill men was essential to maintenance of building carpenters' standards. First vice-president T. M. Guerin observed as early as 1904:

> We should insist upon our outside men paying more attention to our brothers working in the mills and to render them more and efficient assistance. May the time soon arrive when we will better realize the importance of the mill men and of our organization. Show me a city where the mill men have the eight hours and work under fair conditions otherwise, and I will show you a city where the outside men are doing well and control the trade.[35]

Already by 1906 the Brotherhood claimed to have organized 45,000 mill men, more than twice the number belonging to the Amalgamated Wood Workers.[36] Among the concerns that led the Brotherhood to go after these men was the fear of potential competition from them on site work. In an unsigned article that appeared in *The Carpenter* attempting to explain why carpenters' wages were below those of masons, plasterers, plumbers, and iron workers, it was pointed out that these crafts required an apprenticeship and controlled entrance to the trade. On the other hand,

> . . . the carpenter trade, being one comprising many branches, an unlimited number of workers can practically engage in some one or more of them without having to go through much of an apprenticeship, and forthwith, in the public eye, they are carpenters. They are available for a certain low or rough class of work in some of the various branches, at a rate scarcely above laborers' wages.[37]

There was constant complaint from locals that the Wood Workers were unable to enforce adequate standards, and were undermining Brotherhood wages. For example, Thomas Hickey, a New York City carpenter, wrote that in 107 shops controlled by the Brotherhood, wages were $22.00 for a forty-four-hour week, whereas Wood Worker shops were working fifty to sixty hours for earnings as low as $12.00. "All any scab employer needs to do when he has a job where he knows union carpenters will hold the work up, is to send for the officials of the A.W.W. and by paying $1.00 per head for the scabs in his employ secure the label." Hickey ended up by declaring that "the inside man never could and never will gain a permanent union condition without the help of the outside man, and, on the other hand, the inside man is a constant and very often willing menace to the outside man, showing plainly the absolute need for one organization for all woodworkers, inside and out."[38]

The attitude of the Brotherhood was not based on a "personal element," as Deibler asserted; it was rather a reflection of what has been the most important goal of American trade unions: control of the job. As Selig Perlman put it:

> The ideology of the American Federation of Labor . . . was based on a consciousness of limited job opportunities, — a situation which required that the individual, both in his own interest and in that of a group to which he immediately belonged, should not be permitted to occupy any job opportunity except on the condition of observing the "common rule" laid down by his union. The safest way to assure this group control over opportunity . . . was for the union . . . to become the virtual owner and the administrator of the jobs.[39]

Huber advanced this idea clearly when he urged carpenter locals to organize inside men. He warned that "the carpenter, in order to hold what rightfully belongs to him, must control the manufacture of the material—the carpenter of today may be the mill man of tomorrow."[40] American unions could not rely, as they did in Great Britain, on the class consciousness of workers to ensure that the employer would not be able to play off one union against another, or to prevent the introduction of nonunion mem in a strike situation. Dual unionism has always been a cardinal sin in the American labor movement precisely because job control through strong unions had to be substituted for the absent working-class solidarity. McGuire, who retained his basic socialist ideology throughout his life, was not prepared to abandon his belief that in the end, a unified working class would take over the economy by peaceful means — although on the practical level, he acted as though this day were far in the future. His successors were laboring under no such illusion. John Williams, an early general president of the Brotherhood, put the matter well in a retrospective look at the organization's history:

> The worker has come to look upon the United Brotherhood as a vast business institution. Sentiment, which was a potent factor in the determination of my attitude toward it, has given way to a colder calculation of its virtues — to a more practical measuring of its value, and I must say that the contemplation from this viewpoint has caused the Brotherhood to take a firmer hold then ever on my allegiance and affection.[41]

To those who dislike the term "business unionism" and what it connotes, the United Brotherhood deserved condemnation for its long campaign against a fellow labor organization. Christie wrote of a later period in Brotherhood history that "Hutcheson and his lieutenants carefully divested the union of all social overtones and made it a straightforward, cold business proposition."[42] This reveals a lack of understanding not only of the Brotherhood, but of American labor history as well. From their very beginning — and the Carpenters were no exception — American unions were concerned primarily with the economic well-being *of their members*. Broader social concerns were subordinate to this central driving force, and indeed, were conceived of mainly as a means to the same end.

Operating in a hostile environment, with every employer a potential user of readily available strikebreakers, the successful unions concentrated on immediate wage and hour problems. Indeed, they were proud to be labeled business unions. Once the political and economic constraints had eased, as they did in the 1930s, American trade unions could become more "social" in their outlook. But in the early part of the twentieth century, neither the officers of the Brotherhood nor their fellows in the AFL could afford to deviate from the pursuit of fairly narrow economic goals, primarily through the means of collective bargaining, without risking the alienation of their members and the demise of their organizations. They were prepared to use legislation to the same end — for example, their advocacy of restricted immigration and mechanics' lien laws — but in the final analysis, it boiled down to the economic situation of their members.

The view that the disappearance of the Wood Workers meant a loss of membership to the labor movement as a whole is not justified. Christie's figure of 13,000 workers who "were lost in the shuffle" was calculated on the basis of the number of men in locals that catered to mill men exclusively, and does not take into consideration the mill men who belonged to mixed locals, as Christie himself notes.[43] Huber reported in 1912 that 1,100 mills, cabinet shops, and job shops were under Brotherhood contract.[44] There is simply no evidence that any substantial number of men dropped out of the labor movement because of the controversy, or that the Brotherhood failed to pursue the organization of mill hands with vigor.

The Amalgamated Society of Carpenters and Joiners

As already noted, the Amalgamated Society of Carpenters and Joiners began as a branch of the British trade union of the same name. Although it never had a membership of more than 10,000, its pension and unemployment benefits, as well as its effectiveness in finding work for traveling members, earned it the loyalty of those who joined it. Its membership was concentrated in New York, Chicago, and Philadelphia. The Society was chartered by the AFL without any opposition from the Brotherhood when the two unions were still in the early days of their existence.

Prior to the 1898 Brotherhood convention the two organizations enjoyed fairly good relationships, although there was some friction at the local level, particularly in Chicago, where the Society teamed up with the Knights of Labor in opposition to the Brotherhood district council. An agreement in 1895 for joint recognition of working cards did not last long; it was repudiated by the 1898 Brotherhood convention. In October 1901 a committee of the Society met with the Brotherhood GEB to protest local attempts to force their members into the Brotherhood, but the GEB took the position that it could not compel Brotherhood locals to recognize the cards of any other organization.[45] The 1902 Carpenter convention, taking time off from the McGuire affair, resolved that the Amalgamated Society had been acting contrary to the principles of trade unionism, and instructed its delegates to the next AFL convention to induce that body to secure reasonable concessions from the Society under threat of withholding its per capita tax.[46] The AFL convention recommended that a committee of five from each side meet, choose an impartial chairman, and attempt to reach agreement. The AFL executive council had refused to revoke the Society's charter on the ground that the purpose of the AFL "should be to band trade unionists together, and should not be applied to cancel recognition of, or to dismember a body affiliated thereto, which can not truthfully be charged with violating the principles of our movement."[47]

It will be recalled that the executive council was embroiled in the Carpenter–Wood Worker dispute at the same time, and was not inclined to be sympathetic to the Brotherhood. However, when Duffy complained that the council was hounding the Brotherhood, it instructed Gompers to direct the Society to hold an early meeting, with the warning that "any procrastination would be regarded as an evidence of bad faith."[48] This brought the Society to a conference, which was held in New York. The Brotherhood offered its by now standard proposition—merger. The Society rejected this solution on the ground that it would mean loss of the benefits that its members had earned over the years. It submitted a counterproposal to the effect that its locals would join the Brotherhood district councils, to which they would pay a monthly per capita of 1 cent, in return for which Society working cards would be recognized. The Brotherhood rejected this as being a working agreement rather than a merger.[49]

The difficulties were exacerbated by a conflict in New York City. All jobs employing Society men were struck by the Brotherhood, forcing the discharge of the former. The local Building Trades Council demanded that the Society men be reinstated, and when the Brotherhood refused it was suspended from the council, which in turn declared a strike against Carpenter jobs. This move merely resulted in the dissolution of the Building Trades Council. A new United Board of Building Trades was established, with both the Brotherhood and the Society as members. This body in turn refused to support the Brotherhood in a strike against the Amalgamated Society, and the former withdrew.[50] Duffy claimed that the Society had 600 members to 6,750 for the Brotherhood, but wanted half of all the available jobs. The Brotherhood strike, which lasted seven weeks, cost $800,000 in lost wages. The GEB provided the New York locals with $47,000 in strike benefits, while the locals used $25,000 of their own.[51]

The AFL executive council induced the parties to cease hostilities in New York and temporarily honor one another's working cards. At a conference held in August, Adolph Strasser, former president of the Cigar Makers' Union and a leading figure in the formation of the AFL, was chosen as impartial umpire. Strasser submitted a complicated plan of amalgamation that provided for four classes of membership, according to per capita payments, and a schedule of benefits based on the practices of the Society: equalization of funds, sick benefits, unemployment payments, and tool benefits, a diet that was too rich for the Brotherhood, particularly because it would have meant an increase in dues. Gompers wrote to the general secretaries of the American and British branches of the Society, strongly urging that the plan be accepted. One of the conditions of the agreement was that each side deposit a bond of $25,000 to ensure faithful performance of the agreement, which the Society considered excessive. Gompers wrote his British colleagues: "It is needless to call attention not only to the strife, discord and strikes, but to the enormous amounts of money lost in wages as in union funds by reason of the controversies between the two organizations, and we dread to look forward to what will result if the efforts at unity should fail."[52]

Gompers convinced the Society but not the Carpenters, who rejected the agreement in a referendum vote. More conferences were held, and each side submitted its own plan of amalgamation, to be voted on by the respective memberships. Both were defeated, leaving matters as they had been before the Strasser award. The stumbling block was the insistence by the Society that a benefit structure superior to that of the Brotherhood be adopted. A particular bone of contention was the unemployment benefit, of which Huber said, "We did not believe [it] would work successfully in this country."[53]

The Brotherhood attempted to put pressure on the Amalgamated Society by having it expelled from local building trades councils. The AFL executive council directed on several occasions that they be reinstated.[54] The Brotherhood, for its part, accused the Society of providing strikebreakers in a dispute in Washington, D.C., in 1910, and demanded that the AFL revoke its charter.[55] In St. Louis, when 3,000 Carpenters struck, the contractors secured the union labor that the owners of the buildings wanted by signing an agreement with the Society.

Inevitably, the tide of battle swung against the Society. Despite the support of its members, it was not able to stand against the much larger Brotherhood. At the 1911 AFL convention, the adjustment committee recommended that Gompers renew efforts to merge the two organizations, and that if no agreement could be reached by July 1, 1912, the Society charter be revoked. A long debate ensued, with the committee chairman stating: "I want to say right in the beginning that our committee is unanimously of the opinion that there is room for but one organization of one trade in America." Andrew Furuseth, the leader of the West Coast seamen and one of the well-known American trade unionists at the time, made a strong plea against the use of compulsion: "Why, you are about to deny the fundamental principle upon which the labor movement is founded . . . I cannot agree that men are property to be given by the one and taken by the other . . . I cannot agree that this convention or any other power on earth has the right to say to me or to anybody else, 'you shall join this or you shall join that.'"[56]

This argument proved to be of no avail. The committee report was adopted by an overwhelming vote, and the Society was under a mandate to merge. The Brotherhood submitted a new plan of amalgamation, but it was turned down by the Society, which refused to give up its benefit system. On August 1, 1912, the charter of the Amalgamated Society was revoked by Gompers, who characterized the Brotherhood plan as "broad and generous."[57] All unions were notified that the Brotherhood had sole jurisdiction over carpenters. Thus cast adrift, the Society had little choice but to consent to merger.

Having prevailed, the Carpenters showed considerable restraint in the terms offered. Society locals would remain intact; their beneficial system was to be retained and controlled by them, subject to amendment only by those members entitled to the benefits; they were to be exempt from general assessments levied by the Brotherhood; and all Society members were to be entitled to work under Brotherhood contracts without payment of initiation fees.[58] In fact, the Society was to remain a quasi-independent body.

The plan was accepted by the membership of both organizations by votes of roughly 2 to 1. General President Kirby told the 1914 Brotherhood convention that since the Amalgamated Society had a benefit system second to none, "I felt it would be unjust to ask them to surrender the benefits that they had paid for so long . . . By leaving the Amalgamated Society its benefit system, and giving control of the working conditions to the Brotherhood, no hardship was worked upon anyone."[59]

Although the merger was put into effect, the separatist tendencies of the former Society locals led to internal discord for a decade. Some of them were slow in paying their per capita taxes; others resisted working under a Brotherhood charter. To avoid another break, some years later in 1922 General President Hutcheson went to England to advise the former parent organization that a full integration of the Society locals into the Brotherhood was necessary. The British Amalgamated Society, with which the American branch still retained a loose association, decided to turn it over completely to the Brotherhood. When an independence movement manifested itself in 1923, the Brotherhood declared that the 1913 plan had been violated by Society members and was null and void. Members and locals of the former Society were given the option of taking full membership in the Brotherhood either individually or as local unions.[60] Hutcheson reported to the 1924 Brotherhood convention that most Amalgamated members had taken advantage of the offer.

The small Amalgamated Society of Carpenters and Joiners had shown tremendous tenacity in maintaining its identity over a period of half a century, attesting to the strength of benefit unionism at a time when there was no social security legislation. In the end, however, it could not prevail against the determination of the Brotherhood to eliminate dual unionism. As the central administration of the Brotherhood became stronger, it became less willing to tolerate a semi-independent body within the organization. The complete absorption of the Amalgamated Society was another milestone in the drive of the Brotherhood for complete control over all those employed in the woodworking trades.

The United Order of Boxmakers and Sawyers of America

The United Order of Boxmakers and Sawyers, formed in 1898 on the initiative of a local union that had existed in Chicago since 1873, claimed jurisdiction over workers employed in the manufacture of wooden boxes. In 1906 it applied for an AFL charter, but this was denied when the Amalgamated Wood Workers objected. When the latter merged with the Carpenters, the Boxmakers tried again, but were told that the jurisdiction over their industry had passed to the Brotherhood.

A few years later the Brotherhood began putting pressure on the Boxmakers by having the AFL declare that the only box label to be recognized was that of the Carpenters. This action finally led to an agreement in October 1918 whereby the Boxmaker locals were to be chartered by the Brotherhood. There was one obstacle, however, and this had to be covered by the following stipulation:

> That in view of the fact that at the present time there is no provision in the laws of the United Brotherhood of Carpenters and Joiners which accepts women to membership, it is understood and agreed that nothing in this agreement and proposed amalgamation prohibits Box Makers and Sawyers Union, chartered as such by the United Brotherhood of Carpenters and Joiners, from continuing women members as *associate* members of their unions. [italics added]

The women who were thus admitted to the Brotherhood paid no per capita tax to the national union and only such dues as their locals required. Nor were they eligible for benefits.[61] This was the first time women had been admitted to the Brotherhood on any basis; thus the acquisition of the Boxworkers carried with it an unanticipated result.

The International Union of Shipwrights, Joiners, and Caulkers

The Shipwrights union was formed in 1902. It applied for an AFL charter, claiming jurisdiction over "every class of floating structure from canoes to Atlantic liners, wet dock, dry dock and graving dock construction, and everything pertaining to ships and ship building." To the Carpenters' claim of jurisdiction, the Shipwrights replied: "They are house carpenters and joiners. We are shipwrights, joiners and caulkers . . . Ship-carpentry and house carpentry are two distinct trades."[62] The charter was issued with the understanding that any shipwright who worked on buildings would have to join the Brotherhood.

The Brotherhood was not happy about the action of the AFL; the GEB noted that the AFL would have to enforce this decision when Brotherhood locals refused to accept it.[63] In April 1904 a delegation from the Shipwrights met with the GEB to propose an interchange of working cards between the two organizations. In reply the GEB proposed a merger on the ground that it would be unjust to deprive Brotherhood members of death and disability benefits as a result of their joining the Shipwrights.[64] By this time the Brotherhood had recovered from the traumatic events of the 1902 convention, and was firmly set in its resolve to prevent splintering of the craft along industry lines. When the Shipwrights turned down this proposal, the GEB made a constitutional determination "to claim jurisdiction over all Woodworkers in the shipbuilding industry. This interpretation to guide us in our controversy with the Shipwrights."[65]

The Brotherhood continued to issue charters to shipwright locals, but complaints to the AFL were met by the advice to adjust the difficulties amicably.[66] Other locals remained independent. As a result, the Shipwrights union languished, and by 1909 it was down to 1,611 members. It was suspended from the AFL for nonpayment of dues in 1911, and Duffy warned that the Carpenters would take a dim view of its reinstatement. No such request was ever made, and most of the locals were eventually chartered by the Brotherhood.[67]

JURISDICTIONAL DISPUTES

In addition to disputes with other unions that had organized men who were classified as carpenters, the Brotherhood began early in the twentieth century to come into conflict with craft unions that did not lay claim to the carpentry trade but did claim control of certain jobs or work operations over which the Brotherhood had also asserted jurisdiction. In the case of dual union disputes, there was only one remedy as far as the Carpenters were concerned — amalgamation. In jurisdictional disputes, however, the remedy was for the work in question to be performed by members of the Brotherhood rather than by members of other unions.

When unions were established in the nineteenth century, craft lines were either clearly delineated, or the degree of organization was so low that jurisdictional controversery was minimal. As unions grew, as craft lines became blurred, as new technology developed, and as the range of competition among firms expanded, interunion disputes became one of the major problems of the labor movement. Changes in industrial methods, the introduction of machinery, and the development of new materials were among the causes for the proliferation of disputes.

When it came to claiming or defending jurisdiction over particular jobs, the Carpenters were not reticent. Writing in 1914, Whitney, in his classic study of jurisdiction, says of them: "The United Brotherhood of Carpenters has probably had a greater variety of trade jurisdiction disputes than any other union. This has been due partly to the extension of the carpenter's trade in so many directions, and partly to the fact that the most far-reaching and rapid changes in methods and materials in the building industry have centered about the work of the carpenters."[68]

Part of the Carpenters' advantage over other crafts lay in the fact that they had locals in more cities and smaller communities than did their rivals, and could establish customary practices more widely. They also worked directly for general rather than specialty contractors, in the main, giving them greater access to the employers who could most effectively advance their claims.

Over the years there were many internal attempts to define the Brotherhood's jurisdiction, sometimes in the form of constitutional changes, sometimes in statements by union officials. These definitions often became obsolete rather quickly. To outsiders, the rate of obsolescence was proof of the aggressive nature of the Brotherhood. To those who were running the organization, it was simply a question of protecting the livelihood of carpenters in a period of rapid technological change. One of the best early statements of this view was contained in the report of General President Huber to the 1910 convention:

> The disputes generally arise over the erection of certain work which originally belonged to the carpenters, but which through the growth of the building industry has changed form to such an extent that you could not say unless you knew the class of trade which put it up, what trade the work now belonged to. The basic carpenter trade was and is one of the most general and complete trades which a man can learn. It is generally the carpenter foreman who takes care to see that the excavation stakes are properly set; who sees that the proper openings are left; who attends to the scaffolding for the painter, the electrician, the lather and plasterer. In fact, he is usually the superintendent of the job, and on his shoulders falls all the responsibility to see that the work is carried forward promptly and properly.[69]

The 1914 convention called upon the GEB to prepare a detailed trade jurisdiction statement to be incorporated in the constitution. The board found when it approached the task that "owing to the rapid change in the building industry in the last few years, the substitution of one material for another in construction work, as well as the methods of construction, a detailed statement of our claims today may need changing tomorrow or the next day. Therefore, it is a difficult matter for us to draw hard and fast lines governing our work or any particular part of it."[70] Nevertheless, the board went on to define the jurisdiction of the Brotherhood to include "the milling, manufacturing, fashioning, joining, assembling, erecting, fastening or dismantling of wood, hollow metal or fibre, and the erecting and dismantling of machinery, where the skill, knowledge and training of a carpenter are required." The following trade subdivisions were included: carpenters and joiners; ship carpenters, joiners and caulkers; shipwrights and boat builders; railroad carpenters; bridge, dock, and wharf carpenters; stair builders; floor layers; cabinet makers; bench hands; furniture workers; millwrights; car builders; box makers; and reed and rattan workers.[71]

In setting forth the details of the jurisdictional disputes involving the Carpenters during the early years of the century, it seems preferable to use a chronological approach rather than attempting to do it separately by trade. Many of the disputes were going on simultaneously, and the tactics employed by the various parties were not unrelated. The Brotherhood was operating on many fronts; interunion alliances were shifting about; AFL politics at any particular time embraced all of these disputes. One gets a better picture of Brotherhood strategy by examining these developments year by year.

The year 1902 is a good starting point. It marked the ascension to power of a new leadership that was anxious to prove itself vigilant in protecting the economic interests of the carpenters. Mention should be made first, however, of the establishment in 1897 of the National Building Trades Council, which was set up primarily to help settle jurisdictional disputes. It was composed of a number of local building trades councils, not the national unions, and at least in its initial stages, advocated the amalgamation of crafts. The national leadership of the Brotherhood was cool toward it, although district councils were permitted to join. The council was opposed by the AFL, which looked upon it as a potential dual union.

As the AFL began to charter more small craft unions, both on a national basis and as federal locals, the Brotherhood leaders saw the advantage of closer collaboration with other building unions in order to protect themselves. Some of the correspondence at the time reveals the Carpenters' concern. In 1903, for example, Duffy wrote a sharp letter of complaint to Gompers against the issuance of an AFL federal charter to Hardwood Floor Layers' Union No. 9075 in Indianapolis over the objection of the Brotherhood, and demanded that the charter be revoked.[72] Local 9075 had claimed that carpenters could not do the work involved: "There is not one carpenter in twenty that can use a floor scraper, and we, therefore, contend that it is a distinct and separate trade."[73] Gompers agreed that the Carpenters' claims were justified, but he counseled against revocation of the charter on the ground that "this would arouse considerable feeling" on the part of smaller unions. To Duffy this was of little concern; it was his contention that there were thousands of carpenters perfectly capable of laying hardwood floors. The AFL executive council eventually satisfied the Carpenters' demand by awarding them the jurisdictional claim.

Duffy continued to protest the contemplated issuance of charters to small craft groups: locomotive woodworkers, agricultural woodworkers, railway bridge builders, shinglers, and dock wharf builders.[74] It was this kind of activity by the AFL that led to a growing interest on the part of the Brotherhood for an alliance of the larger building trades unions. There was some concern, however, that the existing National Building Trades Council provided for too much local autonomy, and instead of joining it, the Brotherhood took the initiative in bringing about the formation of a new organization, the Structural Building Trades Alliance, in 1903. Huber was one of its principal architects, and he said in justifying its creation that "the time has come when we should act as one of the building trades, and not be pulling apart, creating strikes, ill feelings, among the trades over some petty jurisdictional disputes."[75] The Alliance proceeded to delineate eight basic building trades, and to allocate each of the fields to one union.[76] William J. Spencer, who was elected secretary, wrote that "the Alliance could halt the disintegration of crafts into subdivisions, enabling craftsmen to be replaced by younger and less skilled men"[77] — a paramount concern of the Carpenters.

There was a good deal of discord between local AFL bodies and the Alliance, and while the AFL began to show more caution in issuing new charters, the fear grew that the Alliance might become the focus of an independent labor movement. "You feel that it is a rival body to the A.F. of L.," Duffy wrote to Frank Morrison, the AFL secretary-treasurer, "and you believe such a body superfluous and should not be in existence at all." He accused the AFL of turning building trades disputes over to a committee that was unfamiliar with construction practices. "We were simply ignored in the past, and we decided to protect ourselves." He concluded: "I am surprised to learn that men of wide experience and broad dispositions should allow their prejudices to run away with their good judgment, in the belief that the Structural Building Trades Alliance of America is something organized in opposition to the A.F. of L."[78]

Morrison, in his reply, denied any antagonism toward the Alliance but acknowledged it would be a good idea to move it inside the AFL, citing resolutions to that effect adopted by AFL conventions since 1900.[79] The end result was the creation in 1908 of a Building Trades Department within the AFL. James Kirby, head of the Chicago District Council of Carpenters and a top official of the Alliance, was one of the two members of the organizing committee from the Alliance side, while Carpenter General President Huber and James Duncan represented the AFL. The Carpenters, on both sides of the fence, played a central role in creating the new department.

Kirby, who became the first president of the Building Trades Department, predicted that there were serious difficulties ahead:

What will, perhaps, be the hardest for us to agree upon, will be the defining of the jurisdiction lines of the different crafts; the great trouble is, we have all arranged our jurisdiction claims among ourselves, and then went out with a chip on our shoulder to fight what we are pleased to term "our rights," in our selfishness never considering the other fellow's rights at all.[80]

Instead of eight basic trades, however, the department was initially composed of nineteen building trades unions affiliated with the AFL, some of them quite small. Seeds of future discord were sown by the pattern of voting adopted; instead of weighting convention votes by per capita tax payments, the AFL practice, each delegate was given one vote. Although the larger unions had more delegates, the number of delegates per union did not increase in proportion to membership, with the result that the smaller unions had more power than their size would justify. Kirby warned that "no movement can long be successful . . . that permits of taxation without a just and equal representation," but his warning went unheeded. The Carpenters were soon to regret their acquiescence in this arrangement. They tried to have it changed at subsequent conventions but were always outvoted by the smaller unions.[81]

To retrace our steps a bit, one of the longest disputes in Carpenter history began in 1903 when the GEB signed an agreement with the Wood, Wire and Lather International Union covering lathing and studding work. At first the settlement seemed clear-cut: the Brotherhood agreed not to assert jurisdiction over any iron work, including iron or wire lathing, while the Lathers renounced jurisdiction over any wood work, including shingling, wooden arches, doors or window frames, and wooden studding or furring.[82] But it was not long before trouble developed. When the Brotherhood claimed jurisdiction over the installation of a new type of lathing material, the Lathers protested to the AFL. The agreement had already begun to break down.

Another dispute, which became one of the most famous in the annals of American labor history, began in 1908. What was involved was the right to install inside building trim, which was increasingly being fabricated from metal instead of wood. The International Association of Sheet Metal Workers based its claim on the material involved, while the Carpenters asserted that the installation of trim had always belonged to them. A Carpenter official put it as follows:

> If a competent committee of building tradesmen was to go through the shop and note the method of manufacture and then go to the building and watch the carpenter erect the material it would unquestionably declare that the work belonged to the carpenter and joiner. It is not a question of trying to take something away from the sheet metal worker. It is a matter of holding on to something that is a part of the original basic trade. From the mere fact that some other material has been substituted for wood, to keep abreast of the times, it does not follow that the carpenter must surrender and make a present to some other element of the building trades the improved conditions that we

have fought so earnestly and honestly to obtain and have maintained. I say again that this work belongs to us in and out of the shop and we will never give it up to any other element in the building trades.[83]

The loss of jobs would have been particularly serious in New York, where fire codes required the use of metal trim in tall buildings. The dispute in that city was submitted to an arbitration board under a plan to which all the building trades had subscribed. The board selected Judge N. J. Gaynor, a justice of the New York State Supreme Court (which in New York is the court of first instance) to decide the case, and he awarded the work to the Carpenters.[84] Although this decision had only local application, it was to be cited often by the Carpenters to justify their claims — for example, at the 1910 AFL convention, when the Brotherhood demanded that the decision be honored nationally. At the same convention the Carpenters renewed their dispute with the Lathers, claiming jurisdiction over metal corner-beads and picture molding, and for good measure, of all trim made of wood pulp, straw board, and asbestos.[85]

The Sheet Metal Workers were of the firm belief that they were entitled to jurisdiction over all building trim made of metal, and they brought their case to the second convention of the Building Trades Department, held in Tampa, Florida, in 1909. There a majority of the adjustments committee decided in favor of the Brotherhood, although they followed the Gaynor award by holding that the Sheet Metal Workers should control the factory production of metal trim. The Sheet Metal Workers maintained that they could not organize the factories without controlling the erection of trim at the building site, and were backed by a minority of the committee. The convention vote went in favor of the Sheet Metal Workers by a margin of 35 to 13, with most of the negative votes being cast by Carpenter delegates.[86]

The Carpenters simply ignored this decision and continued to perform the work in question wherever they could. The Building Trades Department was informed that "after learning that the employers in New York City who are doing this work endorsed the decision rendered by Judge Gaynor last year, and after learning that the architects in New York City are distinctly and plainly stipulating in their specifications that this work is to be done by the employing carpenter, and that carpenters must do this work," the Carpenters' Union had reaffirmed its determination to see to it that the work was performed by carpenters.[87]

The Sheet Metal Workers renewed their complaint to the Building Trades Department. As one of their delegates charged, the Brotherhood "ignored the Department. They felt as though they were greater than the Department, and some of them have told you if they do not get what they want they will put the Department on the bum . . . They are helping to make laws. If the laws suit them they will obey them, if they do not they will disobey them."[88] The convention resolved that the Brotherhood would be suspended from the department unless it complied with the Tampa award, whereupon the Brotherhood delegates withdrew from any further participation in the convention.[89]

The department responded by asking the AFL executive council to enforce the award and to revoke the Brotherhood's charter if it refused to comply. The executive council was understandably reluctant to act against one of its largest affiliates, and in its reply, enunciated a philosophy of organization that was to determine its actions for many years to come:

> We present to your consideration the fact which the history of the labor movement demonstrates beyond cavil of doubt, that to revoke charters, suspend organizations, enforce compulsory obedience to edicts and decisions, have invariably led to a rebellious spirit resulting in bitter conflict, relentless antagonisms, and disintegration and dissolution . . . The best, most enduring and advantageous discipline in the organized labor movement is that discipline which comes with time and experience . . . we have great apprehension of the advisability of the suspension or revocation of charters of International Unions based upon the contentions of trade jurisdictional disputes. We submit that such contentions can better be adjusted by the contending parties being in affiliation with the general organizations of labor than by either one or both being in enforced suspension with their charters revoked.[90]

The executive council recommended that the Brotherhood be reinstated in the Building Trades Department, and this was made compulsory by the next AFL convention.[91] The Brotherhood showed some reluctance to return, fearing a repetition of the events that had occurred. But Huber urged reaffiliation, saying that "we belong on the inside, and to make any fight for better conditions and settling jurisdictional disputes, we certainly ought to strive and work toward reaffiliation."[92] After a favorable referendum vote the Carpenters rejoined the department, but nothing had been settled, neither the dispute with the Sheet Metal Workers nor the manner in which department convention votes were to be allocated.

Although the disputes with the Sheet Metal Workers and the Lathers were the most important in terms of the number of jobs involved, they were by no means the only ones that the Carpenters were pursuing at this time. In 1909 the GEB listed the following claims that were in some stage of discussion:

National Association of Heat, Frost, Insulation, and Asbestos Workers: The Brotherhood claimed jurisdiction over cork, asbestos, hairfelt, woolfelt, mineral wool, and asbestos insulation when used in conjunction with carpentry work.

Bridge and Structural Iron Workers: The Brotherhood claimed jurisdiction over all false work; all scaffolding when carpenters' tools were used; the placing of all machines in factories, mills, elevators, and grain elevators; wharves of lumber construction; the setting of seats in all public buildings.

Electrical Workers: The making of all signs constructed of wood, all cutting and channeling when wood was used.

Elevator Constructors: The manufacture and installation of moving stairs and escalators; all overhead work and supports.

Hod Carriers: The erection of all scaffolding when carpenter tools were used; shoring and underpinning of buildings.

Marble Workers: All scaffolds where carpenter tools were used.

Painters: Placing wall molding in position.

Tile Layers: Setting of wooden mantels.[93]

In addition to this substantial menu, there were continuing complaints to the AFL against the granting of charters to specialized contiguous crafts. A group seeking to charter a Shinglers' Union was advised that "if we should consent to allow our work to be taken away from us a little at a time, it would not be long until there would be very little work left for the carpenter. With the modern construction of buildings at the present time in our large cities, much of the carpenter work is done away with. If you could only see these buildings after completion, you would find that the floors were either of cement, tile or mosaic work, slate or marble base, some sort of base blocks, galvanized doors, iron stairs, with slate, marble or iron treads, marble or iron wainscoting, and whatever else possible to do away with wood for the purpose of making it fireproof."[94] The Shinglers did not get a national charter.

Frank Morrison forwarded to Duffy a letter from the president of a local of the International Association of Car Workers, asking whether the Brotherhood claimed any jurisdiction over the manufacture of cars. Duffy replied that all men working on wood and handling edged tools belonged in the Brotherhood, "whether on land or on sea, on the earth or in the air." He told Morrison that he could not find two organizations of Cigar Makers in the AFL; "not even if I use a magnifying glass do I know of two organizations of Granite Cutters," or Machinists, Painters or Hatters. "Put me down once and for all as being totally opposed to dual organizations of any kind."[95] The references to particular unions here were a thrust at Gompers, Duncan (president of the Granite Cutters), and other AFL leaders who had at various times opposed the Carpenters.

It is well to note that when jurisdictional questions were not involved, the Carpenters enjoyed good relations with other unions. The Wisconsin State Federation of Labor was permitted to seek aid directly from Brotherhood locals to assist striking longshoremen and brewery workers. Requests from western locals for sanction of assessments in support of the

Western Federation of Miners were granted. The Brotherhood took the initiative in seeking AFL support for the Iron Workers in their long and bitter strike against the American Bridge Company. A request for a donation to the Moyer-Haywood Defense Fund was turned down, but only on the basis of the fact that a number of Brotherhood locals were already contributing to the fund. The Commercial Telegraphers Union was permitted to issue a circular appeal for assistance to Brotherhood locals.[96] When Gompers and other AFL leaders were threatened with imprisonment for violating an injunction in the Bucks' Stove and Range Company case,[97] the GEB wrote Gompers: "The General Executive Board of the Brotherhood now in session extend to you and Brothers Morrison and Mitchell their undivided support in your fight against the oppressors of the toilers. We are with you in every move you may make to have reversed the sentence imposed upon you." The board gave $2,000 to the AFL to help finance the successful appeal in the case. It endorsed an appeal for an assessment of 10 cents per member to assist striking Iron, Steel, and Tin Workers, as well as one for the Stone Cutters Association. Money was appropriated and locals solicited to help defray the cost of defending John J. McNamara, secretary-treasurer of the Iron Workers, who was indicted for participation in bombing the building of the Los Angeles *Times*. Similar action was taken to raise defense funds for Clarence Darrow, who was indicted for jury bribery in the same case.[98]

The purpose of this catalogue is not to portray the Carpenters as an organization that always leaped to the rescue of brother trade unionists who were in trouble. In fact, the GEB and the general officers were cautious in dispensing union funds for purposes that did not contribute directly to the economic well-being of their membership. At the same time, they felt themselves very much part of the labor movement, and played a not inconsiderable role in general labor affairs. Huber became a vice-president of the AFL in 1905, and was replaced by Duffy in 1913 when Kirby declined the position.

There was one aspect of Brotherhood relationships with other unions that began to cause problems at this early period, and continued to do so in future years. When the secretary of the Building Trades Department forwarded to Duffy a letter from the Painters' Union complaining that the New York City carpenters had failed to quit jobs that had been declared unfair by other trades, Duffy exploded. He declared that it had been the policy of the Painters to let other unions do their organizing for them: "If the Brotherhood of Painters, Decorators, and Paperhangers while in convention would confine themselves to their own business instead of finding fault with others, it would look more becoming. The Painters have no ground for complaint nor can they truthfully and rightfully throw stones at glass houses."[99] In fact, the Carpenters had helped the Painters form a national union and had often provided support.

A similar issue arose with the Stone Cutters when the Carpenters' local in Bedford, Indiana, was charged with remaining at work while other unions were on strike. In a letter to Morrison, Duffy put the dilemma faced by the Carpenters in clear perspective:

> We have been called upon time and again in years gone by to come out on strike in sympathy with other organizations but I notice, and the habit is growing more so than ever, that when we ask other organizations to assist us by taking their men off the work, we are requested not to pay strike pay direct from this office to the men who came off but we are asked to pay the men their days wages . . . Not in one instance to my knowledge have the Carpenters ever asked for one cent from other organizations when they come out in sympathy with them. I know the United Brotherhood of Carpenters and Joiners of America is looked upon as a great, big, powerful labor union with a good treasury, and much is requested and required from us. We have worked hard to build up our organization in the last eight years, devoted our entire time, energy and ability and spent hundreds of thousands of dollars and now that we have accomplished that fact we are called upon to do things that other labor organizations would refuse.[100]

Finally, although the general officers who succeeded McGuire were not as internationally minded as he had been, they were well aware of the existence of the world beyond the borders of the United States and Canada. Local unions were permitted to admit members in good standing from bona fide carpenters' unions in foreign countries without initiation fees. More specifically, an exchange of cards with the Associated Carpenters and Joiners of Great Britain and the Australian Society of Progressive Carpenters was approved.[101] Territorial jurisdiction was extended to include the entire Western Hemisphere for possible organization in the Panama Canal Zone and Mexico.[102]

The place of the United Brotherhood of Carpenters in the constellation of American trade unions is illustrated by the membership data in Table 5.1. Both in 1900 and in 1910, the Brotherhood was the second largest union in the country, after the Mine Workers. Moreover, whereas in 1900 its membership was only 59 percent of that of the Mine Workers, by 1910 the figure had risen to 87 percent. The only building unions among the top ten were the Painters and the Bricklayers, and between them, their membership was only two-thirds of that of the Carpenters in 1910. These figures help explain the success of the Carpenters in jurisdictional controversies.

By 1912 the Brotherhood, to all intents and purposes, had become the only union catering to woodworkers in building construction by absorbing competitors, including those that had been chartered by the AFL. It had staked claims to carpentry work in other industries, as well as to some building where materials other than wood were used. In some

Table 5.1.   Membership in the ten largest trade unions in the United States, 1900 and 1910

	Membership	
Union	1900	1910
United Mine Workers	115,000	231,000
United Brotherhood of Carpenters	68,400	200,700
Brotherhood of Railroad Trainmen	43,200	113,900
Brotherhood of Locomotive Firemen	36,000	69,200
Brotherhood of Locomotive Engineers	35,600	67,400
Brotherhood of Painters and Decorators	28,000	63,500
International Union of Bricklayers	33,400	61,800
International Association of Machinists	22,500	56,900
International Cigar Makers Union	37,100	51,400
International Molders Union	15,000	50,000

Source: Leo Wolman: *The Growth of American Trade Unions* (New York; National Bureau of Economic Research, 1924).

cases, many years were to elapse before the claims were recognized by other organizations. Changing technology and materials continued to pose new challenges; the history of jurisdictional controversy had just begun. Nevertheless, by the end of the first decade of the twentieth century the Brotherhood had established itself as a formidable adversary where its interests were at stake. It may not have been the most popular of unions, particularly among the smaller organizations in the building trades. But there was no doubt about its effectiveness in protecting the jobs of its members — and that, after all, is what they wanted of it.

6

THE EMPLOYER OFFENSIVE AND THE
UNION RESPONSE

The new leadership team that took over in 1902 arrived when economic circumstances seemed fortuitous for union advances. Employment conditions were excellent; unemployment fell from 5 percent in 1900 to 2.6 percent in 1903, rose somewhat in 1904, and then dropped to the almost unbelievably low level of 0.8 percent by 1906. Average hourly earnings of building tradesmen increased by 28.6 percent between 1900 and 1906, while consumer prices rose by only 12.5 percent. Union membership responded at first to these favorable conditions. Brotherhood membership increased from 68,463 in 1900 to 116,205 four years later. But during the next two years, membership growth was restricted to a net of 7,000. Had it not been for a concerted drive by employers in the larger cities to curb labor organization, aimed particularly at the building trades, the rate of growth of membership from 1904 to 1908 would undoubtedly have been much higher.

This campaign began with a lockout against the Chicago building trades early in 1900, in an effort to repudiate wage and other contract gains that had been won the previous year. The contractors attempted to employ nonunion labor, which resulted in a good deal of violence. As the lockout dragged on, one craft after another withdrew from the Building Trades Council and signed agreements. The carpenters held out until February 1901, and when they finally capitulated, the lockout was virtually over. Many of the gains of the 1899 contract were lost.[1]

Events in New York City followed a different pattern, but with somewhat similar results. A jurisdictional dispute between the Brotherhood and the Amalgamated Society of Carpenters (ASC) that broke out in 1903 led to a lockout of 10,000 Brotherhood members. The local Building Trades Council supported the ASC, and although peace was restored by an agreement permitting men from both organizations to work on the same jobs, relations between the Carpenters and the council deteriorated.[2]

At about the same time the major contractors formed the Building Trades Employers' Association and instituted a lockout against the material drivers and lumber handlers, who were seeking higher wages. This brought building in the city to a halt. By the end of April 1903, some 3,800 carpenters were out of work, and Duffy was dispatched to New

York with $16,000 to help.[3] Sixteen of the skilled unions broke away
from the Building Trades Council, and together with the Carpenters,
who were not affiliated with it, offered to handle building materials
whether or not they had been provided by union men. The employers
responded by proposing a plan of arbitration, the purpose of which was
to eliminate sympathetic strikes and reduce the power of the business
agents. The unions rejected the plan, but a few months later they
capitulated and the plan was put into effect. "The employers, by showing
a united front, had undoubtedly carried the day. The power of the
business agent was for a time eliminated and the sympathetic strike
outlawed."[4]

Although the Carpenters signed the agreement, partly to remove the
ASC from the scene, difficulties in enforcing it soon emerged. In July
1904 carpenters were withdrawn from a construction job in violation of
the agreement. When they refused to return to work, they were locked
out for ten days. The same thing happened in August.[5] It was reported in
December that the lockout was still on, and that independent unions
were being organized among the strikebreakers who had been hired.
"From all appearances it is the design of the employers to exterminate the
Brotherhood of Carpenters in New York City."[6]

The employers' terms for settlement of the controversy required
recognition by the Brotherhood of a dual Greater New York Carpenter
and Cabinet Makers Union which they had organized; this was firmly rejected by the GEB, as might be expected. To put pressure on the New
York employers, the Chicago district council struck some jobs of the
George A. Fuller Construction Company, one of the largest in the country, and operating in both cities.[7] The stoppage was finally settled in
April 1905. Huber wired the GEB: "We have won, made a clean sweep.
Strike will be called off today."[8] But it was by no means an unalloyed victory. Brotherhood charters were granted to the independent groups that
had supplied the strikebreakers. The arbitration plan was reinstated, and
remained in effect until 1910. This series of events in New York, from
1903 to 1905, left the building craft unions, the carpenters included, in a
weakened condition.

Chicago and New York were not isolated cases. Dufiy reported to the
1904 convention:

> Never before in the history of our organization have we had to contend with so
> much opposition from our employers as during the past two years . . . They
> organized, combined and affiliated with one another, with the avowed purpose
> and firm determination of putting our local unions out of existence altogether.
> They are fighting us today. Their cry is that they will not be dictated to by
> walking delegates or representatives of organized labor as to how they shall
> run their business. Their policy is to hire whom they please, for what wages
> they please, under what conditions they please.[9]

A Citizens' Industrial Association was established under the leadership of D. M. Parry, president of the National Association of Manufacturers. The new association sponsored so-called local citizens' alliances dedicated to the open shop.[10] W. J. Shields, who was organizing for the Brotherhood at the time, found them to be formidable antagonists, and remarked on

> . . . their splendidly equipped recruiting stations designated as employment bureaus operating in all the principal cities of our country, under the control of their efficient corps of attendants possessed with up-to-date tact . . . with their bull pen institutions to house their hirelings, known as the professional strikebreakers, guarded by their armed squads, titled detectives; creatures who, if justice were done them, would be put behind bars as vagrants and loafers who are no respecters of law and order.[11]

There were lockouts against carpenters in Pittsburgh, Louisville, Houston, and Milwaukee, among other cities. President Charles W. Eliot of Harvard University termed the strikebreaker an "American hero," and urged American employers to allow "no sacrifice of the independent American worker to the labor union." The open-shop drive lasted until 1908, when the influence of the Citizens' Industrial Association declined, partly because of its success in stopping union growth.[12] This employer offensive had a serious impact on the Brotherhood, as both Huber and Duffy noted in their reports to the 1906 convention; the latter termed the two foregoing years "anxious and eventful."

THE GROWTH OF GENERAL EXECUTIVE BOARD AUTHORITY

The GEB had been using its authority to sanction strikes primarily to advise locals on immediate strategy. It now began to establish some general principles for the guidance of the locals. They were forbidden to sign working agreements exclusively with employers who were members of associations, to the neglect of independents, either directly or through a local building trades council. Strike funds could be used only for the relief of strikers and not to pay the salaries of local officials. Locals were told that it was inadvisable to pay full wages to strikers, because "where men have to be paid in full to stand for union principles . . . it is more detrimental than beneficial." A Minneapolis local that had asked whether it was permissible to give employers the right to hire nonunion foremen were told that "if possible, it is desirable that the foremen should be members of the organization, at the same time it is a matter which devolves entirely on the local to decide."

The board reiterated its firm opposition to any formal wage grading of carpenters, considering this practice to be "demoralizing to union principles and the welfare of the trade."[13] This did not mean that all carpenters had to be paid the same rate; employers could and did pay premium rates for special skills on an informal basis. But it did reflect the view that every Brotherhood member had the skill necessary to earn the standard rate, and that to permit grading would have broken down the presumption of full craft competence.

The union label was denied to any shop or mill employing female labor in the manufacture of wood products on the ground that mill work was not suitable for women, and if they were employed, the union scale would be undercut.[14] Local unions were ordered "not to circulate any appeal or circular asking financial aid or calling on the locals in any form to purchase tickets, unless by the approval of the G.E.B."[15] The board wanted to maintain its central authority to sanction and finance strikes.

The attitude of the Brotherhood toward the Industrial Workers of the World (IWW) has already been noted. The Board formalized this by making membership in the IWW cause for expulsion from the Brotherhood. Nevertheless, when Charles H. Moyer and William D. Haywood, who had been involved in the formation of the IWW, were virtually kidnapped by the Colorado authorities and shipped to Idaho, where they were charged with murder, the GEB urged the AFL executive committee to protest.[16]

A Cleveland local informed the GEB that it was forming a private company to engage in building contracting, and asked permission to approach other locals about buying stock. The board refused to sanction this request, and instructed the local to abstain from any official connection with this enterprise. McGuire had been skeptical about worker cooperatives, and his successors were no more friendly to the idea, fearing both that the collapse of such firms would injure the union and that employers would be antagonized by the union competition. But worker savings were encouraged, and the board urged local unions to agitate on behalf of the Postal Savings Banks.[17]

An interesting suggestion by General President Huber to further the expansion of membership in small communities was frustrated by the board. He proposed that in places where there were not enough carpenters in a town to form a local, they whould join the Brotherhood directly at headquarters. The GEB refused to go along, holding that this was not only impracticable, but too great a departure from established policy.[18]

The major events of the 1902 convention have already been discussed, but a few other items are worthy of note. The general president was given greatly augmented power, including the right to appoint organizers.[19] Seven geographical divisions were established, with one member of the GEB to be elected from each—but by vote of the convention as a whole, not by the delegates from each district. Headquarters were finally moved to Indianapolis, reflecting the geographical shift in union membership. Indianapolis had good rail links, and mail could reach it more rapidly than had been the case in Philadelphia. However, the convention turned down by a large majority proposals to raise dues and establish uniform initiation fees. That attitude had not changed.

THE RACIAL PROBLEM

The site of the 1902 convention was Atlanta. An open meeting was held there "for the purpose of endeavoring to organize the colored carpenters of Atlanta." Fifteen people appeared at the meeting, and it was decided to appropriate $200 to appoint an organizer to help establish colored locals in the South.[20] This action by the convention occasioned protests by white locals in the region, and in responding to them, Duffy defined clearly what was to be the policy of the national union on the race question. He warned that unorganized blacks constituted a danger to white wage standards, and continued:

> Our General Constitution does not contain any provision justifying our General President in making any discrimination as to color, race or creed in his appointments. And while it is true that Brother Burgess has been appointed as an organizer for our colored brethren in particular, it would be unwise and in conflict with our laws and principles to debar him from organizing white men of our craft when an opportunity presents itself.
>
> We are banded together in our grand brotherhood for the purpose of elevating the condition of our entire craft, regardless of color, nationality, race or creed. Prejudice on these lines has no standing in the labor movement, and we cannot consistently deny admittance in our organization to any man because he belongs to the African race, which, as we well know, is very numerous in the Southern states . . .
>
> In many instances white men are called on to work with negroes on buildings or in shops, and they do so without raising any objection. Now, if such is the case, why should they object to meeting the negro in their Local Unions and placing them under the control of the organization as a safeguard against unfair competition?[21]

Pointing out that when colored carpenters were refused admission by the carpenters' local in Birmingham, everyone suffered, he added: "Such actions are discreditable to our Brotherhood, and we say that, however difficult the task of the elimination of race prejudice may be, as far as our Brotherhood is concerned the drawing of the color line should be stopped at once and for all time."[22] Not many affirmations of this kind were being made on behalf of organizations, labor or otherwise, in the year 1903, more than a half century before the inauguration of the affirmative action campaign.

This statement was followed by a considerable amount of correspondence, pro and con, that appeared in *The Carpenter*,[23] but the leadership never retreated from this policy. For example, a year later second vice-president Robert Connolly reported that on a visit to Greenville, South Carolina, he addressed a joint meeting of the white and colored locals that was well attended. "The white carpenters of this city seemed to realize the necessity of working in harmony with the colored carpenter in order to better the conditions of both, and I congratulated them on the good sense they displayed."[24] And he observed in a later report:

> Right here I want to say that I believe the color question is the greatest drawback we have to the thorough organization of our craft, and we never will enjoy the success that we deserve until we recognize the fact that a carpenter is a carpenter, and a competitor, and entitled to our unstinted support as such, regardless of his color . . . From Vicksburg I went to Jackson, Miss., where I found our organization almost completely wiped out on account of this senseless color question.[25]

There was, of course, considerable opposition by many southern members to the position taken by the national leadership. A letter from a member of a local in Daytona, Florida, replying to Connolly's observations, asserted that a large majority of his local recognized the right of a contractor to hire blacks, but reserved the right not to work with them. "I am fully convinced that, for the good of our U.B., all local unions, especially those in the South, should be free to meet this negro question according to their environment. Should the U.B. in convention or otherwise attempt by law to encourage a spirit along the lines advocated by our Second Vice President it would kill organized labor here in Daytona deader than African slavery." But Connolly stuck to his guns. "I spent several days in Montgomery, Ala., trying to imbue new life and courage into the members of No. 353, and I believe I met with a fair measure of success. Local No. 353 is a colored union and is badly handicapped from the fact that the white carpenters of that city have allowed their organization to lapse."[26]

However, it was difficult to make headway against the continuing recalcitrance of the southern white membership. Several years later an organizer, W. G. Wilson, reported that he had gone to Tampa, Florida, in an effort to restore harmony between the black and white locals in that city. "The feeling on the race question among the membership of the two local unions being quite bitter, the desired result of my efforts were not perceivable when I left the city. Still, I hope that our brothers in Tampa will be reasonable enough to overcome this trouble, and in the near future form a [joint] district council." Organizer William D. Michler, reporting on a racial conflict in Pine Bluff, Arkansas, warned the white carpenters: "In their own interest it behooves those who enjoy higher living conditions to bend their energies toward the elevation of those who, because of racial or other differences, are so unfortunate as to be forced to accept inferior conditions."[27]

Progress was slow. J. H. Bean, a black organizer, wrote of having met with the members of a white local in Augusta, Georgia, "this possibly being the first time they had a colored organizer in their midst." A year later he continued to complain of the difficulty of organizing the South in the face of racial animosity. Becoming somewhat discouraged, he approached Booker T. Washington, the famous black educator, and asked whether he should continue his organizational work. The answer was: "In my opinion, anything without opposition was a very small thing, and the work of the U. B. was too great to succeed without opposition."[28] The Brotherhood would be justified in being proud of this endorsement by one of the outstanding American blacks.

THE THIRTEENTH GENERAL CONVENTION

The effects of the open-shop campaign were just beginning to be felt when the Carpenters assembled for their thirteenth convention in 1904, but there was a general air of optimism. Some 600 locals had been added during the previous two years. Groups had been chartered in Hawaii and Puerto Rico, in the latter after a membership referendum vote that was overwhelmingly in favor of such an action. District councils had been established in 90 cities. Half-day work on Saturday had become general in the larger cities, and the eight-hour day was in effect in 480 towns and cities, although 791 were still working nine hours, mainly small communities. More strikes had been called between 1902 and 1904 than in any previous two-year period, and most were said to have been successful, supported by almost $190,000 in assistance from the general office.[29]

Charges had been made against Huber and Duffy by several former McGuire adherents who were still on the GEB. They involved the printing of *The Carpenter*; an accusation was made that these two men had accepted gifts of traveling bags as a reward for the grant of contracts.[30] The board ruled that the gifts were accepted simply as tokens of friendship. Nevertheless, in its report to the convention the GEB stated that by assuming supervisory control over printing bids, it had saved the union $1,000 a month. This gave rise to a bitter debate at the convention.

When a delegate accused the board of having failed to report this matter in *The Carpenter*, he received the reply that to have done so "so soon after the exposure that was made at the last convention in regard to the P. J. McGuire matter, would have had a demoralizing effect on our organization." Huber hotly denied any impropriety in accepting the gifts, which turned out to include a gold watch as well as a suitcase. "If as your General President I cannot give and receive tokens of friendship on occasions like that, then I don't want to be your General President. If I had thought for an instant that it could ever be construed into graft, know damn well that Bill Huber never would have accepted them."[31]

The main counter-attack was made by Duffy, who asserted flatly that the whole matter was an aftermath of the McGuire affair: "At every meeting of the Board we had trouble of some kind or another. Members of the Board have told me that it was no pleasure to come to meetings of that body." Board member Wellman had accused him of getting 3 percent interest on Brotherhood money in a local bank, and turning over only 2 percent to the union. The bank confirmed that the rate had been 2 percent: "No, they were not satisfied with that. My word was good for nothing. They appointed a committee—myself included—and I was never so ashamed in my life, but I went with them. We had to go to the other banks where we had money, and where we did not have money, in the City of Indianapolis, to find out if we could not get 3 percent on our money."

Henry Meyers, another board member, took the floor to accuse Duffy of having paid 40 cents apiece for dues books that could have been bought for 11 cents, with a similar overcharge for letterheads. He made this charge, he said, because Duffy had "built up a pile of dirt and filth that I have got to get down into and crawl into." Wellman arose again to state that from the start, Huber and Duffy had attacked his faction on the board for political reasons:

> Every night brothers of this Convention body have been approached on that very line of trying to influence them against the members of the Board because they were Socialists, and the very gang that is doing this is the gang that has been carrying politics into the union . . . Look at the facts, boys, look at the figures and don't be stampeded by hot air. Did you ever hear of any man who is associated in this convention raising objections to any man becoming an officer of this organization because he was a Democrat, or because he was a Republican? Now be fair boys, be fair. Those of you who were at Atlanta [in 1902], do you know or don't you know, that the same tactics were used there? There are fair men that are Socialists, and plenty of them. I respect and honor them for their convictions.

The strength of the opposing parties was indicated by the fact that the convention adopted by a vote of 318 to 117 a resolution exonerating the general officers from all the charges that had been made. Wellman was defeated in his quest for reelection to the board, as were three other members of his faction, while Meyer did not run. Huber and Duffy were reelected by substantial majorities, and thus ended one of the most vitriolic debates in the history of the Brotherhood. The new administration was now firmly in control.

The convention did adopt several constitutional amendments of some interest. The general officers were to be nominated at the convention, but the actual elections were to be referred to the membership for a referendum vote. A similar arrangement was to govern the fixing of their salaries. These changes were approved by the subsequent membership vote, but once again an increase in the per capita tax to 25 cents per member was turned down. This left the national office in a precarious financial situation, for total expenditures for the year exceeded income by $40,000. Death rates had been particularly high, resulting in a heavy drain on the benefit fund, and a good deal of money had been paid out to resist the open shop drive. Therefore, a year after the convention the GEB sent out another proposal for the increase to a referendum vote, and this time the necessary two-thirds approval was secured.[32] A threat to levy an assessment if the per capita increase was not forthcoming apparently did the trick.

THE FOURTEENTH BIENNIAL CONVENTION

At the 1906 convention the Brotherhood celebrated its silver jubilee; twenty-five years had gone by since McGuire had issued his call for an organizing meeting. Huber was able to boast that the Brotherhood was the largest organization of skilled mechanics in the world. Since 1881 the national office had paid out $1.5 million in death and disability benefits, while locals had spent $1.8 million for sick benefits, very large amounts of money for the time.

Some idea of the geographical distribution of Brotherhood membership can be seen from data presented to the convention. Shown in Table 6.1 are the ten largest states by Brotherhood membership, together with the number of locals and members per local. Half the members were concentrated in five states, and only California was a new entrant to the Carpenter strongholds. The size of locals depended largely on the concentration of industry, though Massachusetts stands out as a state in which organization appears to have spread to many smaller communities. There were almost 5,000 Canadian members, but the Canadian percentage of total membership was still small.

Perhaps the most interesting action take by the 1906 convention was the adoption of a political resolution and a proposed political platform to be submitted to a referendum vote. The resolution read:

> Whereas, the American Federation of Labor has decided that it is good policy to use our political power to defeat for office those members of Congress and

Table 6.1. Membership and number of locals, United Brotherhood of Carpenters, for the ten largest states and Canada, 1906

State	Membership	Percentage of total UB membership	Number of locals	Average number of members per local
New York	27,858	16.4	188	148
Illinois	18,051	10.6	167	108
Pennsylvania	16,543	9.7	157	105
California	13,194	7.8	70	188
Massachusetts	11,265	6.6	125	90
New Jersey	10,744	6.3	87	123
Ohio	8,362	4.9	97	86
Missouri	5,855	3.4	44	133
Indiana	3,700	2.2	69	54
Connecticut	3,667	2.2	34	108
Canada	4,792	2.8	62	72
Total U.B. membership	170,192			
Percentage of members in ten largest states		70.1		

Source: United Brotherhood of Carpenters and Joiners of America, *Proceedings*, Fourteenth Biennial Convention, 1906, p. 94.

other officers who have opposed the political demands of the A.F. of L., therefore be it Resolved that the United Brotherhood of Carpenters and Joiners of America . . . hereby endorse such action and request our members to give such policy their undivided support.[33]

This was an endorsement of the AFL's first major venture into nonpartisan politics, marked by the presentation of Labor's Bill of Grievances to the authorities in Washington. In addition, however, the convention adopted a platform that contained a long list of national, state, and local demands, including public ownership of public utilities and transportation; prohibition of child labor; equal pay for women; the abolition of national banks and the issue of money directly by the government; the prohibition of land ownership by aliens; the abolition of all indirect taxes; workmen's compensation laws; and other items that were partly populist in nature and partly the core of what was to become labor's legislative program. However, this program did not receive sufficient endorsement from the locals to be incorporated into the constitution.

This convention marked the beginning of the agitation on the part of the Canadian locals for more autonomy. A request by the Canadian central federation, the Canadian Trades and Labor Congress, that the Brotherhood locals in Canada affiliate with it had been rejected a year earlier by the GEB. However, the board was overruled after the convention received a strong letter from the Canadian Congress pointing out that most of the American skilled craft unions had permitted their members to affiliate with it.[34] Another concession was the designation of Canada, apart from New Brunswick, Nova Scotia, and British Columbia, as a separate geographical district.

Although Duffy was unopposed for reelection, August Swartz was nominated to run against Huber. When Huber's election was certified, the Pittsburgh district council sent out a circular questioning the accuracy of the tabulation. The GEB replied: "That hideous and horrible demon—slander and distrust—must be banished from our midst altogether . . . Such circulars are poisonous and are not issued for the best interests of our organization."[35]

This did not end the dispute between the Pittsburgh council and the national office. At the next convention the Grievance Committee reported that the council was suffering from internal conflict and disorganization, and recommended that the GEB step in to improve matters. After investigation, the board found that some of the locals were in poor financial shape, and recommended consolidation and reorganization under the direct supervision of the national office.[36] Whether or not this action was justified by the financial situation of the Pittsburgh locals is not clear from the rather brief report of the board. However, it did serve to eliminate a source of opposition to the national officers.

THE UNION LABEL AND THE CLOSED SHOP

Prior to 1900 there was no nationwide label to designate Brotherhood-made woodwork. It was used in a few cities, particularly New York. After some discussion at the 1900 convention, the GEB adopted a design the following year and sent it to the U.S. Patent Office for registration, but was informed that trademarks could be registered only for individuals, firms, or corporations. It proved possible, however, to register the emblem in a number of states under state law, and this was done.[37]

The 1902 convention authorized the use of the label by shops or mills employing union labor, paying a minimum of 30 cents an hour, and working an eight-hour day. Each shop was to elect a delegate who controlled the label; only he was authorized to affix it to the product. Each factory had its own number, which was to be stamped on the label. The GEB was given the right to withdraw the label from any shop that violated union regulations. The continued emphasis placed on the use of the label suggests its importance as an organizing device. Huber told the 1906 convention: "Our label is one of the most valuable organizers we have on the road, for it works while our brothers are enjoying the half holiday and only seeks compensation in an increased wage and shorter hours for those who toil."[38]

The union label plus the closed shop placed the Brotherhood in a powerful position. In New York, for example, all carpenters were compelled to strike against a contractor who employed nonunion men or used nonunion material. Manufacturers who did not have the label could be placed on an unfair list, and the master carpenters would refuse to handle their product. In 1909 jobbers in Manhattan were obliged to sign an agreement not to handle nonunion products, subject to a fine of $2,500. In return, the union kept out the products of lower-wage open shops. "So effective has the closed shop been employed in the larger cities that the employer was even deprived of the right to hire men when the union was unable to furnish all the men needed."[39] What the union did in those cases was to issue temporary working permits.

Christie wrote disparagingly of "label unionism" as leading to collusion with employers, a reduction of competition, and higher prices for both materials and buildings.[40] There is little doubt that this was the result of the combination of the union label and the closed shop. But no Brotherhood official lost sleep over it. As far as the union was concerned, higher wages could only be secured through higher prices, and whatever it could do to protect its employers from low-wage competition was all to the good. The men who were now running the Brotherhood prided themselves on their businesslike conduct, which did not necessarily include adherence to the doctrines of Adam Smith. They did not like to be called monopolists because of the possible legal consequences, but they firmly believed in their right to monopolize work for their members. The closed shop was the most effective means of protecting the standards they had gained, and they used it as long as they were permitted to do so by employers and by law. The alternative, a nonunion labor market, did not seem attractive to them, as evidenced by the following account by a business agent in Toronto, where it prevailed:

> We, indeed, have a rough road to travel at this time, and we are up against great odds due, to a great extent, to the enormous influx of carpenters from this country, England, Ireland, Scotland, and the United States. All through the past few years large numbers of carpenters have been flocking to this city, coming here in response to alluring advertisements and inducements held out by employment and immigration agents, their motive being to flood the labor

market, to create a surplus of carpenters, and subsequently cause a reduction of wages that would serve to fill their own pockets.[41]

"STAY AWAY" NOTICES

The desire to maintain a relative scarcity of labor, or at least to maintain an equality of supply and demand at the union wage rate, did come into conflict with one of the original purposes of forming a national union — to augment employment opportunities for the members by making it easier for them to move from one city to another. A dialogue in the pages of *The Carpenter* illustrates the nature of this conflict.

The monthly newspaper of the Brotherhood had for many years carried notices sent in by local unions warning of poor employment conditions in their communities, and urging carpenters to stay away. These were often embellished with charges that employers were sending out advertisements for jobs that were nonexistent in order to flood the labor market. An editorial appearing in *The Carpenter* of January 1906 made the following comment on this practice:

> We are pleased to state that in the course of the past month we have been requested to remove the names of two cities from the dull list in the journal. This is an occurrence so unusual that we can not pass it without particular comment. Almost as a rule, local unions are sending in "stay away" notices, asking the editor to place the name of their locality on the dull list, and after their requests are complied with no notification is sent us of the revival of business or improvement of trade conditions. This is certainly not in conformity with our obligation wherein we promise to use every honorable means to procure employment for brother members . . . many a brother could obtain employment in localities that appear in the dull list of the journal continuously for months and month, upon special request of the respective local unions.[42]

The secretary of a local in Lynchburg, Virginia wrote in to say that the most unpopular feature of *The Carpenter* was "the never-ceasing whine of the various local unions asking traveling carpenters to please keep away from their localities. This is depressing and lowers our confidence in the brotherhood of man." A member of Chicago Local 416 replied that these notices were helpful to traveling carpenters, telling them "what localities are already overflooded with carpenters, and where . . . they would run the risk of being disappointed, of getting stranded and spending railroad fare in vain." A letter from Jacksonville, Florida, complained that some Brotherhood members came there for the winter and worked for nonunion contractors, and urged that the least they could do was to work under union conditions.[43]

This ambivalence about the benefits of labor mobility came up in other contexts as well. Commenting editorially on the establishment of a seasonal unemployment relief fund by several New York locals, *The Carpenter* observed: "Many of us doubt the feasibility of this feature in our U.B., claiming that it cannot be successfully inaugurated nor carried out in our trade on account of the numerous floaters, who, we admit, are difficult to control; yet where there is a will there is a way." An organizer wrote in from Los Angeles repeating the charge that many Brotherhood members from out of town were working under nonunion conditions there. "I believe that a member who will disregard this section of the constitution is so crooked that a corkscrew placed along side of him would look like a straight edge."[44]

As far as the national union was concerned, accepting nonunion work was an unmitigated evil: "This practice of outside members clandestinely working in and defying the D.C. or Local Union of another locality, must be stamped out by all means, and the sooner the better."[45] On the other hand, the GEB time and again reprimanded local unions for refusing to accept the clearance cards of traveling members, pointing out that this was in direct violation of the union's constitution and contributed to enlarging the nonunion labor pool.

The available statistics do not enable us to determine with any degree of precision the volume of carpenter mobility. That it was substantial is suggested by the fact that in 1906, with total membership at 170,000, some 22,000 carpenters transferred by clearance card from one local to another in the period of roughly a year,[46] while from 1906 to 1908, the total was 51,600.[47] However, some of these clearances may represent multiple transfers by the same individuals.

THE FIFTEENTH BIENNIAL CONVENTION

The 1908 convention opened against an unfavorable economic background. Unemployment, almost nonexistent in 1906, had risen to 8.5 percent two years later. The year 1907 had been a period of "fullness and plenty," according to Duffy. Membership had grown by almost 30,000 from the previous year, but 20,000 of the total had been lost by the time of the convention. Huber reported to the convention that the monetary panic that set in during the fall of 1907 led to serious consequences for industrial relations: "In the conflict of industrial life there is no quarter shown when a crisis or business depression similar to the one that now confronts us occurs; the ordinary means and methods of human warfare are abandoned. Agreements, wage scales are set aside, ignored, and every unfair advantage taken of the workmen for the purpose of disrupting and disorganizing the unions."[48] And Duffy added: "With thousands of men out of work, walking the streets for weeks or months, reduced to the last cent, almost begging for food for themselves and their little ones, unable to secure employment, how could it be expected under such trying circumstances for the [local] unions to exist."[49] And indeed, from 1907 to 1908, some 150 locals surrendered their charters.

In his report to the convention Huber made some interesting observations about the internal life of the organization. During the previous two years, seventeen locals had suffered embezzlement of their funds, losing $16,000 in all. In some cases, the funds were restored; in others, the culprits absconded; and in still others, "we caught them, and they are now expiating their crimes in various penitentiaries." He urged locals to abandon the fairly common practice of paying officers a salary based on a percentage of dues income. As an indication of growing organization among factory workers, it was pointed out that of 3,814 mills and cabinet shops in the United States, 982 were employing only Brotherhood men.[50]

The report of the GEB was much livelier than it had been in the past, and contained a good deal more information. For example, it had acted on 105 appeals, most of them involving benefits. It endorsed 264 of 276 requests for strike sanction in 1907, but the requests virtually dried up the following year as a result of the depression. Local 1008 of Brooklyn had written in to request that the Brotherhood constitution be published in the "Jewish language," while a Buffalo local wanted one in Polish, but the convention did not concur, urging instead that each local undertake whatever translations it needed.[51]

A resolution to standardize the ritual for admission of new members into the Brotherhood was proposed and reprinted in the convention proceedings.[52] It is too long to reproduce in full, but a summary will serve the purpose of showing the "brotherhood" origin of the Carpenters' Union. There were still strong elements of the benevolent and fraternal order that characterized the early days of the union.

The proposed ritual provided for the designation of a Conductor to bring in the initiate. As they approach the door, the Warden reports an alarm at the door, and informs the President that Mr. X is seeking admission. The Conductor is asked why, and he replies: "To aid in the betterment of trade conditions, become more skillful in the craft, and to improve his social and moral condition." The warden so informs the President, who asks whether the candidate has answered the proper inquires properly, and when the Warden goes back to the door and receives the proper assurance, he so informs the President.

The Conductor then leads the candidate twice around the room, blindfolded, and members sitting in the room tap with a gavel as he passes. He then stops in front of the altar, faces the President, "removes the hoodwink," and takes the oath of obligation. He is told that he is now tied to the order by a bond that will never break, and once again is blindfolded. He is then introduced to an Apprentice, who explains his job as "assisting the different craftsmen in their work, in distributing material and making myself generally useful." The Apprentice hands him a bunch of shingles to give to "yonder Craftsman." The Craftsman inquires about all the confusion, and is told that the candidate has passed his apprenticeship. He thereupon informs the candidate that he will be entrusted with more important work, "sawing, fitting and placing in position for the erection of a structure." He is asked to saw a board, and then is introduced to a Journeyman—all this accompanied by appropriate rapping of the gavel—who also inquires "why all this disturbance." When told that the candidate has passed the apprentice and craftsman rank, and wishes to become a journeyman, the Journeyman says that "it becomes our duty to saw, plane and form all the material of the structure. It is also part of our duties to receive instructions as well as give them." The candidate is asked to plane a board, then moves, "without alarm," to the Builder. This man grabs the candidate by the shoulder, and exclaims: "What, this man so soon trying to undermine and take away a brother's job? What shall be given unto thee, or what shall be done unto thee, thou false tongue? Brother Conductor, is this in strict accordance with our ritual?"

Three or four members of the admission team adduce arguments pro and con as to the candidate's capabilities, intentions, and so on. The President rises and inquires: "Why this strife among the craft? Give the Conductor time to explain. He that worketh deceit shall not dwell in my house. He that telleth lies shall not tarry in my sight." The Conductor informs the meeting that the candidate means no harm, but comes seeking knowledge. The President asks for charity, and the Team, in unison, declares, "We will have charity." The candidate is told by the Builder: "The station you are now in, that of Builder, is one of great importance. To this position belongs the supervision of every detail in the erection of a structure. But let us not forget that moral structure, of which we alone are builders, so that when it becomes our duty to respond to the Master Builder of the Universe we may be ready to answer the call."

The blindfold is then removed, and the candidate is led to the various stations and presented with the emblems of the order. Each of the initiators, the Apprentice, the Craftsman, the Journeyman, and the Builder, makes a speech. The Journeyman, for example, says:

> My brother, this plane which I present to you is one of the most useful tools of our craft. That you will use it skillfully is our earnest desire. Always keep its edge keen and true and as you use it, as we trust you will, with true craftsman's pride, may it teach you that if you would be perfectly joined to your brother,

your own life must be square and true, and having made it so, you should endeavor to aid your brother to rightly live.

May your life among us be filled with earnest endeavor to smooth the way of life for your brothers, striving to aid them to overcome life's difficulties and discouragements. May the Great Craftsman's life be an inspiration to you and may you so live as to be worthy a place with him in that house not made with hands, eternal in the heavens.

The candidate is then given the password, and the Team sings the initiatory ode:

> Brother, with an earnest greeting
> We now bid you welcome here.
> Wherein labor's cause believing
> We can find some hearts to cheer.
> And we strive to aid our brothers
> In this strife 'gainst odds so great,
> With fraternal love toward others
> We a victory can make.
> And we hope that you will enter
> With whole heart in labor's cause,
> And your strength and mind will center
> In the upholding of our laws.
> For, if victory we would hasten
> We must all united stand
> And each other we can strengthen
> If we labor hand in hand.

The Past President gives a stipulated closing lecture. The candidate is told to study the constitution, to work only on materials bearing the union label, to set an example as a good and faithful workman, to excel in quality and quantity, to make the words "Carpenters' Union" stand for something. Moreover, "if you are always the last man to get a job and the first one to lose one, there must be a reason. It is evident that you are lacking in some important qualification."

The President then sets forth the goals and principles of the Brotherhood: a reduction of hours, discouragement of piecework, encouragement of higher skill standards, death and disability assistance. "This brotherhood guarantees no man employment. Every man must stand on his own merits. But it is both our duty and pleasure to assist you in securing employment, and we expect you to do the rest." The candidate is presented with a Bible, told to be loyal to the American flag, and given an insignia of the craft. Finally, he is urged to "strive to build up the cause of unionism in this or any other place in which you may be located. By careful thought and study prepare yourself to explain to others the advantages of this brotherhood. Be present at all our meetings, attend faithfully to the order of business and take an active part therein."

The convention committee on ritual recommended this procedure as an optional one for local unions to adopt, and also suggested that the general secretary print as many copies as were in demand. However, the committee report was rejected, but since the discussion was not included in the record, we do not know why. At any rate, we do have a record of what was probably a typical initiation ceremony at the time.

The convention had voted to increase dues and to raise officers' salaries, but these proposals were defeated in the subsequent referendum. A detailed plan for tool insurance was also adopted by the convention, since carpenters were required to supply their own tools and theft was a frequent occurrence. This was also rejected in the later vote, *The Carpenter* commenting that "the entire subject is a novelty for the outside carpenters . . . It is not understood and for this reason unpopular with them."[53]

Duffy was nominated once more without opposition, but T. M. Guerin, the first vice-president, was nominated in opposition to Huber. The latter won by a slender margin in the referendum vote, 24,113 to 22,401. The votes of 204 locals were disqualified by the tabulating committee, mainly because they were not taken during the stipulated time period. There was some complaint, but the GEB confirmed the results, since there was no evidence of any bias in the pattern of disqualification.

The closeness of the vote revealed that there was considerable opposition to Huber within the union. Some of this may have come from the former McGuire supporters, but a good part of it appears to have resulted from what often happens in an organization after the departure of a strong leader. McGuire had dominated the carpenters for two decades by virtue of his personal qualities as well as his hold on the affections of the rank and file. The manner in which Huber came in left a great deal of bitterness among his opponents, and he was never able to gain full control of the Brotherhood. In a sense, the Huber regime constituted an interregnum bridging the presidencies of two of the outstanding union leaders in America labor history.

DEVELOPMENTS IN SAN FRANCISCO

The Carpenters played a leading role in the growth of trade unionism in many cities, but nowhere more than in San Francisco. Under the pressure of the Carpenters the Building Trades Council of San Francisco was founded in 1896, with Henry Meyers of Local 22 as president. He was replaced in 1898 by P. H. McCarthy, who had been president of Local 22 and of the Carpenters' district council.

Patrick Henry McCarthy had been born in Ireland; he came to the United States at the age of 17, working in Chicago and then in St. Louis, where he met McGuire. He migrated to San Francisco in 1886 and played an important part in both the labor and political life of that city for a quarter of a century. From 1904 to 1908 he was a member of the Carpenter's GEB; he was elected mayor of San Francisco in 1909 on the Union Labor ticket. "His administration was criticized severely by some people, particularly from the standpoint of the increase in the number of saloons and of the activity in the tenderloin area of the city." Defeated for reelection in 1911, he remained president of the Building Trades Council until 1922, when he became a contractor. He career was summarized as follows:

> He died in 1933. P. H. McCarthy's ability was beyond question; he was an able organizer and administrator. However, it was said that he was selfish and domineering; that power went to his head; that he was determined to "rule or ruin." While questions were raised from time to time as to the disposal of funds of the Building Trades Council, there was never any evidence of the use of bribery or the levying of "strike insurance" by the officials connected with the Building Trades Council or any of its affiliated unions.[54]

In 1901 a conflict developed between the Building Trades Council and the San Francisco Labor Council, the local body of the AFL. Many of the building crafts had withdrawn from the Labor Council, in part because of McCarthy. "The ambitions of P. H. McCarthy may have explained to some degree this attitude of the Building Trades Council toward the central labor body. McCarthy was merely a delegate in the Labor Council and it is quite possible that this subordinate position hurt his vanity."[55] The immediate cause of hostilities was the refusal of the Labor Council to join with the Building Trades Council in the boycott of an entrepreneur who was using nonunion fixtures in his establishments. The Building Trades Council notified all the building unions affiliated with the Labor Council that they too would have to withdraw or lose their Building Trades affiliation. A number of them did, but Carpenters' Local 483 refused to do so, partly out of resentment of McCarthy's domination of the Building Trades Council, whereupon it was suspended from the latter. About 250 members of Local 22, who sympathized with Local 483, formed a new Local 1082. These two bodies joined forces with Local 616 (stair builders) and 304 (Germans) to fight McCarthy and the Building Trades Council.

At first the GEB came down squarely on the side of these four locals. In October 1901 it ruled that "local unions may be affiliated in central bodies without being persecuted by other locals which may be affiliated with a Building Trades Council," and directed that Local 483 should not be injured in any way by other locals.[56] This had come in response to the refusal of the Building Trades Council to recognize the cards of Local 483, and a personal appearance before the GEB by Frank Morrison,

AFL secretary-treasurer, with the request that something be done to settle the dispute.

This failed to achieve any results, and a few months later the GEB issued a stronger statement. "Local 483 or any other local may belong to the Labor Council if they so desire, and this Organization will not stand for any discrimination against members of the UB who may belong to a local union affiliated with the Labor Council so long as said union complies with the working rules governing the UB in their affiliation with the BTC."[57] Again, McCarthy refused to yield.

Local 483 attempted to secure an injunction against the Building Trades Council to prevent interference with its members, but was turned down. The Builders' Protective Association tried to settle the controversy, but when its efforts came to naught, it came down on the side of the Building Trades Council. Finally, the GEB suspended four locals loyal to McCarthy, including Local 22, and dissolved the district council.[58]

The AFL executive council stepped in and proposed a scheme whereby the San Francisco unions would be divided into five departments, each of which was to coordinate the work of its affiliates, the central council to be merely advisory. It also recommended that the suspension of the Carpenter locals be lifted.[59] The Building Trades Council accepted this plan, for it would have been given dominance over the building crafts, but the Labor Council refused to go along after a negative referendum vote.[60]

A special committee was appointed at the 1902 Brotherhood convention to seek a solution. It managed to secure an agreement whereby the suspended locals were immediately reinstated, while a new district council was to be elected to replace the old one.[61] The four anti-McCarthy locals were permitted to affiliate with the Building Trades Council without resigning from the Labor Council. However, the controversy between the Building Trades Council and the AFL Labor Council continued until 1910 when McCarthy brought an end to the controversy, possibly because he was interested in solidifying his support for an upcoming mayoralty campaign.

Although McCarthy was obliged to back down in the internal Carpenter conflict, the real basis of his power was the Building Trades Council. After a lockout in 1900, which the building unions won with the help of $40,000 supplied by the Carpenters, the employers agreed to purchase only union-made lumber or other material. "The score of years following the 1900 lockout was a period of carefully regulated employer-worker relations. The Council, under the leadership of P. H. McCarthy, not only controlled the industry 'from the foundation to the roof,' but it also had complete supervision over employers and employees and their organizations that were parties to the industry."[62]

A letter to *The Carpenter* from the president of Local 815 in nearby Haywood complained that high initiation fees and dues were keeping men out of the union.[63] McCarthy took issue in a reply in which he claimed that there were very few nonunion men left in San Francisco, and that carpenters were quite willing to pay the $20.00 initiation fee. He continued:

> Organized labor the country over, in all its departments, in order to be successful, must be conducted in accordance with improved business methods. When the labor movement is constructed on strictly business lines, to the exclusion of sympathy, prejudice, passion or hate, then and not until then will it be a success . . . organized labor must be financed, because while it had done a wonderful job for the wage earner, it has much more to do, and because of what it has done, the non-union man has no right to object to payment of the initiation and dues of today, and he will not object if he knows, as he certainly must, that to do so means loss to him.[64]

Christie, in his history of the Carpenters, referred disparagingly to McCarthy as "politician and labor boss extraordinary," "czar of the trades," who tied up the city with monopolistic combinations.[65] Compare this with the report of Carpenter organizer M. C. Hughes, who wrote in 1908: "In mentioning organized labor in San Francisco I feel that with it I must mention the name of P. H. McCarthy for a greater friend to unionism does not exist. He has spared no pains to further its interests; and there is no doubt that the conditions existing today in San Francisco are largely due to his efforts."[66] Perlman and Taft refer to the McCarthy era as the "the labor barony on the Pacific Coast," but concede that he installed the union shop, the protection of employers from low-wage competition, and the elimination of sympathetic strikes and of favoritism among contractors.[67] The open-shop drive of the Citizens' Alliance made no headway in San Francisco. Or finally, compare the situation in San Francisco with that reported by organizer E. Rosendahl from Los Angeles in 1909:

> The city was at one time fairly well organized as far as the carpenters were concerned, but through poor management, such as giving the bosses about fifteen minutes notice of a raise in wages, and at the same time demanding a half-holiday on Saturday and the closed shop, and not allowing contractors to finish work contracted for under the old scale, and many other bad moves, the unions here are almost runined.[68]

P. H. McCarthy was the prototype of an old-time business unionist. His methods may have left something to be desired, but they were in keeping with the times. He was not disposed to tolerate opposition. He did make a major contribution toward building a strong labor movement in San Francisco at a time when employers were doing their best to eliminate unionism throughout the country.[69] He continued to play a major role in San Francisco labor affairs until 1922, when he was forced to resign as president of the Building Trades Council after a successful employer counterattack.

THE SIXTEENTH BIENNIAL CONVENTION

Prior to the 1910 convention, tensions began building up between the majority of the GEB, led by William G. Schardt, who was also president of the Chicago Federation of Labor, and the general officers. At its meeting of April 1909, the board ruled that all purchases of supplies and orders for printing, which had previously been handled by the general officers, were to be approved by the board in advance, and that the same held for any financial aid given to district councils or local unions to help them fight injunction suits.[70] Huber's reaction was later reported by board member R. E. L. Connolly:

> It was rather an innovation to me; I was under the impression that we were coordinate officers of this organization. We were asked to come in to the General President's office; we went in there, and we had just about got seated — Brother Duffy and Brother Neale were also present — when the fireworks began. We were asked why we "butted in" on the printing matter, and we were given to pretty plainly understand that it was none of our business . . . From that day forward, brothers, the war was on.[71]

Connelly claimed that the board's intervention had led to reduced printing costs.

The next dispute arose a year later, when the St. Louis district council filed charges against Huber for "maladministration, gross neglect in the discharge of his duties, and for oppression in office."[72] The GEB and the two vice-presidents set themselves up as a trial court and proceeded to hear the case, as provided in the constitution. The board decided, on the basis of the hearing, that the charges were not sustained, and sent their decision out to a referendum vote, which went in Huber's favor by a margin of better than 3 to 1. Immediately following the report of the tabulating committee, *The Carpenter* carried a letter from the members of the committee calling for an amendment to the constitution to prevent any further "ill-advised" referendum of this nature.[73] Although he was cleared, Huber had to undergo the indignity of having his administrative abilities and integrity questioned.

Huber reacted by urging the next convention to change the constitutional provisions relating to the status of GEB members. As it was, they had to come to the general office four times a year, at the expense of spending a great deal of time away from their regular jobs. He recommended that they be made full-time, salaried officials, and placed under the jurisdiction of the General President.[74]

Huber's opponents countered by proposing a constitutional amendment making organizers, who were under the president's control, ineligible to be convention delegates, arguing that "organizers were in the convention for political purposes only . . . they should be on the road working for the best interests of the organization in general . . . The organizers had a certain amount of influence in the convention and they used that influence in behalf of certain candidates for office." The amendment was adopted and approved by referendum. The convention then proceeded to adopt Huber's suggestion that between its quarterly meetings, the members of the GEB should act as full-time organizers under the control of the general president, but this was overwhelmingly defeated in the subsequent referendum.[75]

The hostility to Huber was also reflected in a hotly debated resolution that would have required all organizers to be suspended immediately and not reappointed until after the balloting for general officers. The rationale was as follows: "The use of our organizers for electioneering purposes is contrary to the intent and purposes of the United Brotherhood . . . and is a power never intended by our members to be placed in the power of any one man to perpetuate in office any man or set of men, nor to thwart an honest effort to amend our laws."[76] The resolution was defeated by a roll call vote of 234 to 122, but many bitter charges and countercharges were made in the debate.

Although internal politics occupied a good portion of the convention's time, a number of other issues were considered. Duffy reported that 1909 had been a bad year for the Carpenters, but 1910 a good one. Membership climbed above the 200,000 mark for the first time during the latter year, an increase of 22,000 over a two-year period. General Treasurer Neale, in his report, observed that "at no time in our history have we been able to present such a favorable financial report as at present, notwithstanding the great financial panic which we have of late so successfully passed through."[77] A board of trustees had been created to manage the real estate and property of the union, and its first acquisition was the general headquarters building in Indianapolis, which was built at a cost of $72,000. The contractor became ill and defaulted, so the general office supervised the completion of the building. This was the first national headquarters building in the United States to be owned by a union.

The convention voted to increase the salary of the general president to $2,500 a year. At the time, Gompers was being paid $5,000, as were the presidents of the Street Car Workers and Railroad Telegraphers, while the Bricklayers' president was earning $3,000. The increase was defeated in the referendum, as was a dues increase that the convention had approved. The members continued to be very conservative where additional funds for the national office were involved. A similar fate befell a plan to create a fund that would have financed an annual pension of $150 a year for any member over 60 years of age who had been a Brotherhood member for twenty-five years, and could no longer earn over 75 percent of the average wage in the district in which he worked.[78]

That there were still some socialists in the Brotherhood was evidenced by the introduction of a resolution endorsing the Socialist Party and calling upon members to study the doctrines and principles of socialism. Another resolution called for the abolition of the capitalist system and its replacement by collective ownership of the means of production. Both were rejected by the resolutions committee on the ground that it was not the function of the Brotherhood to support political parties or to interfere with members' political views, and the committee was sustained without debate. The socialists had few remaining supporters.[79]

Another resolution urged that the creation of state councils of carpenters be encouraged since those in existence had been helpful in securing beneficial legislation. Despite committee nonconcurrence, the resolution was approved.[80] As already noted, the national office had been hesitant to back organizations at the state level, fearing that they might lead to a dispersion of power. The convention also resolved to appoint a delegate to attend the next convention of the Canadian Trades and Labor Congress.

Women were not made eligible for Brotherhood membership until some years later. However, a Ladies' Auxiliary had been established in Indianapolis, and it was proposed that the union give official guidance to such organizations by chartering them. "It is a well known fact that in places where men are on strike no more loyal support is given the men on strike than that given by the wives and mothers and sisters of the striking men." The resolution was enacted, and thus was created an adjunct to the formal organization of the Brotherhood that played a significant role in its subsequent history. The convention also came down strongly on the side of women's suffrage, a burning political issue at the time.

William G. Schardt, the leader of the opposition to Huber on the GEB, was nominated for the presidency, along with Huber and Harry Payne. In its report on the results, the tabulating committee stated that it had found many irregularities in the ballots, and the committee was split on many cases of disqualification. However, illegal ballots were not confined to any one candidate. The committee concluded that the method of election employed was clumsy and costly, and would eventually wreck the organization. It recommended that officers be elected at the convention, since the membership at large was in the final analysis obliged to rely on the convention delegates for guidance.[81] In any event, Huber was reelected with a bare majoritiy over the combined totals of his two opponents, with Schardt getting the bulk of the anti-Huber vote.

THE GEB VERSUS THE GENERAL OFFICERS

Shortly after the convention had ended, Schardt called a meeting of the GEB and charged that there were discrepancies in Duffy's books. Duffy described the background to the charges in a letter to Morrison:

> The election of National Officers just closed was something that I am ashamed of. Scurrilous, defamatory letters went the rounds of our organization from one end of the land to the other . . . What with the Socialist element on our Board and the Chairman of the said Board running for General President against Brother Huber, you can readily realize what sort of a time we had . . . As far as the election is concerned we had a peculiar tabulating committee. They threw out votes right and left, for cause or no cause. If President Huber had a square deal he would have had at least a ten-thousand majority. The committee took upon itself powers and authority never intended for it. On both reports we win.[82]

An accounting firm was engaged by the board to audit the books. It found that with respect to the stock records, "the clerical work of keeping the accounts had been very poorly done;" that many shipments to locals for which payment had been received had never been entered in the stock ledgers; and that the membership tally as reported by the general secretary was off by about 500. After an investigation of eight weeks that cost $7,000, Duffy was given a clean bill of health as far as any fraud was concerned, but it was found that there was some sloppiness in the manner in which the accounts had been kept.[83] After it was all over, Duffy once more wrote to Morrison about the matter:

> The whole affair from beginning to end was nothing more or less than a scheme, cunningly devised, in order to whip Wm. D. Huber over my shoulders. Our Constitution specifies that he as General President shall supervise the entire interests of the U.B. and the desire of his enemies, or in other words his principal opponent, was to show that there was something wrong at the General Office . . . [But] the bubble burst and there is nothing left. I do not know of any National or International officer who has been put to the test so severely as your humble servant, nor do I know of one whose accounts would come out as correct.[84]

Schardt, who did not stand for reelection to the GEB because of his presidential candidacy, retired from the board at the meeting at which the auditor's report was received and discussed. The minutes record that "the G.E.B. presented the retiring chairman with an emblem ring as a slight token of our friendship for him as a man and appreciation of his services on the G.E.B., as a member and as chairman."[85] Huber and Duffy must have breathed a sigh of relief when he went, but a majority of the board, including R. E. Connolly, the new chairman, were still in the opposition to the general officers. Suggestive of the deteriorating relationships that prevailed between the board and the officers was the fact that while from 1908 to 1910, three-quarters of the appeals from Huber's disciplinary decisions were rejected by the board, the figure fell to 50 percent from 1910 to 1912. By way of comparison, Kirby, Huber's successor, reported that from February 1, 1913, to July 1, 1914, he was sustained by the board in thirty-nine of forty-five appeals, with three reversed, two referred back, and one not considered.[86]

THE LABOR INJUNCTION

Employers who were engaged in the campaign to eliminate unionism found in the labor injunction a powerful legal weapon. Before they were curbed by legislation, courts of equity, both federal and state, possessed almost limitless power to order that specified actions be taken or not be taken. The Carpenters, like other unions, found themselves increasingly subject to lawsuits, and were obliged to spend large sums of money defending themselves.

Leading the assault was an attorney named Walter Gordon Merritt, one of the founders of the Anti-Boycott Association, an organization dedicated to fighting unions. This organization was behind the famous lawsuits of *Loewe* v. *Lawlor*, in which the Hatter's Union was obliged to pay $234,000 — most of which was raised by the AFL through an assessment — for violating the Sherman Act, and *Bucks' Stove and Range Co.* v. *AF of L*, in which Gompers was sentenced to 30 days' imprisonment for not preventing an AFL boycott (the sentence was reversed on a technicality).

The employer efforts did not bypass the Brotherhood. In his report to the 1908 convention, Huber listed twenty-four cases in which affiliates of the Brotherhood had been sued. A few of them involved efforts by expelled members to secure reinstatement, but most were employer suits for damages or restraining orders. Huber commented:

> The injunctive writ is most beneficial, but by judicial usurpation it has become dangerous not only to the bulwarks of our American citizens (the wage workers) but a menace to the proper respect, esteem, and confidence which we, as citizens, should and must have in the judiciary if we expect our republican form of government to maintain and endure . . . All the protection seems to be for the man with the corrugated forehead, and none for the man with the calloused hands.[87]

New York City became the focal point of the injunctive onslaught against the Carpenters. In January 1912 the GEB compiled a list of a dozen cases against the New York district council, to defend which the council had paid out $20,000 in legal fees, half reimbursed by the national office.[88] A few of these cases made legal history, and are worth re-counting for the light they throw on some of the hurdles faced by the Carpenters.

In April 1910 Irving & Casson filed suit in federal court against the New York district council and the general officers of the Brotherhood, with Walter Gordon Merritt as the plaintiff's attorney. The firm sought to restrain the Carpenters from interfering with some decorative work that it was performing in constructing the Cathedral of St. John the Divine. The union had pulled all its members off the job because the material being used by Irving & Casson was of nonunion manufacture, having been produced at the company's Boston shop that was operating on a nonunion basis.[89] A preliminary injunction was issued, prohibiting the union from threatening architects and builders and from calling out men in other trades in support of the stoppage. But it did not prohibit the union from striking the plaintiff and fining carpenters if they worked on nonunion trim.[90] Commenting editorially on the decision, *The Carpenter* noted that the judge had deleted from the order those parts that would have been most destructive of labor's rights. "We can communicate the facts about the mill to each other and can apparently advise our own members not to handle the non-union trim."[91] The record does not show whether Irving & Casson were able to finish the job.

The most important case of all was *Paine Lumber* v. *Neal*, which was eventually decided by the U.S. Supreme Court and represented a renewal of the attack that had failed in the Irving & Casson case. The plaintiffs were eight manufacturers of wood trim in southern and western states, who claimed that because of a combination in restraint of trade between the employers and the unions in New York City, all wood trim made in a nonunion shop was barred from the city. The Brotherhood charged that Paine Lumber, one of the plaintiffs, was operating a shop in Oshkosh, Wisconsin, with 400–500 men and 200 girls; the men were being paid $18.00 a week, the girls, 6 to 9 cents per hour.[92] A preliminary injunction was granted by a federal judge and affirmed by the circuit court of appeals. When the case went to trial, however, the district court held that the union had the right to fine any member who refused to stop work on nonunion trim; that the union constitution's bar of nonunion trim was reasonable, since it was for the benefit of the members; that there was a right to strike against the trim, as well as to use the union's paper to warn members. This decision was handed down on November 6, 1913, and affirmed by the circuit court of appeals.[93]

The case was hailed as "a sweeping victory for organized labor . . . it will go down in history as one of the greatest achievements of organized workmen during the past fifty years . . . If the carpenters had lost . . . they would in effect be driven to work by court order to undo their fellow members in mills."[94] But though the injunction was lifted, the Anti-Boycott Association appealed the case to the U.S. Supreme Court, which handed down a decision in favor of the Brotherhood in June 1917, almost seven years after the original action had been filed.[95] The case had been argued before the Supreme Court on May 3 and 4, 1915, but Chief Justice Hughes resigned from the court to run for the presidency of the United States in 1916, so the case had to be reargued in October 1916. The Supreme Court held that a private party could not secure an injunction under the Sherman Anti-Trust Act, the basis of the original suit, but would have to go through the Attorney General of the United States. Duffy made the following comment on the majority decision: "The majority opinion by Justice Holmes does not justify the old adage that a judge of mature years is a man from whom the milk of human kindness has been squeezed, whose ears and eyes are shut to present conditions and who plods through the ruts made by other plodders."[96]

In the end, the Carpenters won most of the cases that had been brought against them by the Anti-Boycott Association. However, during the rainstorm of lawsuits that fell on the union, large sums of money that could have been used to better purpose were spent on legal fees, and if things had turned out differently, the Brotherhood would have faced financial disaster. General President Kirby, in his first report to the GEB, observed that "no one subject affects our organization so generally as the litigation now pending in the courts of Denver, Chicago, and New York City."[97] Fortunately, the decision in the Paine Lumber case and in several similar ones lifted this threat for the time being, although it arose once more in an even more virulent form a quarter of a century later.

THE SEVENTEENTH BIENNIAL CONVENTION

The 1912 convention opened with an address by Samuel Gompers in which he eulogized Peter J. McGuire: "I knew him, and knowing, loved P. J. McGuire, the founder of your Brotherhood. It was he, too, as much as any other man in America, who did so much to give the life and the spirit to the organized labor movement of our continent in the American Federation of Labor." However, with an eye to the jurisdictional conflict in which the Brotherhood was involved, he continued: "There is one thing I want to say to you with whatever impressiveness I can utter it — that great power brings with it great responsibilities and the exercise of great care . . . There are rights to which the working people of other trades are entitled and which you, in your great power, should bear in mind and keep before you."[98]

The years from 1910 to 1912 had not been particularly good ones for the Brotherhood. Membership dropped by 5,000 over the period, and the general office had paid out $242,000 in aid of strikes, by no means all of them successful. In his report to the convention, Duffy condemned ill-timed strikes. "The haphazard fashion of entering into them, whether appropriate or not, whether trade is good or bad, whether the men are organized or not, without due deliberation and careful consideration, should be stopped . . . We have advised our unions time and again that unless they see their way clear to victory they should not enter into any of these movements at all, as we realize that lost strikes are detrimental and a drawback to our organization." On the other hand, Huber spoke of the largest Brotherhood strike in 1911, involving 3,000 furniture workers in Grand Rapids, Michigan, in positive terms; the strike had been lost, but he felt that the workers would have prevailed had the general office been in a position to continue financing it. The level of strike benefits, $4.00 a week, was too low, but even this level was endangered by the weak financial position of the national office.

Huber raised an issue that was again causing problems in some of the high-wage cities like New York. Trim bearing the union label was put on the unfair list if it derived from mills that paid 30 to 40 cents an hour less than the city rates, but that nonetheless were working under union contract. He urged that such trim be accepted lest it endanger the working standards that had been secured in the lower-wage rural areas.[99] The city locals had acted to protect their employers, and had not taken kindly to previous efforts by the national office to secure the admission into their jurisdictions of economically damaging material.

Right from the start, the political struggle that had dominated the 1910 convention erupted once again. The immediate issue was the attempt of the Huber forces to seat several national organizers as delegates, a practice that had been barred by a constitutional amendment adopted by the previous convention and approved overwhelmingly by the membership vote. A local union had challenged the validity of this amendment as being in conflict with other portions of the constitution, and Huber had sustained the appeal. After a long discussion, in which Huber and Duffy argued in behalf of seating the organizers, the opposition prevailed. Huber had committed a serious tactical error in attempting to override a clear expression of membership opinion.[100]

A great deal of convention time was devoted to the controversy that had taken place the previous year between the general officers and the GEB over the condition of the union's accounts. Duffy complained: "Again I am on trial . . . Things are not altogether smooth underneath the surface; we are not working in harmony in this organization. We are split and divided, and it is a mystery to me how it is that we grow and prosper under these conditions." He asserted that when the auditors had cleared him, his opponents had said: "'We'll find something on him; we'll get other accountants.' Why? Because it was politics, and now is the time to cut them out and get down to our laws." He laid the blame for these attacks on board members Schardt, Connolly, and Wahlquist.

Schardt defended the board's action as justified by the fact that there were erasures in the stock books as well as other irregularities. "When Brother Duffy casts slurs upon me, I am here today to defend myself. I expect my name to come before this convention as a candidate [for the presidency]." Connolly remarked not without some justification that "it is a feature of Brother Duffy's explanations to inject extraneous matters into his arguments and have your mind going in several directions at one time." Wahlquist, referring back to earlier events, charged that "every time we have endeavored to improve the book-keeping in this organization we have had to crucify an Executive Board to do so." Board member Bauscher, who was generally supportive of Duffy, conceded nonetheless that "highly improper methods had been used in inserting some of the entries in the stockbook in order to balance it and make the figures conform to the stock on hand. Now, we did not find much, true, I will agree to that." Another Duffy supporter, Post, expressed unhappiness about the fact that "since our Des Moines convention there has not been enough handshaking in our general office among our general officers in order to have things go as they should go . . . If I were to tell you some of the remarks that were passed against your General Secretary it would make you shiver."[101] It must have been a very rough two years in the general office of the Brotherhood.

Fireworks erupted once more during the debate on a resolution calling upon the tabulating committee for the last referendum to explain why it had thrown out the votes of 7,000 members. The chairman of the tabulating committee, Thomas Ryan of New York, a strong opponent of Huber, defended what had been done, and added: "Let me tell this convention, and let me tell Brother Huber, that if it were not for me he would have been counted out." Huber replied with a bitter attack on Ryan. He "accused me of being drunk on my return from St. Louis, and said things were crooked in the general office." Huber read letters and affidavits to the effect that Ryan was slandering him, and concluded: "Why did they not turn the ballots over to the General Secretary? I will tell you why: they thought we were all crooks; in fact, I was told to my face that we were rotten from the top to the bottom, and one man would have went out the window, as old as Bill Huber is, if that expression had not been withdrawn."[102]

It must have become clear to the delegates during the course of the debates that this situation could not be permitted to continue if the Brotherhood were to make progress toward the achievement of its primary objectives. Indeed, this squabbling at the top may have been responsible for the failure of the organization to make progress during the previous two years, despite the relatively favorable economic conditions that prevailed. At any rate, a far-reaching change in the governing structure of the organization was made by the convention, and subsequently ratified by the membership. The GEB was enlarged to include as full voting members the general president, first general vice-president, general secretary, and general treasurer, with the general president to be the chairman. All GEB members were to be full-time salaried officials, devoting their entire time to the interests of the Brotherhood under the supervision of the general president. By thus bringing the board members under the direction of the general president, a continuation of the controversy between the general officers and the board members became much less likely, even though the board members continued to be elected by the convention.

The position of the general president was further strengthened by the defeat, in a referendum, of a proposal that would have required organizers to be appointed in the districts in which they resided, a proposal that would have made their assignment less flexible. Also defeated was a set of rules laying down a procedure for recalling general officers. And the membership finally approved an increase in dues to a minimum of 75 cents a month for beneficial members and 50 cents for nonbeneficial members.

Apart from the constitutional changes, the convention considered several additonal items of interest. There was a preliminary report on the feasibility of establishing a home for retirees and victims of tuberculosis, a disease that affected many carpenters at the time,[103] reflecting the conditions under which they worked. A resolution that would have made it mandatory for local unions to affiliate with state councils of Carpenters was defeated, despite committee support. A proposal that one page of *The Carpenter* be opened up prior to an election for discussion of the issues by candidates for office was defeated on the ground that "*The Carpenter* should not be used for the political purpose of any candidate for national positions in our Brotherhood." Proposals to establish a sick benefit fund and an unemployment benefit fund went down to defeat, the former because it was felt to be a matter for local action, the latter because "we are not in a position financially to take this matter up and deal with it in a proper manner."

Two resolutions introduced by Local 309 of New York City were approved by the convention, rather surprisingly. The first, calling the unemployed "part of the reserve army of capitalism," proposed that the Brotherhood "propagate among our members the abolishment of the present wage system and the establishment of a cooperative commonwealth, where the problem of unemployment, with all the accompanying misery, will be banished from the human race." The resolution also called for discussion of social issues at local meetings. The resolutions committee recommended against adoption, but after some discussion, which was not printed in the record, the convention endorsed it.

The second resolution noted that the president of R. H. Macy and Company, a New York department store that had refused to employ union carpenters, was a member of the National Civic Federation, an organization of liberal businessmen, labor leaders (including Samuel Gompers), and public figures, which had fought the open shop and attempted to mediate several major labor disputes. The resolution termed the Civic Federation "an organization founded . . . for the purpose of establishing peace and harmony between capital and labor; and Whereas, we are absolutely opposed to such harmony between capital and labor as that here above mentioned; therefore, be it Resolved: That the officers of our organization shall sever their connection with the Civic Federation."[104] In 1907 the GEB had authorized the general president to attend a meeting of the Civic Federation,[105] since its activities appeared to be very much in line with the ideological bent of the Brotherhood leadership. However, by 1912 the Civic Federation had virtually ceased to exist.

It is not easy to understand why the convention approved the first of these resolutions, but it may be that the heated national presidential campaign taking place at the time of the convention had something to do with it. Former President Theodore Roosevelt was running on a third-party ticket against William Howard Taft, the Republican nominee, and Woodrow Wilson, the Democratic nominee. A month before the convention, Duffy had written to Gompers asking whether he had "handed over to Wilson the vote of the organized wage workers of our country as represented in the American Federation of Labor. Some individuals resent this action on your part without consultation first. Now understand me, Sam, I am not finding fault nor am I kicking; I am simply trying to place matters before you as they appear." He told Gompers that there was a move afoot to hold a conference outside the AFL to endorse Roosevelt, although Duffy himself had defended Wilson's record within the Brotherhood.[106]

It might also be added that some remnants of socialist influence still remained within the Brotherhood, although Duffy's earlier characterization of the opposition to him and Huber as "socialistic" appears to have been an exaggeration. In this respect, the Brotherhood was by no means unique. At the 1912 AFL convention Max Hayes, a socialist printer, ran against Gompers for the presidency and polled 30 percent of the votes. Among the delegates voting for him were the representatives of the Brewery Workers, the Bakers, the Machinists (whose president, William H. Johnston, was an avowed socialist), the Painters, the Tailors, the Printers, and most of the Coal Miners.[107]

When the time came for the nomination of officers, Huber declined to run again. He told the convention: "The salary you pay your general officers is insufficient. I have been a pauper for twelve years, that is long enough for me. I have been on the road about two-thirds of my time and am allowed only $1.50 per day spending money and I have to meet Presidents, Congressmen, from the highest to the lowest man in the ditch." At this time Huber was 60 years old, and the duties of office plus the constant infighting within the top leadership had taken their toll. His continuing popularity was attested to by the fact that the convention took the unprecedented step of voting him $2,500 in recognition of his services. James Kirby was nominated by the Huber faction to replace him, while the opposition nominated William Schardt. R. E. L. Connolly, another GEB member and opposition leader, ran against Duffy. Kirby received 35,119 votes to 26,107 for Schardt, while Duffy won by 36,773 to 23,516.[108] Wahlquist was defeated for reelection to the board, which meant that the leading oppositionists were all gone from the national leadership of the Brotherhood.

William L. Hutcheson was elected second vice-president against two opponents. Hutcheson was born in Bay City, Michigan, in 1874. After six and a half years of school, he went to work in local sawmills. He spent several years as an itinerant carpenter, a common event in the lives of young men at the time; he then returned to Michigan and joined Local 334 of Saginaw in 1902, and became a full-time business agent four years later. His first and only setback on the union ladder was in 1910, when, as a delegate to the national convention, he was nominated for the post of GEB member from his district and lost to John H. Potts, an older man from Cincinnatti. He was more successful in 1912, when he ran for the second vice-presidency. This post did not then carry with it membership on the GEB, and it would have been difficult to predict that within a few years, he was to occupy one of the most important offices in the American labor movement.

The 1912 convention marked the close of an era for the Carpenters' Union. Factionalism at the national level had come to an end, though it continued to linger in a few localities. The concept of the union as an organization whose role was primarily to improve the material conditions of its members, eschewing politics, was firmly established. The governing structure had been rationalized, and there was no longer any question of where executive authority lay. Everything boded well for a successful Kirby administration.

7

CONSOLIDATING THE NATIONAL UNION

James Kirby, a millwright by trade, was the second full-time president of the Brotherhood of Carpenters and Joiners. Born in Illinois in 1865, he joined Local 199 in Chicago in 1898, represented the local in the Chicago district council, and eventually rose to the council presidency. He played a major role in the national organizations of building trades: first as vice-president and president of the Building Trades Alliance, then as first president of the AFL Building Trades Department, from 1908 to 1910. When he assumed the general presidency of the Brotherhood in 1912, it was a badly split organization. At his untimely death in 1915, he left it united. Although his tenure of office was short, he played an important role in the development of the Brotherhood.

The Kirby era was marked by generally favorable economic conditions. Kirby reported in 1914 that business was good until October 1913, and then after a dull winter, things picked up. This was reflected in an initial jump of 23,000 (9 percent) in Brotherhood membership from 1912 to 1913. But the outbreak of World War I had a depressive effect on the economy, with average unemployment rising from 4.4 percent in 1913 to 8.0 percent in 1914, and again to 9.7 percent in 1915, higher than any previous twentieth-century rate. Brotherhood membership fell by 6,000 in 1914 and by another 18,000 in 1915, putting the total back to the 1912 level. Even a new national administration, not handicapped by factional controversy, proved unable to advance in the face of an unfavorable turn in the business cycle.

THE APPRENTICESHIP SYSTEM

The apprenticeship system had been installed before unions were organized. Both masters and journeymen were interested in securing an adequate flow of competent workmen. Employers often took on a larger number of apprentices than the journeymen would have liked in order to take advantage of their lower wages. The relationship of journeyman to apprentice was often close. For example, early apprenticeship rules adopted by the carpenters' union of Tacoma, Washington, specified that "nothing in this article shall be construed to prevent any minor son working with his father, and under his instructions, said father being a member of this union."[1]

The Brotherhood was concerned about apprenticeship from its start, with mixed emotions. It wanted to make certain that a limited number of well-trained journeymen entered the trade each year, with a level of skill sufficiently high that they could perform all tasks required of a well-rounded carpenter. There was concern lest the training be done in vocational schools, for "the products of the industrial schools, having cost but a nominal sum, would be placed on the market in competition with those products exclusively the work of journeymen. This would tend to a still further reduction in wages."[2] The solution was a union-regulated system, and to that end the 1888 convention recommended that a four-year indenture be required, and that specific limits in the apprentice-journeyman ratio be installed.[3] However, because of the diversity of local conditions, no attempt was made to create a national system.

Once it became evident to union employers that they could not use the formal apprenticeship system to limit wage increases, they tended to lose interest in it. Smaller contractors found it difficult to provide steady employment for young men. Larger employers began to look upon apprenticeship as a means of restricting entry to the trade, with a consequent tendency toward higher wages.

Almost every convention dealt with the apprenticeship problem, but found it difficult to resolve. In 1909 the GEB recommended that no one over the age of 21 years be admitted as an apprentice, and that there be a minimum training term of three years.[4] A proposal calling for regularization of existing programs and the establishment of evening training schools run by local unions was adopted by the 1910 convention, but was defeated in a referendum vote.[5] In 1912, however, a resolution was adopted, and subsequently ratified, setting a four-year term for an apprenticeship initiated between the ages of 17 and 22 years, with the number of apprentices and the wage levels to be set by each district council for its locality. Any apprentice breaking the agreement was to be barred from Brotherhood membership.[6]

This plan was merely advisory, not mandatory, and there was a good deal of variation among localities in the specific schemes adopted. The New York unions allowed one apprentice for every ten journeymen, with every employer entitled to a minimum of one. In Pittsburgh, anyone under 25 years of age was permitted to enter the program. In San Francisco, special apprentice working cards were issued, with the following limitations on numbers: two apprentices for every eight journeymen; three for thirteen; five for twenty; six for twenty-eight; and ten for forty-eight, with ten the maximum. These liberal ratios reflected the relative scarcity of skilled labor in the West.

The Chicago system, which was often recommended as a national model, was based on an agreement between the district council and the Carpenter Contractors Association. All apprentices were to be laid off during the first three months of each year to enable them to attend a special school. Their regular wages were to be paid during that period, and school attendance was compulsory, with fines for unexcused absences.[7]

It was established as national policy that all apprentices were to join the union as semibeneficial members immediately upon entering the program, and to remain in that status until satisfactory completion of their training.[8]

Duffy was a staunch supporter of industrial training in the public schools as a basis for improving skill levels. He told the 1916 convention: "We wanted to get away from that education, that part of it at least that is no good to our boys and girls, and will be of no use to them when they get out into the world to earn their living. We said to the school authorities: 'Cut out the refuse.' What do the children want with the dead languages, or with ancient history. Give them a good practical education, prepare them for the trades they want to follow, and we will give them the practical end of it."[9] At this time Latin was widely taught in the high schools, as well as Roman history. Duffy was no great believer in the value of humanistic education.

THE EIGHTEENTH GENERAL CONVENTION

In April 1913, between the seventeenth and eighteenth conventions, an event occurred that was to have great significance for the future of the Brotherhood. According to a constitutional amendment adopted in 1912, the first vice-president was to be in residence at the general office headquarters in Indianapolis. Arthur A. Quin, who had been elected to that post, reported that when he arrived there, he discovered that his eyesight was not up to the demands that the office placed on him.[10] This may have been an excuse for his unwillingness to leave New Jersey, where he was president of the State Federation of Labor and had great influence in the labor movement.[11] Whatever the reason, William L. Hutcheson became first vice-president, while Quin assumed the office of second vice-president, which permitted him to remain in New Jersey. Thus when the 1914 convention opened, Hutcheson was installed as second in command, a rapid promotion for a man who had been elected to his first national office only two years earlier.

It is of interest to compare Hutcheson's rise to national office with that of the man who was to be his principal opponent in AFL politics for several decades, and with whom he maintained friendly relations despite the divergent views they held on some issues. John L. Lewis was a legislative agent for Illinois District 12 of the United Mine Workers, and in 1917 became statistician for the national union, his first national post. When President White of the Mine Workers joined the wartime Fuel Administration, vice-president Frank Hayes became president and appointed Lewis acting vice-president. Hayes resigned in 1919 and Lewis succeeded him as president, with only two years in national office behind him.

The two men knew one another very well. For many years the Miners, like the Carpenters, had their headquarters in Indianapolis, and for a time they rented space in the Carpenters' headquarters building. In 1921, when Lewis ran for the presidency of the AFL against Gompers, he received the vote of the entire Carpenter delegation, headed by Hutcheson. The two men shared much the same political and trade union philosophies, although the nature of their constituencies brought them into conflict for a relatively brief but important period in the 1930s.

Kirby's report to the 1914 convention contained some interesting remarks that foreshadowed the expanded role that the national union was later to play in collective bargaining. He started with a strong statement on the importance of observing both the letter and spirit of agreements, and added: "I do not believe in evasive and underhanded tactics. I believe such actions despicable, and no matter if our employers should adopt those methods it does not justify trade unionists in doing so." He noted that the general office had assisted in negotiating agreements, but was never a party to one. Then he continued:

> It would, however, be to the best interests of the Brotherhood if the General President, in conjunction with the General Executive Board, had the authority to sign agreements with employers doing an interstate or international business. This would tend to insure the employment of union carpenters, at the same time it would protect the employer from unnecessary interruption of his work. An agreement of that character could not set about to fix the wage scale or working conditions, but it could provide that union men must be employed and the prevailing rate of wages paid.[12]

The problem of the traveling employer had not yet become acute, but it had already been dealt with by other unions that were, like the Carpenters, mainly concerned with local product markets. Ulman cites the Theatrical Stage Employees, the Bill Posters, and the Bridge and Structural Iron Workers as having established national rates applying only to work "done by members travelling from city to city in continuous employment of one firm or individual."[13] The Carpenters were by no means alone in wrestling with local autonomy versus the necessity of providing some standardization of wages and working rules for the emerging contracting firms operating on a national basis.

The 1914 convention adopted some constitutional changes that served to augment further the powers of the national union. On Kirby's recommendation, a resolution was introduced to give the general president, in consultation with the GEB, authority to consolidate local unions where too many small locals existed in the same community. Duffy pointed out, for example, that there were 67 locals in New York City, 13 of which had fewer than 50 members, and 20 fewer than 100 members. This amendment was opposed from the floor on the ground that it put too much power in the hands of the general president, but the convention passed it. The general president was also empowered to "take possession for examination of all books, papers, and financial accounts of any L.U., D.C., S.C., or P.C. summarily when he may deem it necessary, and the same shall remain in his possession until a complete report has been made and filed."[14]

Also in line with Kirby's suggestion, the GEB was authorized to enter into agreements with employers, provided the terms conformed to the trade rules of the district where the work was to be done.[15] This opened the way for what was later to be an important national union function: dealing directly with large contractors operating on a national basis.[16] An attempt to curb the authority of the general president to appoint organizers, and instead to have them elected by the locals in the districts in which they were to work, went down to defeat after a long debate in which the centralization of power was an explicit issue. All the constitutional amendments relating to expansion of the central authority favored by the national office were carried in the subsequent referenda. The intense factionalism that frustrated previous efforts to move in this direction appeared to have dissipated.

The convention sustained a ruling of the GEB requiring a 51 percent affirmative local vote before it would sanction a strike. A committee report putting the convention on record "as favoring public ownership and control of public utilities and machinery associated with the presenting to the laborers of our country the necessities of life" was carried. A populist spirit still lived on among the Carpenters.

Local 309 of New York proposed once again a political resolution that made the only reference to the war that had broken out in Europe a month earlier. The war was attributed to commercialism, militarism, and imperialism, "spelled in one word 'Capitalism'"; the convention was called upon to condemn every form of militarism in the United States, and the general officers were instructed to inaugurate or participate in efforts to bring about peace. The Committee on Resolutions recommended that the subject matter of the resolution be referred to the AFL, and the convention agreed, despite the Marxist flavor of the wording.[17]

In his report to the convention, Vice-President Hutcheson published detailed statistics on the degree of organization and the level of wages prevailing in woodworking factories, the data having been secured by a questionnaire sent to all locals. They showed 17,345 union men as against 37,490 nonunion.[18] Incomplete though the data must have been, they indicated that although a great many of the mills were not organized, the mill men were becoming a significant bloc within the Brotherhood, which had a total membership of 261,000 at the time.

The ubiquitous Local 309 of New York presented a resolution to the effect that since the Brotherhood was rapidly becoming an industrial organization, in order to avoid jurisdictional quarrels the delegates to the AFL convention should be directed to "propagate to their best ability the principle of industrial form of organization."[19] The craftsmen who constituted the bulk of the membership were not enthusiastic about this idea, particularly if it might lead to free entry of the mill men into building construction jobs.

In a statement defining the Brotherhood's jurisdiction, the Committee on Constitution added "auxiliary unions" to the list of bodies that might be chartered. When questioned about the meaning of this term, the committee chairman said it was "very elastic," and could embrace both ladies' auxiliaries and former locals of the Amalgamated Wood Workers "who were not fully qualified for membership in the Brotherhood." A motion to insert the word "ladies" before "auxiliary unions" brought forth a fuller explanation of the committee's intent:

> The object and intent of the committee . . . was to give us a chance to handle the men that we cannot handle otherwise; to put them off to one side under the banner of the United Brotherhood, with no voice in the doings of the Local Union or the Brotherhood. It is simply a side issue, and the ladies the same. We would give them a charter from the Brotherhood, but it does not give them the right of voice in our District Councils or anywhere else. But they are under our authority and must obey the mandates of our General Constitution.[20]

By way of example, the case of Boston hardwood finishers was cited; they had come into the Brotherhood when the Painters refused to accept them. Another example given was cue and billiard ball makers, who joined the Brotherhood when Brunswick-Balke-Collender was organized. A motion to restrict the "auxiliary" concept to "ladies," which would have meant a first step in the direction of industrial unionism by rejecting secondary status for non-craftsmen, was defeated by a vote of 145 to 115.

Christie interpreted this vote as clear evidence that the Brotherhood leaders "did not intend to unionize indiscriminately the whole industry. Rather, they claimed jurisdiction as the paramount trade in the industry in order to police their industry effectively. There is a vast difference between policing and unionizing an industry."[21] There is something in this view, but it was not as clear-cut a decision as he makes it out to be. The Brotherhood was still overwhelmingly an organization of skilled craftsmen, and there was a long way to go before the limits of the craft jurisdiction were reached. Organization was neither cheap nor easy, and the resources that were available had to be devoted largely to signing up the construction men. It was necessary to claim jurisdiction over the mill hands, particularly in the larger cities, to regulate the competition of the semiskilled machine workers when technological advance enabled them to take over work that had been traditionally done by carpenters. The Carpenters' Union was seeking to protect the jobs of its members, and the long dispute with the Amalgamated Wood Workers, involving mutual charges of scabbing and strikebreaking, had convinced them that this could be accomplished only by absorbing the semiskilled men who were directly involved in the controversy.

There was nothing to stop the Carpenters from adopting an industrial union stance, and the closeness of the vote over the definition of "auxiliary union" indicates that there was considerable support for such a policy within the union. But Hutcheson's circular inquiry had come up with a figure of 1,673 nonunion mills as against 1,172 unionized. Going after these mills, many of them small, may not have seemed like a good investment to the Brotherhood leaders, particularly if their location or the type of work they were engaged in did not threaten the primary carpenter jurisdiction. It was not a question of "policing" the manufacture of wood, but rather one of preventing the erosion of job opportunities for carpenters. The Carpenters were doing what a good business union should have been doing: allocating their resources in the best way to achieve their objective function — the maximization of jobs for skilled craftsmen. Peter J. McGuire may have chosen another road and sought to organize all the semiskilled mill workers, but his conception of unionism no longer dominated the United Brotherhood.

In marked contrast to what had happened at earlier conventions, the incumbent general officers and all but two GEB members were renominated without opposition. The antagonism between the general officers and the GEB was a thing of the past.

CLEARANCE CARDS

The centralizing tendencies of the convention were reinforced by subsequent GEB action on clearance cards.[22] In an editorial headed "The Stay-Away List Growing," *The Carpenter* noted that even when economic conditions improved, no deletions from the list came in. As it had done in the past, it urged local unions to exercise great care in keeping any Brotherhood members out of their labor markets:

> It should be borne in mind that the traveling brother is not always favored of fortune. As a general rule his travels in search of work may in many cases be traced to ill-luck or depression in the trade in his home town. To lock the door against him when there is no need of it is a flagrant injustice and violates the great fraternal principle which is one of the cornerstones of our organization.[23]

A year later the journal carried a letter from a local in Ann Arbor complaining that being on the stay-away list did more harm than good, since it kept union men away and permitted nonunion men to secure available work.[24] Complaints continued to come in that clearance cards were being withheld arbitrarily, and the GEB was obliged to act.

Under the constitutional provisions in effect at the time, the application of a traveling Brotherhood member for a clearance card had to be voted on at a local meeting of the host union, and a favorable majority vote was necessary for admission to the local. The GEB ruled that no local had the right to pass a resolution or send out notices to the effect that it would not accept clearance cards. "The clearance card system is the means whereby we preserve unimpaired the lines of industrial intercourse between our members, and any breakdown in it correspondingly weakens the spirit of fraternity which should exist in our organization."[25] This did not solve the problem completely, since the local members could always vote against the admission of travelers, and the constitution was later amended to make the issuance of clearance cards compulsory unless there was a strike or lockout in the locality concerned. In the meantime *The Carpenter* discontinued the stay-away list, pursuant to a GEB decision, in order to discourage local exclusionary policies.[26]

To promote solidarity among members, *The Carpenter* published a communication from John Quinn, a member of Local 714 of Queens, New York, in which he set forth ten commandments for the Brotherhood. It provides an interesting insight into some problems faced by local unions at the time:

1. Thou shalt be a union carpenter, a member of the U.B.
2. Thou shalt not belong to any other organization. . . .
3. Thou shalt keep whole the Saturday half-holiday and all other holidays. . . .
4. Honor and respect thy officers. . . .
5. Thou shalt not become "boisterous" in the meetings of thy local union and want to lick anyone who may disagree with thy opinions. . . .
6. Thou shalt not commit offenses against the laws of the U.B. . . .
7. Thou shalt not steal from the boss. . . . Show him it pays to employ U.B. members.
8. Thou shalt be charitable toward fellow members. Thou shalt not try to gain favor with the foreman by pointing out their shortcomings. . . . Thou shalt not be a "boss's stool pigeon" for in his heart he shall despise thee.
9. Thou shalt not be envious of thy fellow member if he should happen to be working while thou art on the sidewalk. He may need the money as much as thyself. . . .
10. Thou shalt not covet thy fellow member's good fortune, and if he should happen to own his little home . . . he may have scraped all his life for the few dollars he has in that home, and will keep on scratching the remainder of his days paying the interest on the mortgage.[27]

Not a bad code of conduct, even for today!

FRANK DUFFY AND THE AFL

In 1913 Duffy was elected eighth vice-president of the AFL, and in that connection, some interesting correspondence between him and Gompers has been preserved. When Gompers asked him to use AFL letterheads in dealing with AFL matters, Duffy replied that it had become second nature with him to think of himself as general secretary of the Brotherhood:

> I would rather hold that position than any other in our Brotherhood. No doubt I could have been General President long ago if I had only said the word, in fact at our Niagara Falls convention in 1905 that proposition was made to me to which I replied,"I am better suited for the position of General Secretary than that of General President . . . At that I may not be the right person for the job. I am a carpenter and I received a good every-day elementary school education; beyond that I got my education as best I could, I may not be as bright, as clever or as forcible as others, but whether I am or not I know I hold the confidence and esteem of the members of our organization."[28]

He told Gompers that he had accepted the AFL vice-presidency only when Kirby declined it, and warned that "underneath an apparently fighting exterior" he was extremely sensitive in regard to his nationality, religion, politics, education, and family affairs. Gompers replied that "there is no one who appreciates and values your ability and your devotion to the labor movement more than I do. I always count upon your sympathetic cooperation and support in all the work for the advancement and protection of the rights and interests of the men and women of labor."[29] Relations between the two men were not always so cordial, however, particularly when the jurisdictional interests of the Carpenters were involved.

The sensitivity of which Duffy warned was reflected not long after in another letter to Gompers. Duffy told about a visit to the United States in 1912 of Carl Legien, president of the International Federation of Labor, at the invitation of the AFL. In a tactless article later written about the visit, Legien castigated Duffy as "a fanatical Catholic who will not allow Socialistic matters to be printed in the official monthly journal, The Carpenter"; asserted that most of the membership of the Brotherhood held opinions at variance with Duffy's; and that Duffy was regularly elected to office only through the aid of his political machine. "We are neither Democrats, Republicans, Progressives, Prohibitionists, Socialists, Single Taxers, or any other party; we are first, last and all the time trade unionists, and not politicians," wrote Duffy. He continued:

> I have no machine. I am elected for a term of two years at a time. I became General Secretary on July 24th, 1901, and from that date up to the present

time I had opposition only on two occasions, the Socialists not even protesting when I was reelected time after time unanimously. Can you call that a machine? As far as my religion is concerned, that is my own personal business, and not that of Mr. Legien or of any one else. To tell you that I am sore at the actions of Mr. Legien is but putting it mildly.[30]

Duffy was never shy in approaching the AFL when the interests of the Brotherhood were at stake. In 1915 Local 64 had been expelled from the AFL United Trades and Labor Assembly of Louisville, Kentucky. The Assembly had come down firmly against the Prohibitionist movement that was gaining strength at the time, on the ground that prohibition of the sale of alcoholic beverages would cause unemployment and "deprive the working classes of one of the luxuries within their reach." Kentucky, it should be said, was the major whiskey-producing state in the country. Local 64 opposed this position on the ground that the whiskey companies were using nonunion labor in their construction and repair work. "Of the two evils we prefer Prohibition instead of the open-shop or non-union policy practiced by the Whiskey interests."[31] As soon as Duffy heard of what had happened, he wrote Gompers demanding that the delegates of Local 64 be seated by the Labor Assembly "without further parley or delay." Gompers replied immediately that he had written an "urgent letter" to the Louisville Assembly directing that the Carpenter local be reinstated.[32] The Whiskey Rebellion was crushed.

Another interchange of letters between Duffy and Gompers suggests that the views of the Brotherhood leadership on social legislation were in close accord with those held by the AFL. Gompers wrote asking for Duffy's views on a current proposal to enact an eight-hour-day law for private industry. Gompers expressed his strong opposition to the degree of government intrusion upon industrial relations that this would have meant: "The question involved is one of concern to all liberty-loving citizens and I venture to ask you [for an expression of opinion] in the interests of justice, liberty and humanity . . .I am seeking the judgment of yourself and others who are interested in the maintenance of the fundamental principles of liberty."[33]

Duffy did not disappoint Gompers. In his reply, he went back into history to make the point that the Magna Charta was exacted by the barons to give them more power, and if they could have foreseen that it would have been used eventually to emancipate their serfs, they might not have been so enthusiastic about it. The unintended consequences of an eight-hour law might be the enactment of wage and other restrictive legislation:

It would be an unwise move for labor ever to sanction any such legislation. I am of the opinion that more can be accomplished by strong, self-governed trade unions, for the benefit of the wage workers, than by legislative enactments. You cannot legislate and make prosperity, any more than you can legislate and make a happy and contented people. This can only be accomplished through economic trade organizations.[34]

Finally, there was the "Mexican question." In several communications to Gompers, Duffy objected to Gompers' calling for a conference between the AFL and the Mexican labor movement, on the grounds that the Mexican government was both socialist and anti-Catholic, and that the Mexican unions were paper organizations controlled by the government. He insisted that before Gompers committed the AFL to international conferences, the AFL executive council should be consulted. "The United Brotherhood claims Mexico under its jurisdiction, but so far we have not a single local union in that Country and you know we spend more money for organizing purposes than any other organization affiliated with the American Federation of Labor."[35]

THE NEW YORK AFFAIR

James Kirby died on October 8, 1915, and was succeeded automatically by William L. Hutcheson. It was not long before the Brotherhood was given an example of the style of leadership he was to provide.

We have already seen that the New York carpenters had been among the first to organize, and that although they had eventually affiliated with the Brotherhood, they had maintained a tradition of independence. When Hutcheson became president there were seventy-three local unions in the city, with 17,000 men among them; 40 percent of the locals were organized on an ethnic basis.[36] We have also seen that building trades industrial relations in New York were not of the best; arbitration had been forced on the locals after they had lost a dispute in 1903, and the Carpenters in particular had been the focus of the city's anti-boycott group.

The origin of the controversy between the New York locals and the national office was a demand by the district council late in 1915 for a wage increase; there had been no rise in wages for several years. Secretary Elbridge Neal of the New York district council met with the GEB and requested strike sanction in support of a demand for an increase of 50 cents per day in four boroughs and 60 cents in Manhattan. The board agreed, subject to the conditions that the district council raise strike funds of its own and that an organizing campaign be initiated immediately.[37]

The Employers' Association offered a 30 cent a day increase to go into effect on October 1, 1916, after the peak of the building season, while the council had demanded that the increase take effect on May 1. Prior to May 1, however, the GEB adopted the following resolution: "In approving trade movements the Board rules that in case of failure to reach an agreement with the employers the D.C. or L.U. is directed not to call its members out on strike until the General President can send a representative to assist in bringing about a better understanding and settlement."[38] Neal was informed in a letter from Duffy that he was not to call a strike until this resolution had been complied with. It was charged later that he deliberately concealed the letter from the locals until after a strike had been called.[39] Be that as it may, 17,000 men walked out on May 1, without the required consultation. The small contractors quickly gave in, and somewhere between 10,000 and 14,000 men returned to work at an increase of 50 cents a day, subject to a scale revision when a final settlement was reached with the large contractors — a common arrangement in construction bargaining then as now.

Hutcheson arrived in New York a few days later and claimed that he had consulted with the officials of the district council, who admitted that "the best offer they could secure was one-half of the increase they were asking. They further admitted to us that they were helpless, if it came to a fight with the Building Trades Employers' Association, to properly protect and maintain the jurisdiction of the United Brotherhood." He met with the Employers' Association, together with two GEB members, and reached a settlement providing for 25 cents a day effective July 1 and another 25 cents a day effective September 1.[40]

Apart from the dates of the increase, the Manhattan carpenters objected to the fact that they had not gotten the extra 10 cents that they had demanded. The local leadership was outraged by the fact that the agreement had been made without consulting them, and refused to abide by the settlement. They were sustained in this action by an overwhelming vote of the membership. The response of the national office was quick to come: on May 12, sixty-one locals were suspended by the GEB.[41]

The board also chartered a new local, which, along with three locals that remained loyal to the national union, formed a new district council. Robert Brindell, the head of Local 1456, a dock builders' organization, was the guiding spirit behind the new council. At this juncture, the old district council made a serious tactical error; it applied for and secured a temporary injunction against Hutcheson, restraining him from taking over its property and records. This later gave Hutcheson the opportunity to castigate the council in no uncertain terms: "I can only express contempt and loathing for anyone who will resort to the civil courts and use the weapon that has been used by those opposed to organized labor in an endeavor to crush them and stop their progess, namely, the injunction."[42]

This was not the only mistake. The council employed as its attorney in the injunction proceedings Morris Hillquit, a prominent New York labor lawyer who was also head of the New York Socialist Party. This enabled the loyalist locals to claim that "the supporters and advocates of revolutionary socialism were the cause of the present trouble in the New York District."[43] In fact, Hillquit was not a revolutionary socialist, and while there were undoubtedly socialist carpenters in New York, there was no evidence that socialism had anything to do with the action of the district council. From the point of view of the true New York radicals, the leaders of the district council were anything but socialists:

> They are politicians of the worst sort. Natuarally, there are also exceptions, but the majority of them are bound fast to their jobs, and they lost every sympathy for responsibility to their union brothers. And still, this Council became leaders in this case. It may be that they were driven to it by jealousy. What! A stranger comes to their town and takes away from them such a wholesome bite! They could not stand it.[44]

Another bone of contention was the charge made by Hutcheson's opponents that the $70,000 the Brotherhood had spent financing a seven-month dock workers' strike in 1915 and 1916 was really intended to buy Brindell's support for the new national administration. This charge was aired at the 1916 convention, and Duffy responded as follows:

> Yes, Brother Cook, we did pay that $70,000, every cent of it, and if you come down to the hotel I will show you an accounting for it. There isn't anything wrong about it. We defended these men, we will defend them again, and if any of you get in the same fix and another organization tries to put you out of business the Executive Board will come to your defence and defend you to the last ditch.[45]

On July 13 a committee representing the old district council appeared before the GEB and was told that if the Hutcheson agreement were accepted, the suspended locals would be rechartered. The response of the committee was that it lacked authority to accept this proposition, whereupon the board agreed that it would meet in New York with a committee that was empowered to reach agreement. Two days later a formal appeal against suspension was filed with the board, which decided that "inasmuch as the said D.C. did not comply with the laws of the U.B. before taking their case to the courts the appeal cannot be considered." Hutcheson and three members of the board returned to New York to meet with a committee, but the board reported that the controversy could not be resolved "owing to the fact that the committee representing the suspended Local Unions could not agree among themselves, although they had full power to act in this matter."[46]

This was the situation when the 1916 Brotherhood convention opened in Fort Worth on September 16. The suspended New York locals sent full delegations. They were not admitted to the hall, even as visitors. Hutcheson declared that "no man or body of men with an injunction standing against this organization are entitled to a seat in this hall during our deliberations." Robert Brindell spoke on behalf of a claimed 5,000 New York members who remained loyal to the Brotherhood. A committee was appointed to examine the dispute, and when it was to make its report, some delegates demanded that the New York delegation be permitted to attend. The record reads: "Cries of 'Let New York in to hear this trial.'"[47] When a motion was made to this effect, Hutcheson ruled it out of order. "Delegate Bundy: I appeal from your ruling. President Hutcheson: You can't on a motion of that character. The delegates will please come to order. I am not going to keep repeating that but I am going to have order." He continued to refuse to accept a motion or any appeal from his ruling.

The committee recommended to the convention that the suspended locals be rechartered, but combined into twenty-five locals, and that they become affiliated with a new district council formed by agents of the general president. The debate was long and vitriolic. A Massachusetts delegate challenged the ability of Hutcheson to go to New York and understand the tangled situation there. The committee was criticized for not having presented any evidence to substantiate its recommendations. A motion to hear the entire committee report, including the case made by the New York locals, was brought to a roll call, and was defeated by only 39 votes.

Faced with this convention decision, the New York locals capitulated. They were obliged to pay all back per capita taxes in order to be rechartered. The district council was reorganized; the rebellion was over.

Looking at these events sixty-five years later, and considering the record alone, one can make several points. The first is that Hutcheson, as presiding officer, acted in a high-handed manner. He not only used the power of the chair to rule legally seated delegates out of order, but he refused to entertain appeals from his rulings. He failed to adhere scrupulously to the Brotherhood rules of order. On the other hand, Hutcheson had full constitutional authority to intervene in the New York situation, to reach a settlement with the contractors, and to call off an unwise strike. The failure of the New York locals initially to follow constitutional procedures gave to the national union the power to reorganize them and justified the convention in refusing to seat their delegates. This affair accomplished something that would have had to come about in any event: strengthening the national union at the expense of the district councils and the locals. The nature of the industry was changing from a purely local one to one in which local actions could have serious repercussions on other areas. The role of large interstate contracting firms was growing; prefabrication was to become a national problem; the federal government was becoming a major factor if only because of its growing demand for construction.

This is not to say, however, that local autonomy was destroyed by the New York affair. On the contrary, most collective bargaining continued on a local basis, and both wages and working conditions varied with local circumstances. What was established was that a district council could no longer set up a local barony and tell the national union to stay away.

It is interesting once again to look at the parallelism of events in the only other contemporary union that could match the Brotherhood in size and influence—the United Mine Workers. When John L. Lewis took over the presidency in 1919—four years after Hutcheson—he was faced with a revolt against national union policies by the 90,000-strong Illinois District 12. The charters of twenty-four locals were revoked, and the delegates from the expelled locals were denied seats at the 1919 convention. In 1920 Alexander Howat, the head of Kansas District 14, refused to discipline thirty-three locals that Lewis had suspended because of an unauthorized strike. Howat was expelled and the district reorganized, and eighty-nine locals loyal to Howat had their charters revoked. Howat sought an injunction but failed to get it.

Howat came to the 1922 Mine Workers' convention to protest. "Lewis argued that any member who applied to a court for an injunction before exhausting all the avenues of relief within the union was automatically expelled. The forces in the convention were almost evenly divided. Under the circumstances Lewis felt obliged to put to a vote Howat's appeal from the decision of the chair, notwithstanding his own contention. On a roll call Lewis was upheld by a vote of 2073 to 1955."[48] When factionalism continued in other mining areas, Lewis responded by putting a number of districts into receiverships. The end result was, as in the case of the Carpenters, a more powerful national union and the same circumscription of local autonomy.

The New York affair did have one very unfortunate consequence, however: it made Robert Brindell the most powerful building trades leader in New York. Canadian born, Brindell came to New York in 1905 and went to work as a dock builder. He joined the Independent Dock Union, which was chartered as a federal local by the AFL in 1907 but was suspended in 1910 for failure to pay its dues. The local explained its failure by the fact that it had lost all its records in a fire, and applied for reaffiliation, but both the Iron Workers and the Carpenters objected on jurisdictional grounds, so the local group remained independent. Brindell became its business agent in 1912.

In 1910 employees of the New York City Department of Docks and Ferries, a separate group of dock builders, had become the Municipal Dock Builders Union No. 13041. The AFL organizer who secured their charter said that they had 25 members, with a potential of 300 who were employed by the city.[49]

In 1914 representatives of the Carpenters met with the officers of the independent dock builders, including Brindell, with Samuel Gompers there to represent the AFL. It was argued that the dock builders should be chartered by the Carpenters' Union on the basis of its jurisdictional claims.[50] A Carpenter charter was issued, as Local 1456, but the Municipal Dock Workers Federal Local 13041 refused to go along. Under pressure from the Carpenters, Gompers ordered them to affiliate by June 15, 1915, or lose their charter. When they persisted their AFL charter was revoked, but the Bridge and Structural Iron Workers, who also claimed jurisdiction over dock builders, took them in as its Local Union 177.[51]

The AFL executive council appointed a committee to look into the matter. It reported that Iron Workers Local 177 had entered into a three-year agreement with the Contracting Dock Builders Association that purported to cover all dock building in the city. The committee recommended that there be only one local of dock builders in the city, and that Carpenter Local 1456 be designated for that role.[52] The president of the Iron Workers replied to Gompers: "Why should we obey a decision or rather recommendations of a committee so constituted and biased when their report is favorable to an organization that has never in its history observed or obeyed any decision handed down against it."[53] The Brotherhood clearly did not enjoy universal popularity.

The Brotherhood brought the controversy to the 1916 AFL convention, and secured the passage of a resolution directing the Iron Workers to turn the Municipal Dock Workers over to the Carpenters, on pain of suspension. The Iron Workers were incensed; their AFL delegates claimed that they were not given a hearing.[54] They revoked the charter of Local 177, but turned right around and chartered the same members as a subdivision of Local 189 of Jersey City, across the Hudson River. Duffy labeled this as "deception, trickery, and a farce" and demanded their suspension.[55] When Gompers wrote the Iron Workers to this effect, the president of the organization replied: "Must we withdraw the sub-charter of Local No. 189A and tell the men who compose that body that they are not wanted in our International Association? . . . What are we to tell the 150 men who are now in that local who have been members of our Association for a number of years prior to our controversy with the Brotherhood of Carpenters?"[56]

The AFL executive council had given the Iron Workers until July 1, 1917, to comply, but it was reluctant to suspend the organization over this relatively minor dispute. But Duffy would not brook any delay: "Now Brother Morrison, I may as well be plain with you and tell you that we do not propose to stand for this game of delay. This whole matter has been thrashed out and rehashed so often that for the life of me I cannot understand why there should be any further dilly-dallying."[57]

Gompers wrote to the Iron Workers' president, urging him to comply with the AFL convention decision: "The Executive Council and I, as its President, join in an appeal, not only to your country patriotism, but to your labor patriotism, to deal with this subject in a high-minded and broad spirited manner."[58] But this did not prove sufficient, and the Iron Workers were suspended. However, their charter was reinstated the day before the opening of the 1917 AFL convention when they disbanded Local 189A.[59]

The Carpenters had won complete control of the New York dock builders against a formidable antagonist. But there was still the problem of Robert Brindell, who had been promoted to the presidency of the New York district council. His subsequent career did not bring credit upon the Brotherhood. By 1919 he had become head of the New York Building Trades Council, which included all the building trades except the Painters, Bricklayers, and Plasterers. An investigating committee of the New York state legislature chronicled some of his exploits as follows:

1. The Starboard Realty Co. paid $25,000 to Brindell's representatives to have a strike called off.
2. In 1920 Joseph Paterno was putting up two large buildings. Brindell threatened to call a strike on the ground that the steel erectors were non-union, but refrained from doing so after being paid $3,000.
3. Anthony Paterno was obliged to pay $3,000 to keep a job going even though he was employing union men.
4. Allen Hershkovitz paid Brindell $25,000 to keep the construction of a loft building strike-free. Joseph Goldblatt paid $2,000 for a similar favor.
5. Meyer S. Blumberg was putting up an apartment house, and employed plasterers who were union men but non-members of the Building Trades Council. One hundred and twenty-five men walked off the job, but after Brindell received $5,000, the same plasterers went back to work.[60]

These were by no means the only examples of extortion. Builders discovered that work went much more smoothly when they accepted contractors accredited by Brindell. "On his part, Brindell would recommend to the builder the contractor who was ready to pay him the largest bribe for the recommendation." Not only small contractors but even large ones such as the George A. Fuller Construction Company and the Thompson Starrett Company discovered that "if they wanted to avoid labor troubles they must never award a contract for the demolition of an old building except to a contractor designated or approved by Brindell and that the 'rake-off' must be satisfactory to him."[61]

In the spring of 1920 Hugh F. Robertson, a prominent builder, was engaged in the construction of the Cunard Building, at a cost of $35 to $40 million. A strike was called on the ground that the steel contractors were using nonunion men, but it was called off in fifteen minutes when Robertson went to see Brindell, who suggested that it would be a good idea for Robertson to secure "strike insurance." An agreement on a sum of $50,000 was reached, and at intervals Robertson would meet Brindell, drive him around the block, and pay installments in cash. The job was completed without a strike. It is estimated that through such arrangements, Brindell collected over a million dollars.[62]

In 1921 Brindell was convicted of extortion and sentenced to serve from five to ten years in prison. He was paroled after having served almost four years. The AFL reorganized the Building Trades Council, and Brindell disappeared from the scene.[63]

In his history of the Carpenters, Christie charged that "Brindell was sponsored by Hutcheson for his own purposes. Further, Hutcheson continued to back him in 1919 and 1920, long after it was common knowledge that Brindell was a thief."[64] On the first point, there is no question. Hutcheson used Brindell in his fight against the New York locals. There is no evidence, however, either that Brindell's depredations were "common knowledge" prior to the Lockwood Committee hearings, or that the general office of the Brotherhood was aware of what was going on in New York. Brindell, moreover, was head of the Building Trades Council, and subject to AFL discipline in that capacity.

JURISDICTION (2)

The return of the Brotherhood to the AFL Building Trades Department in 1912 was not of long duration. The issue of voting strength had not been settled, and a heated debate took place at the 1912 meeting of the department. The president of the Painters argued that "the men who are in control of the smaller organizations ought to have something to say on how the organization should be run . . . It does not stand to reason that all the building trades of any city in this country want to be dictated to by one or two organizations." To which a Brotherhood delegate responded: "We do not want to control this body. We want a little protection, that is all. We have had experience in city after city where half a dozen trades that would not form one-third of the carpenters by trade votes in local councils could put our men on the street."[65] The convention referred the whole issue back to the AFL executive council.

Trouble soon developed in two areas: the erection of metal trim, which the Sheet Metal Workers claimed, and the installation of machinery, which the Machinists said belonged to them. Early in 1913 the AFL executive council agreed to inform all state and local bodies that the

Building Trades Department had awarded jurisdiction over metal trim to the Sheet Metal Workers,[66] a decision that did not make the Carpenters happy. At the same time, the AFL Metal Trades Department charged, on behalf of the Machinists, that Brotherhood men had replaced machinists in the performance of millwright work in several firms that the Machinists were attempting to organize. To this, Huber replied: "I have always contended and always will contend that millwrighting is a part of carpenter work."[67]

The millwright controversy also arose in connection with the construction of breweries in St. Louis. An arbitrator appointed by the Building Trades Department awarded the machinery installation work to the Machinists, a decision that was upheld by the department convention.[68]

It was not only these disputes that led to renewal of Brotherhood unhappiness with the department. Duffy developed another theme in a letter to the department's secretary-treasurer:

> The questions of jurisdiction between trades affiliated with the Department are not to be settled by local building trades councils, and local building trades councils should not under any consideration take part in jurisdictional disputes; that is, if the Lathers and Iron Workers have a jurisdictional dispute the other trades should not be called upon to quit work on that account . . . it is unfair to a good union employer to stop his work on account of a jurisdictional dispute between two trades . . . You can rest assured that if something is not done to relieve the Carpenters from the unjust yoke placed upon them by local building trades councils, it will be an impossibility to keep us affiliated with the department.[69]

A few months later, when he was informed that the executive council of the department had refused to recommend changes in the convention voting pattern, Duffy replied: "Fair, flattering and flowery promises were made to us if we would only again affiliate with the Building Trades Department, and the reason for those promises as you well know was to rebuild your department which was then on its last legs as far as authority and actual work was concerned. Without the Brotherhood of Carpenters the Building Trades Department was a failure."[70]

Duffy stoked up the fire by informing Secretary Spencer of the department that on the initiative of the Carpenters' Cincinnati district council, its local unions were being asked to endorse the proposition that the Brotherhood withdraw from the department once again. "Let this be a warning to you in the future that business must be attended to in a business manner and in an up-to-date fashion, and that the method used by the Building Trades Department in staving off from time to time questions of moment and importance will not be tolerated by the rank and file." But Spencer would not be bullied; he replied that the Cincinnati council had never been cooperative with the other building trades, and warned that to withdraw from the department implied simultaneous withdrawal from the AFL as well.[71]

Gompers backed Spencer. He wrote Kirby that withdrawal from the Building Trades Department would invalidate affiliation with the AFL, "due to the fact that the regulations governing the Department require that a union to be in good standing with the American Federation of Labor must of necessity be a member of the Department to which it is eligible."[72] This appeal fell on deaf ears, particularly in view of the fact that the 1913 Building Trades convention went against the Brotherhood on every issue. The Brotherhood was instructed to obey previous decisions awarding metal trim to the Sheet Metal Workers. The claim of the Machinists to millwright work was sustained, by a 36 to 21 vote. And although the Carpenters had 41 percent of the total membership represented in the department, its plea for an increase in its voting strength was turned down.[73]

In February 1914 Duffy informed Gompers that a Brotherhood membership poll had resulted in a decision, by a 3 to 1 margin, to withdraw from the Department.[74] Despite Gompers' warning, the AFL did not act against the Carpenters. To suspend from membership one of its largest affiliates would have been difficult. Indeed, when the Louisville Trades and Labor Assembly refused to seat Brotherhood delegates, it was informed that the Brotherhood was still in the AFL. When the Building Trades Department asked that the Brotherhood's charter be revoked, the AFL executive council not only refused, but urged the Sheet Metal Workers to enter into further negotiations with the Brotherhood.[75]

It is important to note that the Carpenters not only had size on their side in this dispute with the Sheet Metal Workers, but economic logic as well. The Sheet Metal Workers would not furnish men to general contractors for the installation of trim, but required them to subcontract the work to the specialty contractors with whom the Sheet Metal Workers dealt. This would have been uneconomical and impractical, compared with the more direct avenue offered by the Carpenters. Having the general contractors in their corner was of significant help to the Carpenters.

The reaction of the Carpenters was to enter into an alliance with the Bricklayers, who were not in the AFL at the time,[76] thus buttressing their determination not to be drawn into local disputes as well as protecting themselves from pressure by the Building Trades members. The agreement provided that the locals of both organizations would not strike without first conferring with their respective national offices. Kirby explained its rationale as follows:

> Union principles do not depend upon the unionizing of one particular job. The protection of union interests on future jobs being provided for by

peaceful means is worth far more than all the sympathetic strikes that were ever entered into to force union conditions of employement on some particular operation . . . Only as a last resort will we enter into strikes, or ask those with whom we are now affiliated in protection of our mutual interests to do so.[77]

A considerable portion of the time of the 1914 AFL convention was devoted to the Carpenters' problems, and the convention appointed a three-man committee to deal with them. Included in its membership were Matthew Woll, who was regarded as Gompers' crown prince, and William Green, the future AFL president. This committee predictably recommended further negotiations, to which the Sheet Metal Workers reluctantly agreed.[78]

Continued negotiations proved fruitless. Gompers and representatives of the two organizations visited a number of shops; Gompers reported that in his opinion, the work in question belonged neither to the Carpenters nor to the Sheet Metal Workers exclusively. He suggested that the Brotherhood relinquish the *manufacture* of hollow metal trim doors and sashes, but that the *erection* of this material should be shared by the two organizations, with interchangeable working cards. This solution was recommended by the 1915 AFL convention; the Brotherhood was agreeable, but the Sheet Metal Workers, fearing that they would be swallowed up by the larger organization, were opposed. The convention upheld the latter by the close vote of 84 to 81.[79]

The Convention Committee on Adjustment, at the behest of the Machinists and other unions, determined that the Carpenters should be instructed to discontinue all jurisdictional claims for work that had been conceded to other organizations by AFL conventions or by the AFL executive council, on pain of suspension. A Machinist delegate stated that the Carpenters "just laid aside everything that was done and treat with contempt all that we may resolve to do," a comment that was not far off the mark. Gompers remarked that he had "looked askance for a considerable period of time at the constant extension of claims of jurisdiction by the Brotherhood of Carpenters and Joiners. Many of their claims to jurisdiction are unwarranted. But they are not the only offenders." However, he opposed suspension, calling it "a policy of mutual throatcutting." Andrew Furuseth of the West Coast Seamen, who had been having his problems with P. H. McCarthy in San Francisco, said that the Carpenters "have about two hundred thousand heads and only one immense ravenous stomach to digest everything they can put into it." He warned them that "the rank and file of the Carpenters throughout this country [will] suffer because of the crazy ambitions and the unbridled greed that exists in your claim for jurisdiction," concluding: "You Machinists, don't be sissies; when you are attacked, defend yourselves!

And so I say to the other unions: If these men haven't got any sense or any consciousness of decency, teach it to them."[80]

The convention decided not to let the controversy be fought out on the picket lines. Instead, it resolved that Gompers shoud appoint a committee to attend the next Brotherhood convention to try to get the Brotherhood to recede from its positions. It called on the Brotherhood to rejoin the Building Trades Department, and in the meantime, "not to place in force or operation . . . the claims that they have set forth in their journal."[81]

When the Brotherhood delegates appeared at the 1915 Building Trades Department convention, they received a much more positive reception. The convention annulled by a vote of 35 to 5 the so-called Tampa decision that had awarded jurisdiction over metal trim to the Sheet Metal Workers, and called for further negotiations. A number of delegates abstained from voting rather than opposing the Carpenters directly. By a vote of 42 to 23, the Machinists and Boilermakers were unseated on the ground that they were not building tradesmen. A voting pattern was instituted which in effect gave delegates whose organization was paying per capita on more than 128,000 members the right to cast two roll call votes each. The Brotherhood was the only union that qualified under this new rule. Moreover, the executive council of the department was reduced to seven members, replacing the previous practice of including a representative from each trade. Hutcheson was elected a vice-president of the department.[82]

Why did the opposition to the Brotherhood collapse? It is not possible to answer this question precisely, but several possibilities may be suggested. For one, it must have been clear by 1915 that the AFL had neither the authority nor the willingness to challenge the jurisdictional claims of one of its largest affiliates.[83] To do so would have meant to risk splintering the labor movement. Gompers, for all the prestige he had acquired after thirty years as AFL president, was well aware that national leaders like Hutcheson were prepared to challenge him. The whole theory of the movement was against compulsion. Duffy put it this way:

> When we get down to hard facts the American Federation of Labor does not approve or disapprove jurisdiction claims at any time. All that they do is to publish them without assuming any responsibility for them. Therefore, if an affiliated organization extends its jurisdiction claims the American Federation of Labor has no power to prevent it from doing so . . . We reserve the right to say what our jurisdiction claims shall cover, and we don't propose that they shall be curtailed, altered or amended through any other agency.[84]

The Brotherhood's alliance with the Bricklayers was probably a second factor. Jurisdictional strikes against the Carpenters involved the risk of retaliation by these two large organizations. Moreover, the Bricklayers had a loose alliance with the Operating Engineers and the Plasterers, and there was always the possibility of a rival building trades organization being formed. The trades that were not directly in conflict with the Carpenters must have felt that there was no longer any point in confronting the Carpenters on behalf of the Sheet Metal Workers and the Machinists, particularly when the latter were only peripherally involved in the construction industry.

Shortly after the Carpenters' success at the Building Trades convention, Duffy called Gompers' attention to the fact that the Machinists, "who seemed to take pleasure and delight in lambasting the Carpenters for their unwarranted claim of jurisdiction without first having that claim approved by the American Federation of Labor," had submitted to their membership a vote on the proposition that in localities where no locals of any other metal trade existed, the Machinists would take all metal tradesmen into mixed locals, regardless of jurisdiction. "What right did the delegates of the Machinists have to open their mouths against the Carpenters?"[85] Even though the proposition was approved by the Machinists' membership, their officers decided that it was not a good idea, and the plan was dropped.[86]

When the Sheet Metal Workers protested to the AFL executive council against the overturn of the Tampa award by the Building Trades Department, the committee denied the appeal on the ground that "deliberative bodies in accordance with customary procedures have a right to change decisions previously rendered."[87] The executive council reported to the 1916 AFL convention that the Sheet Metal delegation that Gompers was supposed to take with him to the Brotherhood convention to urge a compromise of claims had been canceled because preliminary conferences had yielded no progress. Conferences at the convention were no more sucessful, and the convention simply recommended that the presidents of both organizations meet again by April 1, 1917. A similar solution was advised in the case of the Carpenters versus the Machinists, with the Iron Workers demanding that they should be included as well. "The Carpenters have enlarged their claims greatly, they have put these things into their claims and spread them broadcast."[88] But by October of the next year, the AFL executive council was told that the Carpenters and Sheet Metal Workers were further apart than they had been a year earlier.[89] As for the Machinists, five conferences were held with them, also without result.

Although these were the most spectacular jurisdictional controversies in which the Carpenters were involved during the years 1912 to 1917, several others might be mentioned. A dispute arose with the United Mine Workers over the erection of buildings in and around coal mines. A tentative agreement was reached providing that all carpenters employed in shafts or in the mines themselves were to belong to the Mine Workers, while those engaged in building breakers, houses, or other structures

were to be Brotherhood members. After some initial hesitation, involving an interpretation of the famous Scranton Declaration, the agreement
went into effect.[90]

A controversy developed with the Brewery Workers over the repair of
boxes used in the breweries. It was the contention of the Brewery
Workers that the men who cleaned the boxes should repair them, while
the Brotherhood insisted that if there were enough work to keep one or
more men steadily employed, the work should belong to the Carpenters.
A settlement was reached giving the Brotherhood all building and repair
work in breweries, as well as the manufacture of bar fixtures and beer
boxes. The Brewery Workers retained the right to repair the boxes.[91]
This agreement was canceled two years later because of alleged violations
by the Brewers, and a more explicit one substituted for it. The Brewers
agreed to do their best to see that only Carpenters performed repair work
in breweries and saloons, and that they would refuse to place bottles in
cases not bearing the Brotherhood label. More important, they agreed to
remain neutral in any jurisdictional disputes between the Carpenters and
other trades; what was at stake here was millwright work in breweries.[92]

An agreement reached between the Carpenters and the Slate and Tile
Roofers Union did not last long. That organization was to have jurisdiction over the laying of asbestos and asphalt shingles above the eaves line,
with the Brotherhood to work on the side of the building. This was
another example of the impact of changes in materials on jurisdiction – the substitution of asbestos and asphalt for wood.[93] Hutcheson explained that the Brotherhood had agreed to this compromise at an AFL
convention because there was a good deal of sympathy for the small
Roofers Union, and Carpenter intransigence would have solidified the
delegates against it on other more important matters.[94] But there must
have been some dissatisfaction with this concession, for a year later the
Brotherhood canceled the agreement. The Roofers took their case to the
AFL executive council, which simply refered the matter to the Building
Trades Department. The executive council of that department decided
that any union had the right to abrogate an agreement. "Your Council
recognizes that innovations are at hand every day. Some mind conceives
something new in building erection and today there are innumerable new
ideas in regard to shingles. With every new idea comes a jurisdiction
dispute, therefore the Council recommends that the next convention of
the BTD shall endeavor to bring about an amalgamation of the Slate and
Tile Roofers with the next organization in kind."[95] This organization was
not the Carpenters, however, but the Composition Roofers, Damp, and
Waterproof Workers, with whom the Roofers merged the following
year.

The dock worker controversy that played a role in the New York affair
also had some important jurisdictional results for the Brotherhood.

When the 1916 AFL convention ordered the Iron Workers to give up the dock local they had chartered against Brindell's local, Hutcheson was asked whether he interpreted that decision as giving him jurisdiction over pile driving in general. He replied: "I don't know if it would be proper for me to interpret that at this time, but I will say this: that insofar as dock and pier work is concerned, it was conceded by the representatives of the iron workers as well as the contractor at the time of the hearing that it would be a very important proposition not to separate pile driving from the dock, pier and wharf building."[96] Shortly before the convention, Duffy had protested to the AFL that the Iron Workers were adding the term "pile drivers" to their title.[97] Gompers advised the Iron Workers that the change of name was in violation of the AFL constitution in that it implied an extension of jurisdiction without approval of the AFL convention, and the Iron Workers agreed to drop the change.[98] This did not mean, however, that they waived jurisdiction; they claimed that if they were to do that, a third of their members would be obliged to join the Brotherhood. The Carpenters proposed that all pile driving for docks and piers be allocated to them; they were willing to cede pile driving for bridges and viaducts where the superstructure was built by the Iron Workers. The Iron Workers would not agree. However, after having been forced by their suspension from the AFL to give up any claim to dock builders in the New York area, they finally ceded pile driving to the Carpenters, who also managed to add to the agreement the setting and framing of iron braces on walls.[99]

In 1912 the Shingle Weavers' Union was given jurisdiction by the AFL over all men working in and around sawmills and logging camps, and in 1914 was given permission to change its name to the International Union of Timber Workers — all without any objection by the Brotherhood. The reason for this uncharacteristic reticence appeared in a resolution adopted by the GEB indicating that "since many of these men are laborers, loggers, drivers, packers, filers, engineers, firemen, etc., as well as men working in the sawmills," the possibility of acquiring this work was deferred until an investigation was undertaken as to whether those eligible could be admitted to the Brotherhood.[100] The influence of the Industrial Workers of the World among the western loggers, as well as unsuitability of many of those working in logging camps for Brotherhood membership, apparently dissuaded it from claiming jurisdiction until two decades later.

One final footnote on jurisdiction: in 1916 an AFL organizer wrote to Secretary Frank Morrison asking whether ship caulkers in Baltimore, most of them black men whom the white ship carpenters would not accept in their local, could be chartered as an AFL federal local. Morrison sent this inquiry to Duffy, who replied:

The men engaged in this class of work properly belong to our organization, and nine-tenths of such work is done by the ship carpenters. That being the case, we feel that the issuance of a charter to these men by the American Federation of Labor would only cause trouble and misunderstanding later on, and it strikes me that the best thing to do is to organize the colored men in a separate local union if the feeling is such that they cannot be taken into the ship carpenters Union already existing.[101]

THE NINETEENTH GENERAL CONVENTION

A good deal of the time of the 1916 convention was devoted to an airing of the charges and countercharges arising out of the New York affair, as well as discussion of the Brotherhood's relationships with the Building Trades Department. A much smaller number of constitutional changes were considered than was customary, and only a few were adopted. Increases voted in monthly dues and per capita tax were defeated in a later referendum. However, ladies' auxiliaries were incorporated into the formal structure of the Brotherhood.

Harry L. Cook of Cincinnati, who had been one of the principal opponents of Hutcheson in the New York affair and had spoken on behalf of the New York locals at the convention, was nominated to run against Hutcheson for the presidency. Hutcheson won handily, his position strengthened by his victory in New York. A resolution proposed by Local 104 of Dayton, Ohio, at the end of 1917, to the effect that conventions should be held at four-year rather than two-year intervals, received the required number of twenty-five local endorsements from twenty-five states to put the matter to a referendum vote. Although this was a controversial proposal, nothing appeared in the pages of *The Carpenter*, pro or con. The proposal was carried by a vote of 49,734 to 22,680, so the next convention was put off until 1920.[102] Thus Hutcheson had four years in which to consolidate the power he gained at the 1916 convention, the first over which he presided, and he used them skillfully. He never again faced any serious challenge to his leadership.

The centralization of authority in the national union went on apace. The GEB instituted a rule requiring that in the event of strikes or lockouts, a list of local members affected would have to be submitted to the general office before financial aid would be allowed. Moreover, an initial waiting period of two weeks was imposed before strike benefits would be paid.[103] The central strike fund was a weapon in the hands of the general office, providing it with leverage in the control of local trade movements, in addition to the constitutional power of the general president that required consultation with him or his representative before a work stoppage could be initiated. The strike philosophy that was to dom-

inate the Brotherhood for the duration of the Hutcheson regime was
clearly stated by him at the outset of his presidency:

> We have found that when a General officer or deputy arrives on the ground
> previous to the men being called out, in many instances an agreement could be
> made, and the reason is very simple. Such officer or deputy is far removed
> from the scene of the conflict and he has no personal interest to serve . . . He
> can approach the task before him in a calm, dispassionate manner, and usually
> knows and has at his command statistics showing conditions in cities and
> towns of a similar size to that in which the trade movement is contemplated
> and is ready with argument to convince the employers of the justness of the de-
> mands of our members. These means, thus employed by our General officers,
> have tended to prevent many ill-advised movements and to increase the in-
> terest in our organization and to teach the employers that we are not a body of
> irresponsible men who do not keep agreements made, but that we are an
> organization of mechanics who believe in justness and fairness.[104]

It has been argued that the structural changes initiated under Huber,
and continued under Kirby and Hutcheson, impaired the local autonomy
and democracy that characterized the Brotherhood during the first
twenty years of its existence, and moved the union toward greater central
control and more businesslike management. This is undoubtedly true; by
1917 the general office was able to exercise a degree of control over the
conduct of its local affiliates that had never been contemplated when the
organization was formed. The ambitions of individuals certainly contributed to the centralizing process, but it is unlikely that the new institutions would have been adopted in the first place, or would have continued to function effectively, if they had not met the needs of the time.
The business union approach exemplified by the Hutcheson statement
just quoted was ideally suited to the hostile environment in which the
labor movement functioned during much of twentieth-century America.
Labor movements founded on other philosophies rose and fell—the Industrial Workers of the World provide a good example for the years with
which we have been dealing.

Why unions like the United Brotherhood did not evolve toward the
pattern of the major European labor movements is a long and complicated story that cannot be developed here. Suffice it to say that some
of the basic ingredients that made for characteristic western European
labor organization were not present in the United States in anything like
the same degree: a class-conscious, homogeneous, blue-collar labor
force; a fluid political party system; a rate of economic growth that
limited the economic gains that could be made through collective
bargaining; a less secure and confident entrepreneurial class than that of
the United States.[105] Why should the Carpenters, or any other group of
workers, have supported a socialist solution, for example, when their
businesslike organizations gained for them substantial annual increases

in living standards and improvements in their working conditions? That they did not, and instead built unions that embraced capitalism and cooperated in the improvement of its productivity, attests to the good sense of American workers and to the major contribution they have made to the stability and progress of American society.

8

WORLD WAR I AND ITS AFTERMATH

The entry of the United States into the first World War on April 6, 1917, brought the Brotherhood into close contact with the federal government on a wide range of matters for the first time. Its previous bargaining experience had been almost entirely with the private sector. What happened during World War I was significant in determining the future attitude of the Brotherhood, and particularly that of its president, William L. Hutcheson, toward the appropriate role of government in the labor market.

The initial reaction of the Brotherhood was to "offer the services of our organization as mechanics to the government of the United States in whatever manner they may be most needed. Our organization consists of patriotic American citizens, ready and willing at all times to do their duty when called upon. If the Government officials will only let us know in reasonable time when our men are required for industrial service at any particular point designated by the Government, we will cooperate with those officials in supplying the men required."[1] In offering to act as an employment service for carpenters, the Brotherhood hoped and expected that union men would be employed on war-related work whenever they were available. A Labor Day editorial in *The Carpenter* called upon the government to sign collective agreements with the unions, indicating that cooperation was a two-way street:

> Doubtless the trade unionists of the country will reiterate their loyalty to the nation and to the ideals for which our government stands this forthcoming Labor Day, and that is as it should be. But there is something else we should not forget to do. We should in addition affirm in no uncertain voice the ideals and purposes of the organized labor movement and let our opponents know that the workers of the nation while united in the prosecution of the war stand just as resolutely as ever for their rights and privileges and that no infringement of them will be tolerated.[2]

This message was addressed not only to the government, but to the employers and the AFL as well. The day after the United States declared war, Gompers called a meeting of labor, management, and public representatives which created a special committee with himself as chairman and William B. Wilson, Secretary of Labor, as secretary. The committee adopted a resolution "advising that neither employers nor employees shall endeavor to take advantage of the country's necessities to change existing standards." The Brotherhood was by no means the only union to express concern about this policy. Daniel Tobin, president of the Teamsters, told Gompers that he had no authority to make such a statement, and warned that his organization would "continue to fight and struggle, even to the extent of striking for better conditions, war or no war." Gompers denied that a no-strike pledge was intended, and indicated that all he had in mind was the protection of existing standards.[3] Gompers, however, had no immediate constituency for which he had to bargain, and appeared to the more cautious national union leaders to be giving too much away without adequate recompense.

One of the first orders of business was to build cantonments to assemble and train an army. The War Department began to let contracts without any attention to the union status of the contractors involved. Moreover, some employers were attempting to use the war emergency as an excuse for weakening labor standards on the ground that the shortage of labor made union provisions burdensome and unnecessary. Gompers was obliged to take some positive action to meet the growing restlessness in the ranks of organized labor.

THE BAKER-GOMPERS AGREEMENT

On June 19, 1917, Newton D. Baker, the Secretary of War, entered into an agreement with Gompers that became the basis of the government's construction labor policy. The agreement read:

> For the adjustment and control of wages, hours and conditions of labor in the construction cantonments there shall be created an Adjustment Commission of three persons, appointed by the Secretary of War; one to represent the Army, one the Public, and one Labor; the last to be nominated by Samuel Gompers, member of the Advisory Commission of the Council of National Defense, and President of the American Federation of Labor.
>
> As basic standards with reference to each cantonment, such Commission shall use the union scale of wages, hours and conditions in force on June 19, 1917, in the locality where such cantonments are situated. Consideration shall be given to special circumstances, if any, arising after said date which may require particular advances in wages or changes in other standards. Adjustments in wages, hours or conditions made by such boards are to be treated as binding by all parties.[4]

General G. H. Garlington was appointed to the commission to represent the Army; John R. Alpine, president of the Plumbers' Union, to represent labor; and Walter Lippmann, the journalist, to represent the public. This was potentially a pathbreaking agreement, foreshadowing the later Davis-Bacon Act. But the bloom was taken off it when on the very next day, a representative of the War Department informed the AFL that the agreement did not commit the government to the closed shop, and Gompers acknowledged that this was an accurate interpretation.[5]

Unaware of Gompers' concession, Hutcheson demanded that the closed shop be enforced in government contracts. He wrote Gompers that "we feel that when firms . . . are given contracts by our Government and permitted to employ non-union men therein even though they may be paying the wage and working the hours established by our organization, that it tends to jeopardize and retard, and even tear down the conditions and standards as established."[6] When he learned of Gompers' interpretation, he told him that the officers of the Carpenters had kept their members on the job because they were of "the opinion that the understanding arrived at between yourself pertaining to hours, wages and overtime rates and conditions as in effect on June 1st would be made applicable and would mean the recognition of the union or closed shop so long as that condition prevailed in the locality where the work was under course of construction. If this is not recognized by the Government officials we will have to assume the same attitude towards the Government work that we would in reference to the work of any contractor or builder, which I am frank to admit would mean the ceasing of work to enforce our conditions."[7]

Gompers' reply was diplomatic but nonetheless firm: "If you desire to continue to assume the position stated in your letter, you must do so with the realization that practically all the other trades engaged in various capacities for the Government have assumed an entirely different position. It is not my purpose to dictate any mode of procedure, but only to lay before you the facts as they exist with the hope that mature judgment will be exercised in dealing with the problems which concern our organization and our Government in this crisis."[8]

Gompers was clearly annoyed with Hutcheson. He felt with considerable justification that the government had made a substantial concession in agreeing to enforce union wages and hours on its construction work, and in fact asked Hutcheson whether "anyone can fail to understand that the Government of the United States, representing all the people of the United States, cannot enter into an agreement to employ exclusively members of any one organization."[9] Hutcheson was one who failed to understand; the closed or union shop had been fought for and secured by the Carpenters in the private sector, and it seemed perfectly logical that union men should refuse to work with nonunionists in public construction as well. The Carpenters would not endorse the Baker-Gompers agreement, and continued to work on government construction jobs under the conditions that prevailed locally. Nevertheless, there were few tieups during the war; the Carpenters acquiesced in the agreement in fact, despite their unhappiness with it.

THE SHIPBUILDING LABOR ADJUSTMENT BOARD

A crash program of shipbuilding was instituted to carry the men and supplies needed to sustain the American Expeditionary Force in Europe. Not only ships but shipyards as well had to be built. Edward D. Hurley, who was chairman of the United States Shipping Board, signed an agreement similar to the Baker-Gompers agreement with Gompers and the presidents of the AFL Building and Metal Trades Department. Pursuant to the agreement, the Shipbuilding Labor Adjustment Board was set up with Everett V. Macy of the National Civic Federation as chairman.

Hutcheson reported to the GEB that on August 15, 1917, he received a communication from Gompers outlining this agreement. The Navy Department, represented by Assistant Secretary Franklin D. Roosevelt, was also a party to the agreement. Hutcheson refused to sign it, "believing that as an organization we are entitled to some consideration in the way of maintaining unto ourselves the conditions that we had established, and it has been clearly shown and demonstrated by some few employers doing work for the government that they will not overlook an opportunity to take advantage of the workmen and ignore whenever possible the conditions established in the various localities and districts of our organization."[10] However, GEB member Guerin did sign the agreement; apparently Hutcheson felt that this did not bind the Brotherhood, since he had not endorsed it or secured GEB approval.

The Adjustment Board paid an early visit to the Pacific Coast, where a good deal of labor unrest had occurred. The unions there agreed to accept a 31 percent wage increase, which paralleled the rise in the cost of living since July 1916, the last time a wage adjustment had been made. However, the board took an action that was bound to cause trouble: it stipulated that carpenters who moved into shipbuilding were to be paid at a lower rate than that earned by regular ship carpenters, presumably because of their lack of shipbuilding experience. Hutcheson would have none of it; he notified Brotherhood members that they should accept nothing less than the minimum shipyard wage, regardless of their previous occupation.[11] The Adjustment Board persisted, carrying the dual wage structure initiated on the Pacific Coast to the rest of the rest of the country.

This policy, plus a general board ineptness in handling industrial relations, led to a proliferation of labor disputes. After some preliminary discussion with the Shipping Board, Hutcheson submitted to the board a memorandum outlining the conditions the Carpenters wanted the shipyards to observe. It called for an eight-hour day; Saturday half-holidays during the summer months; double pay for overtime; access to all shipyards by Brotherhood representatives; an elaborate grievance mechanism; a uniform rate for all carpenters working in shipyards; and all carpentry labor to be furnished by the Brotherhood through Department of Labor employment offices.[12] If this had been accepted, it would have been one of the best agreements ever gained by the Brotherhood.

After the government representatives on the board had conferred with Hurley, the answer came back, "nothing doing." Hutcheson then met directly with Hurley and told him that considerable unrest had manifested itself, particularly in the ports of New York and Baltimore. Hurley's response was that the Brotherhood should sign the agreement that had been reached half a year earlier. Hurley and Hutcheson did not hit it off; according to the latter, "Mr. Hurley again demonstrated his ability in evasiveness."[13]

Hutcheson and Duffy, together with AFL Secretary Morrison, then proceeded to meet with the Adjustment Board, and proposed that a representative of the Brotherhood with full voting power sit on the board when carpenters' wages and conditions were being considered. The board chairman, in a letter to Gompers, responded as follows:

> The matter has been discussed by our Board and we have unanimously voted that the Board cannot extend privileges to one craft that has not been extended to others. You are a party to the Agreement under which we are working, and if any changes are to be made therein we believe that they should be made by those who created the Board. To accede to Mr. Hutcheson's request that the functions of the Board should be limited to questions involving hours and wages would leave this organization free to establish closed shops, etc., which privilege has been waived by the organizations signing the agreement. To accede to his demand that a representative of his organization should sit on our Board would result in the Board being composed of four members, two of whom would represent labor as against one for the Navy and Fleet Corporation.
>
> These changes are so fundamental that they could only be adopted by those creating the present Board. It is our belief that should it be necessary to grant these concessions to the Carpenters' organization, it would be better to ask for a separate Board, so that all organizations coming before our Board may be given equal consideration.[14]

A few days later, while negotiations were still going on, the ship carpenters in New York went on strike, followed by those in Baltimore. Hurley responded immediately by sending a telegram to Hutcheson, which, after noting the grim war casualty reports, asked: "Do you realize that you are adding to the fearful danger our soldiers already face, the danger of starvation and the danger of slaughter if food and ammunition are not sent over in ships and many ships at once? Do you think the fathers and mothers whose sons are making this sacrifice will sit patiently by and permit this paralyzing of the lifeline between us and the Western front to go on?"[15] He urged that the strike be called off, and ended by warning that "those whose sons are now giving their blood that you and I and our children may be safe and free will not long permit either you or me to invite destruction of heroic lives and disaster to a great world cause." This telegram was given to the press; it unloosed a barrage of criticism against the Carpenters reminiscent of that directed against John L. Lewis when he took the coal miners out on strike during World War II. As Duffy noted: "The Press berated them for being unpatriotic and unprincipled; even Billy Sunday, when he hit the sawdust trail in New York City, took a slam at them as being traitors and slackers. General President Hutcheson was looked upon as a dangerous man, unworthy of being at the head of such an organization."[16]

Hutcheson was not intimidated. He wrote Hurley that the Brotherhood was composed of patriotic and loyal citizens who were prepared to return to work on the basis of his proposal to the Adjustment Board. Hurley's response was a long letter, conveyed to Hutcheson by way of the press, recalling the manner in which the Adjustment Board had been set up and pointing out that Hutcheson was the only union president who had refused to sign the original agreement. It concluded: "Will you help now —when every day's delay means the slaughter of our boys?"[17] Hutcheson responded in kind through the press, stating that although the shipyard strikes did not have his approval, the Brotherhood had never been a party to the original agreement. With this avenue of communication closed, Hutcheson sent a telegram to President Wilson, asking for an opportunity to lay the entire matter before him. Wilson's reply said, in part:

> I feel it is my duty to call attention to the fact that the strike of the carpenters in the shipyards is in marked and painful contrast to the labor in other trades and places. Ships are absolutely necessary for the winning of this war. No one can strike a deadlier blow at the safety of the Nation and of its forces on the other side than by interfering with or obstructing the ship building program . . . I must say to you very frankly that it is your duty to leave to [the Shipbuilding Wage Adjustment Board] the solution of your present difficulties with your employers and to advise the men whom you represent to return at once to work.[18]

On February 19, after Wilson had refused to meet with Hutcheson, the strikers returned to work on the promise that their grievances would be given immediate consideration by the Adjustment Board. After several days of negotiation between the Carpenter officers and Hurley plus the board members, the latter came up with a long statement to the effect that the original agreement could not be modified without bringing in all the signatories. On March 5 Gompers convened a meeting to discuss the entire matter, at which Franklin D. Roosevelt and a representative of the Shipping Board were present. Both the Shipping Board representative and the Metal Trades representative expressed their satisfaction with the original agreement. Faced with the opposition not only of the government but of the other trade unions as well, Hutcheson decided that "it was useless to continue the conference further." But before leaving, he told the conference that "we reserved unto ourselves the right as citizens of the United States and the constitution of our country which gives us the right to say whom we would or would not work with and that we proposed to retain that right of citizenship so long as there was permitted to exist between us and our Government a profiteer who was paid a percentage on labor performed by our members."[19]

And there the matter rested for the duration of the war. The board never yielded on the closed shop demand, and although the Carpenters refused to sign the agreement, they continued to work. Hutcheson had failed to achieve his goal.

It has been said that Hutcheson resented Roosevelt's attitude during these negotiations, and that this was responsible for the former's opposition to the New Deal. Hutcheson's attitude toward the role of government in labor affairs may also have been colored by this bruising first encounter with the government. It is likely, however, that Hutcheson's philosophy was already firmly established when he assumed the presidency of the Brotherhood, and that his subsequent caution about government labor market intervention, which he shared with Gompers and other contempoaries in the AFL leadership, would not have been much different even in the absence of his wartime experience. In fact, some of Hutcheson's colleagues in the AFL subsequently indicated their regret that they did not take a firmer stand on the closed shop. President William Bowen of the Bricklayers, after reporting that his membership had been cut in half by a combination of the military draft and the open shop, stated:

> I hold no brief for Bill Hutcheson; he is not only big in stature but he is big enough mentally to take his own part; but I want to have it clearly understood that the time he was under fire men high in the affairs of organized labor in America ought to have made it clear why he made the stand . . . I am opposed to the efforts of those who condemned Hutcheson's efforts to make a condition prevail. I take my stand with Hutcheson. Those who would distort what the man attempted to do, either did not fully understand or they designedly took advantage of a situation that might make them stand good with somebody else.[20]

THE NATIONAL WAR LABOR BOARD

Early in 1918, almost a year after the United States entered the war, the need began to be felt for something more far-reaching than the Baker-Gompers agreement. That it took the federal government so long to recognize that a comprehensive labor policy was essential to the effective prosecution of the war can be put down to inexperience; it responded much more rapidly a quarter of a century later.

A War Labor Conference Board, with five representatives from the AFL and five from the National Industrial Conference Board, a management group, was set up in February 1918 to work out a more general plan. Hutcheson, despite his difficulties with the government, was one of the AFL nominees. The employers selected former President William H. Taft as a public member, while the labor group selected Frank P. Walsh, a lawyer who had been prominent in industrial relations. This board reached an agreement that guaranteed workers the right to organize and bargain collectively through organizations of their own choosing, without employer interference—the first time that basic principle had been recognized by the federal government. Prevailing working conditions, including the union shop where it existed, were to be continued. The basic eight-hour day was to be enforced where the law required it. The right of all workers to a living wage was recognized, and in fixing wages, "minimum rates of pay shall be established which will ensure the subsistence of the worker and his family in health and reasonable comfort."[21]

To handle disputes, a National War Labor Board was created with the same distribution of membership as the original Conference group. Mediation and conciliation were to be employed first, with the appointment of arbitrators as a last resort. As Hutcheson explained to his members, the labor representatives who helped frame this agreement were of the view that there should be no wartime strikes:

> This, however, would not mean that as a last resort we could not strike, but it signifies our willingness to agree that all methods of mediation and conciliation should be exhausted before resorting to ceasing work to enforce that which we deem is just and right . . . I do not intend to convey the thought that we should in any manner deviate from the principles of our organization that we have followed so closely in the past. But I feel we can attain the same end and bring about the same condition by following the avenue of adjustment provided for by the National War Labor Board.[22]

Hutcheson's optimism may have been due to the fact that he was appointed a labor member of the board, together with the presidents of the Seamen, Miners, Garment Workers, and Machinists. The GEB was a bit more cautious. When it wrote President Wilson agreeing to the establishment of the board, it added: "We hope that all departments of the government will cooperate with this Board, so that basic conditions under which labor shall be employed on government work, direct or through contractors, may be established and thereby bring about contentment which has not existed up to this time."[23]

The National War Labor Board began its operations on April 8, 1918, and continued in existence for sixteen months; after the Armistice, only disputes jointly submitted by the parties were considered. Some 1,251 controversies in all came before the board, and awards or recommendations were made in 490. The Carpenters were among the most frequent clients of the board, and supported it strongly, which was not true of many unions. Thus, when several locals asked the GEB to sanction strikes for higher wages in October 1918, it ruled that "any arrangement for an advanced wage has to be sanctioned by the Government representatives."[24] Employers were not as cooperative, since many of them believed that the War Labor Board was encouraging unionism.[25]

During the war the German language section of *The Carpenter*, which had been a feature since 1881, was terminated. This was undoubtedly due to the anti-German feeling generated by the war, although the decline of the German-born membership would have led to a similar result not too far in the future. However, a proposed constitutional amendment to revoke the charters of all locals conducting their business in German or Hungarian was defeated in a referendum vote.[26]

The war years were good ones for organized labor. There was a scarcity of labor, and the favorable attitude of the government, particularly after the establishment of the National War Labor Board, led to a growth in membership. AFL membership rose from 2,072,000 in 1916 to 2,726,000 in 1918. The Carpenters did considerably better; average membership increased from 212,800 to 321,700 during the same period. Part of this, of course, was the great wartime need for carpentry, combined with the Brotherhood's efforts to maintain the closed shop.

BUSINESS AS USUAL

The Armistice was signed on November 11, 1918. On December 9, the largest lockout in Brotherhood history began in New York City. The next summer, the carpenters of Chicago went on strike. The relative labor peace that had characterized wartime America was at an end.

The renewed labor strife was a consequence of both employment and wage developments. While unemployment rose from the extremely low wartime rate of 1.4 percent in 1918 to only 2.3 percent in 1919, there were major shifts among industries. Shipbuilding, which had grown very rapidly in 1917, declined with equal rapidity after the war. A letter from the secretary of the Puget Sound district council of Carpenters tells the story: "Since the signing of the Armistice the shipyard work has been torn to pieces as far as the carpenter is concerned. For the last ten days our men have been laid off by the hundreds and more are being laid off every day. Even housing propositions are being cancelled right along." He noted that there was a demand for only 2,000 of the 6,000 carpenters in the Seattle area.[27]

The construction industry was not sufficiently buoyant to provide jobs for all the displaced shipyard workers. Although the value of residential building contracts rose by 75 percent from 1918 to 1919, the rest of the industry was fairly sluggish, particularly the military sector.

The dispute in New York exemplified the problem of adjusting wages in the transition from war to peace. Shortly before the end of hostilities, the New York district council began discussions with the Building Trades Employers' Association on the renewal of the agreement that was to expire at the end of the year. As a result of the 1916 shakeup in the union, the seventy-four New York locals had been consolidated into twenty-two, and restored to full autonomy on July 1, 1917. The Employers' Association decided that they would not raise wages in 1919, and published the same pay scale that had been put into effect on September 1, 1916. Independent contractors had agreed to wage increases, with the result that carpenters working for some subcontractors were earning more than those employed by the associated contractors. As a result, employees of some of the latter stopped work, and the Employers' Association locked out the rest.

The Brotherhood responded by striking jobs that were being done in other parts of the country by contractors who belonged to the New York Employers' Association. The deadlock continued until March 13, 1919, when it was agreed to turn the matter over to an arbitrator. He awarded an increase that represented a compromise between the association offer and the union demand.[28] This strike and lockout involved a greater number of Brotherhood members than any previous stoppage in the history of the union.

In Chicago the union asked for an increase of wages and went on strike early in the summer of 1919 to back its demand. Members of the Building Contractors' Association of Chicago responded a few weeks later by closing their material yards, which made it impossible for independent contractors who had granted the raise to continue to operate. The Chicago district council went to court and charged the Contractors' Association with conspiracy to restrain trade. Even after a wage settlement was reached, the union prosecuted its claim, and secured a judgment against the association. In general, there were more requests for strike sanction during the four-year period 1916–1920 than in any previous four years in Brotherhood history; most of the strikes occurred in 1919 and 1920.[29]

The Carpenters would not have been opposed to continuing to operate under a peacetime version of the War Labor Board, despite their aversion to governmental intervention:

Those who welcomed the rulings of the War Labor Board as the truest and most conscious effort which this country has made to formulate a body of just principles to govern industrial relations will regret that in a time of world upheaval like the present, when labor at last is conscious of its power, is demanding its "place in the sun," we are reverting so blindly to the old warfare between labor and capital, whereby might makes right, living standards are slashed, and the workers are penalized for the industrial stagnation which has followed the war.[30]

There may have been nostalgia for the War Labor Board, but there was none for the Baker-Gompers agreement, which the Carpenters had never accepted. Hutcheson took the initiative at the 1919 Building Trades Department Convention by introducing a resolution seeking its annulment, and was delegated to pursue the matter at the next AFL convention, which he did successfully. The agreement was declared null and void.[31]

THE INDUSTRIAL CONFERENCE

Early in 1919 President Wilson called an industrial conference in order to find "some common ground of agreement and action with regard to the future conduct of industry . . . and to obtain the combined judgment of representative employers, representative employees, and representatives of the general public conversant with these matters." He asked the AFL to appoint fifteen members to meet with like numbers of businessmen and public figures. There were divided views within the ranks of labor on the advisability of participating in this conference, not least because Judge Elbert Gary, the president of U.S. Steel and one of the most anti-union employers in the country, was among the employer delegates.

Gompers asked Duffy to attend the conference in his capacity as a member of the AFL executive council, but the latter declined when he was told that not all the council members had agreed to be present. He wrote Gompers:

I am not looking for much good, if any, to come from the conference. You know the committee has neither the power, the right or authority to tie down any organization to anything agreed to nor has the A.F. of L. . . . As far as the Carpenters are concerned, we don't propose to be hampered, limited or restrained in any manner from carrying on our legitimate work as formerly or from improving our working conditions, reducing our hours of toil and increasing our wages whenever the opportunity presents itself.[32]

Gompers then asked that Hutcheson or another officer of the Brotherhood be designated as a delegate. Hutcheson replied: "The matter of appointment of delegates has been under consideration for several weeks, and why at this late date you should suggest that I serve is more than I can understand. In view of the circumstances I would not feel justified in considering same, neither do Secretary Duffy and myself as Executive officers of the Brotherhood feel like suggesting anyone alse to represent our organization."[33] Gompers sent back a conciliatory telegram expressing regret that his earlier communication to Duffy had not adequately conveyed his desire to invite Hutcheson to attend the conference, and urging him once more to come.[34] Hutcheson simply wired back: "Due to circumstances and occurrences I fail to see wherein I would be justified in accepting appointment to the Industrial Conference now being held."[35]

Hutcheson's acerbic replies reflected his distrust of Gompers because of the latter's wartime initiatives. The large national unions had always been concerned lest the AFL assume authority that would impinge on their autonomy, and the leaders of the largest unions — the Carpenters and Miners — seemed to be suspicious of the role that the AF of L appeared to be assuming in national affairs. This may help to explain the Brotherhood's support of John L. Lewis as a candidate running against Gompers in 1921, although personal factors also entered into the picture.

In any event, the industrial conference ended in failure. The employers refused to vote in favor of a resolution introduced by Gompers that read as follows: "The right of wage earners to organize without discrimination, to bargain collectively, to be represented by representatives of their own choosing in negotiations, and adjustments with employers in respect to wages, hours of work, and relations and conditions of employment is recognized."[36] A second conference, consisting only of public representatives, was equally barren of results. American employers were not prepared to concede the necessity of living with trade unions. Two more decades, together with strong government pressure, were required to secure their grudging consent to what Gompers asked them to give voluntarily in 1919.

THE BOGALUSA INCIDENT

A foretaste of what was in store for the Carpenters, as well as other American unions, came on November 29, 1919, when four officers and members of the Carpenters' Local 2203 in Bogalusa, Louisiana, were murdered while they were in the offices of the Central Labor Council (see Appendix B). The Brotherhood was attempting to organize the Great Southern Lumber Company, which operated one of the largest sawmills in the world. The local had more than 300 members, and was making headway. "There was no strike, lockout or other labor trouble . . . and the killing of these four men was for no other purpose than to destroy, disrupt and disorganize our ranks for the reason they might be distasteful to the principal industry in the town and bringing men from peonage to a realization of Americanism."[37]

Local members gave evidence before a grand jury, which failed to find a true bill. A complaint was filed with a local judge, who brought thirteen men into court and required them to post bail of $40,000 each. This bail was supplied by the Great Southern Lumber Company.

The GEB immediately appropriated $5,000 to help defray the costs of bringing the murderers to justice, and solicited local unions to make additional contributions. It asked the AFL to press for a Federal investigation of the crime. The GEB noted in September that there had been no trial as yet, nor any further move toward either state or federal action. This was an example, an extreme one perhaps, of the obstacles to the progress of trade unionism in the South.[38]

POLITICAL DEVELOPMENTS

The Bolshevik triumph in Russia and the spread of European communism intensified the antipathy of Brotherhood officers to left-wing agitation. The GEB sent out a circular to all locals in May 1919, warning members not to affiliate with organizations called the Council of Workers, Soldiers and Sailors, or The One Big Union (a Canadian version of the IWW) on pain of expulsion from the Brotherhood.[39] An article that appeared in *The Carpenter* in June, probably written by Duffy, who had become the Brotherhood's expert on left-wing movements, denounced communism in no uncertain terms:

> There is no comparison whatever between unionism and bolshevism; they are as far apart, in every way, as the poles of the earth; in fact, they do not tread the same paths . . . Bolshevism . . . seeks to tear down, to destroy, to annihilate. To do away with every bulwark of modern business and society. It is antagonistic to every principle fostered by unionism. It is a dream of degenerates, the pack cry of inhuman brutes, begot in ignorance and nurtured in anarchy . . . They should be dealt with and stamped out like so much vermin; treated as a pestilence whose tentacles stretch out to crush and stifle the good that is in humanity.[40]

Local 1834 of Detroit adopted a resolution protesting a circular issued by the AFL Seattle Labor Council calling for a conference to bring about a general strike to prevent the manufacture and shipment of munitions to aid Poland in fighting Soviet communists. Duffy passed it on to Gompers with some indignation. Gompers replied that either "there is a misunderstanding or someone is maliciously inciting the Poles against the American Federation of Labor," and asserted that in fact the AFL itself "has not taken any action against the Poles, but on the other hand it has condemned the bolshevists for endeavoring to foist on Poland a Bolshevist government." Satisfied, Duffy wrote Local 1834 that the AFL "will stand with any country against bolshevism."[41]

In preparation for the 1920 election campaign, the AFL set up a National Non-Partisan Committee to work for the election of congressmen who were favorable to the cause of labor. Daniel Tobin, then treasurer of the AFL, appeared before the GEB to ask for a contribution of 1 cent per member in support of the committee's work. The GEB cautiously endorsed the program, but felt that the increase in the per capita tax voted to the AFL at its 1919 convention was sufficient to meet what it regarded as the sole legitimate action of the committee: "to secure the records of legislators favorable to the workers and those opposed, for distribution to the public as soon as possible."[42] It refused any additional contribution.

There was no question of where the Brotherhood stood on political involvement. When the Indiana State Federation of Labor endorsed the newly formed Farmer Labor Party, Duffy immediately protested to Gompers: "This is in violation of the laws of the American Federation of Labor and I therefore call it to your attention. No doubt you are aware of the fact that party politics is debarred in our organization and you can therefore realize that if our Local Unions sever their connection with the Indiana State Federation of Labor, we will not be to blame."[43]

Although the 1920 Carpenters' convention was held only a month before the elections, there was no discussion of politics at the convention. However, Hutcheson began his personal lifelong support of the Republican Party at the national level by working on behalf of Warren G. Harding. Harding invited him to participate in the Labor Day program in Marion, Ohio, and Hutcheson agreed to do so.[44] He was a not infrequent visitor to the White House during Harding's incumbency, although there is no evidence that he had any influence on the administration's attitude toward labor matters.

THE TWENTIETH GENERAL CONVENTION

The 1920 convention, the first in four years, was held in Indianapolis on the basis of a referendum vote stipulating that all future conventions were to be in that headquarters city. Membership was at an all-time high; there was a net increase of 37 percent in the number of local unions since 1916. The number of mill men among the members had risen by 83 percent, double the percentage of overall membership growth during the same period, and stood at 11 percent of total membership. Although unemployment had risen somewhat from the extremely low figures of the immediate postwar period, construction activity was brisk. The outlook for the future seemed good.

There were a few clouds on the horizon, however. A concerted anti-union offensive among employers had just begun, and the GEB devoted a major portion of its report to an attack on this movement. "The term 'open shop' has been coined by employers of labor who are opposed to all labor unions and was brought into general use in an attempt to deceive the public . . . It is a place of employment where the employer exercises the absolute right of freedom of contract between himself and the individual workman, the employer reserving to himself the privilege of employing such persons as he sees fit to perform the labor that he wants them to work at . . . The 'open shop' is a denial of the right of individual liberty."[45]

Hutcheson reported on a perennial problem—the traveling member. This became particularly acute during strikes, of which there had been a great many since the end of the war. Many locals and district councils levied strike assessments, and to avoid payment, some of those affected by the strikes took clearance cards from their locals and moved to other localities where work was going on. Others left the strike area to help avoid a drain on local strike funds. To prevent the former but encourage the latter, Hutcheson recommended constitutional changes that would make it optional for striking locals to issue clearance cards and would permit members with clearance cards to work in a new locality without having to transfer their membership, on the payment of a working fee no greater than the dues prevailing there. However, the convention took no action on this proposal.[46]

Another familiar issue was brought up in the form of a resolution to elect GEB members by district rather than at large. The main arguments advanced in favor of the change were that it would lead to a more democratic choice of representatives and that those elected were likely to be more competent, since their qualifications would be better known to the membership. The counterarguments were that such a change would be divisive and would prevent optimal reassignment of board members from district to district. Despite a strong speech by Hutcheson in opposition to the resolution, it carried.[47] However, it lost in the subsequent referendum.

The convention voted to establish a pension fund through an increase of 25 cents per month in the per capita tax, and to pay pensions of $20.00 a month to members over 65 years of age with twenty-five years of continuous membership. It also agreed on an unusual proposal: to levy an assessment of $1.00 per member three times a year to set up a wood products factory, the profits to be used to promote the Brotherhood label. Both proposals were defeated in the referendum vote, reflecting continuing caution on dues. However, a constitutional amendment requiring that all Brotherhood business be recorded in English did succeed.

Several locals submitted the following resolution to the convention:

"In view of the fact that in many localities where men of color desire to become members of the United Brotherhood, and debarred from the lack of consent of existing locals in said locality, be it Resolved, That in any locality where there are a sufficient number that is eligible to membership in the United Brotherhood, charters may be granted without the consent of the existing union." The resolution was defeated when Hutcheson stated that "the color line was never recognized in dealing with applications for charters, or in any other connection in the work of the organization, and there were cases where charters had been granted over the objections of existing Local Unions and District Councils." He reminded the convention that the constitution drew no line insofar as creed, color, or nationality were concerned.[48]

The color question came up again in connection with the dedication of a bronze tablet in memory of Gabriel Edmonston, the first general president of the Brotherhood. Duffy declared that Edmonston had opposed the introduction within the union of any issues relating to religion, nationality, or politics. And on color:

> Gabe Edmonston was a Southerner, but he said that the colored men of the craft must be taken care of as well as the white men, because if the colored men are not organized, if they are not getting the same privileges and benefits, then when the white men come out on strike for better working conditions, the gap will be wide open for the colored men to go in and take their places. From that day to this we have taken the stand that men of the trade, no matter what their color, are eligible to membership in our organization, and in the conventions of the past, in this connection, we have colored men sitting as delegates, and they are entitled to all the rights, benefits, and privileges of the white carpenters.[49]

These words were not merely for public consumption. A few months earlier Frank Morrison had written to Duffy inquiring whether a carpenters' local in the Canal Zone admitted colored men. Duffy replied that he did not know, but that qualified colored carpenters were fully eligible to join. That was not the problem: "The trouble in the Canal Zone is that they employ the white men to do the fine work and the colored men to do the rough work. The white man is paid the scale of wages as established by our union; the colored man is paid about one third the wages of the white man. This causes much friction and ill-feeling."[50]

The incumbent general officers were reelected without opposition. Everything appeared to be in order, and there was general confidence that the progress of the past decade would continue. What was not foreseen was the strength of the antiunion drive mounted by the employers, on top of a severe economic decline that set in shortly after the convention had adjourned, raising unemployment to the highest level since the beginning of the century. The next convention was a different story.

9

THE NINETEEN-TWENTIES:
A DECADE OF STAGNATION

The years 1921 to 1929 are puzzling ones for the labor historian. Although there was a serious economic recession in 1921, it was not of long duration. Recovery had set in by 1922, and from 1922 to 1929 the gross national product rose by the remarkable average of 5 percent a year. Unemployment peaked at 11.9 percent in 1921, then fell to a low of 1.9 percent in 1926; the average unemployment rate for the years 1922 to 1929 was only 4.2 percent. Construction activity was buoyant, particularly for the years 1923 to 1927 inclusive. Prices remained remarkably stable; there was almost no change in the consumer price index during these years.

After a decline from 1921 to 1922, union wage rates in the building trades rose steadily until 1929, increasing more than 5 percent a year. Wage rates for union carpenters went up by the same amount. All of this added up to a substantial improvement in real living standards for American workers in general and carpenters in particular.

Normally, favorable economic conditions have provided a good background for trade union progress. Yet the 1920s were an exception to this general rule. Membership in unions affiliated with the American Federation of Labor fell from 4.1 million in 1920 to 2.9 million in 1928. Some of the largest unions in the country were badly hit; the Machinists, for example, went down from 331,000 members in 1920 to 74,500 in 1928, the Coal Miners from 394,000 to 325,000. The Carpenters, by comparison, did relatively well: membership fell from 400,000 in 1920 to 346,000 in 1928. In 1920 the Carpenters had 9.8 percent of total AFL membership, but their share rose to 11.9 percent of the total by 1928. During most of these years, the United Brotherhood was the largest union in the country.

The reasons for this divergence between economic growth and union growth are still not clear. Employer resistance, in the form of the open-shop drive that commenced in 1920, was undoubtedly a factor—but then, American employers had hardly welcomed unions earlier. Shifts in the distribution of employment may also have contributed to the failure of unions to progress. Although total nonagricultural employment rose by 14.5 percent from 1920 to 1929, the increase was primarily in the service sector (other than transportation), where unionism had little more that a toehold. Employment in mining actually fell by 12 percent over the period, while total employment in manufacturing remained almost unchanged. There were some shifts within manufacturing in favor of newer industries that trade unions had been unable to enter, for example, automobiles and electrical manufacturing. But the employment factor was not operative for construction; employment in contract construction rose by 48 percent from 1921 to 1929 (1920 employment was abnormally low due to a major building recession).[1]

A close look at what was happening to the Carpenters' Union during these years may help provide some insight into one of the major paradoxes of American labor history.

THE AMERICAN PLAN

Beginning in 1920, open-shop associations began to spring up all over the country. New York State had at least fifty associations, Illinois had forty-six, Michigan more than twenty-five. A national convention held in January 1921 adopted the name "American Plan" for the open-shop campaign. "The postwar drive to liquidate labor's war-time achievements was on the entire industrial front. In the highly organized trades wage deflation and weakening union control were its twin objectives. In a poorly unionized industry, like the textile, the attack was mainly on wages and hours."[2]

Early in 1921 Hutcheson warned the Brotherhood members that employers were seeking to reduce wages. He advised members not to "become excited, or permit themselves to be stampeded into the acceptance of a wage reduction, but in each instance give calm, cool and deliberate consideration to circumstances and conditions with which they are surrounded."[3] But the anti-union drive continued, and it became clear that it was no temporary phenomenon. Hutcheson summarized the events of the next three years in his opening address to the 1924 convention:

> In many localities there have been attempts made to disrupt not only the morale of the membership of our organization but the conditions that have been established. These attempts were made by different forces and from different angles. In some localities, the endeavor was made by the employers, backed by the Manufacturers' Associations, the Chambers of Commerce, etc. These attacks took various forms; the one that was the most catching was where they attempted to put into effect what they were pleased to term the 'American Plan' of employment, but which really meant putting into effect a condition whereby men of our organization could not procure or secure employment other than under the rules and conditions as laid down by those who were advocating that system of employment.[4]

The other "forces" to which Hutcheson referred were unions impinging on the jurisdiction of the Brotherhood and left-wing movements.

THE LANDIS AWARD

A wage movement involving all the building trades of Chicago began in January 1921.[5] The contractors demanded a wage cut, and at their request, the matter was referred to a referendum vote of the building tradesmen. Every craft rejected the demand. On May 1 the contractors instituted a lockout, which continued until June 7, when most of the trades agreed to submit the dispute to arbitration by federal judge Kenesaw M. Landis. Only the wage question was submitted, but Landis announced to the parties that he could not arbitrate wages without getting into working rules. The unions were dismayed by this, but because of disclosures of corruption in the building trades by an Illinois legislative committee earlier in the year, they felt obliged to agree.

The Carpenters and the Painters refused to participate in the arbitration. The award, which was rendered on September 21, 1921, was a shock to the participants. Some of the wage rates fixed were below those that had been offered by the contractors. Working rules, including the union shop, were modified. "The Award became the fighting issue of the open shop movement in Chicago."[6]

The Building Trades Council accepted the award, despite some wildcat strikes, but the Carpenters refused to go along and led a group of unions in opposition to it. A Citizens' Committee to Enforce the Landis Award was formed, and the open shop was declared in carpentry and several other crafts. The Carpenters were able to sign agreements on the old wage scale with independent contractors, but the Citizens' Committee instituted a boycott against the independents. The banks refused to lend money to contractors who did not observe the conditions laid down in the award. Nonunion carpenters were imported from other cities. The Carpenters sought an injunction against these activities, but their petition was denied.

A number of international union presidents came to Chicago, and with the cooperation of the Building Trades Department (from which the Carpenters had withdrawn) reorganized the Chicago Building Trades Council to exclude the anti-Landis unions. Twenty-two crafts joined the new council, but thirteen, led by the Carpenters, formed their own council.[7]

The contract under which the Landis Award had been rendered expired in May 1923, and the pre-Landis wage scale was reestablished. By October *The Carpenter* claimed that the Citizens' Committee was on the way out. Several large contracting firms broke away from it: "Stubborn resistance of those unions in the building trades that refused to accept the Landis Award . . . is what whipped the big contractors into line and resulted in the final downfall of the Citizens' Committee. The immediate cause of the decision of the three big firms was pressure put on them on a national scale by the International officers of the United Brotherhood of Carpenters and Joiners of America."[8]

In June 1924 the president of the pro-Landis Building Trades Council asked Hutcheson to come to Chicago to negotiate with the larger contractors there, and he reached an agreement with them that essentially terminated the effects of the Landis Award. The Citizens' Committee lingered on for another two years, but the union shop was reinstated in Chicago by 1926.[9]

Thus the American Plan was defeated in Chicago after five years of struggle, but the cost to the Carpenters was great. There were not only direct strike costs, but also the efforts that went into fighting the employers and the pro-Landis unions. The Chicago episode illustrates what employer resistance meant for the labor movement from 1921 to 1926. Money and time that might have gone into organizing new members were expended in preserving existing strength. The wage settlement that Hutcheson obtained in 1924 was what the Carpenters might have obtained in 1921 were it not for the activities of the Citizens' Committee. The Carpenters had eventually won a victory, but it was not an easy one.

DEFEAT IN SAN FRANCISCO

From 1900 to 1920, industrial relations in the San Francisco building industry were controlled by the Building Trades Council under the leadership of P. H. McCarthy. In 1920, the American Plan arrived. The Builders' Exchange, which had been a loose organization of employers, was reorganized and assumed wide authority to bargain on behalf of the contractors. A strike occurred in October 1920 when the unions refused to accept a wage cut ordered by the Exchange. The unions agreed to submit the matter to arbitration, but as in Chicago, the arbitration award was completely unsatisfactory to the membership, and McCarthy was forced to repudiate it. Also as in Chicago, the Carpenters led a few other crafts in refusing to participate in the arbitration proceedings from the start. In San Francisco, however, the Building Trades Council remained united in opposing the award.[10]

The Builders' Exchange responded by announcing that it intended to enforce the award, and instituted a lockout on May 9, 1921, in an effort to do so. Some of the contractors were not enthusiastic about the lockout, but the Builders' Exchange controlled the materials suppliers.

The Building Trades Council finally accepted the award in June, but the Builders' Exchange decided nonetheless to continue operating on an open-shop basis, and the lockout continued. A Citizens' Committee was formed to help the Builders, staffed by the Industrial Relations Committee of the Chamber of Commerce. This group proposed a new agreement that called for a 7.5 percent cut in wages, as well as a mixed union and nonunion shop arrangement. The officers of the Building Trades Council, fearing further losses, were in favor of accepting it, but it was turned down by a wide margin in a referendum vote. Carpenters' Local 22, which was McCarthy's own union, appointed a committee of five to work with other locals outside the council, which had lost the confidence of the members.

A conference of all the building trades was called on July 19, 1921, and 400 delegates attended. A Conference Committee of the Allied Building Trades Unions was formed; R. E. Currie of Carpenters' Local 1082 was elected president. When the Builders' Exchange refused to make any concessions, a general strike was called for August 3, 1921. The strike was endorsed by the Building Trades Council, although not by the San Francisco AFL Labor Council. Of the ten leaders of the strike committee, five were heads of Carpenters' local unions. Hutcheson, perhaps remembering the New York experience, and concerned about radical influence, threatened to suspend the charters of locals participating in the strike. They nonetheless went ahead, and the construction industry was tied up.

The Conference Committee decided to establish a new body to circumvent the existing labor leadership in the Bay Area, and the Rank and File Federation of Workers in the Bay District came into being, again with local Carpenter officials in the lead. Many locals, including some outside the building trades, joined it, and refused to withdraw even when ordered to do so by their international unions. However, they could make no progress against the resistance of the employers, and the strike was called off on August 27. Although the workers continued to refuse to agree to the terms that had been proposed by the Chamber of Commerce Committee, they were forced to go back to work on that basis, and the open shop came to San Francisco.[11]

As a consequence of their failure to call off the strike and disassociate themselves from the Rank and File Federation, five locals of building carpenters, as well as Millmen's Local 42, a Pile Drivers local, and the Dock Builders, lost their charters. Individual leaders of the Rank and File Federation were expelled from membership.[12] They attempted to set up a building construction branch of the Rank and File Federation on an industrial rather than a craft basis, with a constitution in which the preamble spoke of the necessity of organization along class lines. Many of the members, however, were less concerned with socialism than with the removal of P. H. McCarthy, whom they distrusted. "Consequently, when the latter resigned the presidency of the Building Trades Council on January 12,1922, many of the workers thought that the principal reason for the existence of the Construction Branch was gone, and the organization gradually dissolved. By the middle of 1922, many of the former members of the Rank and File Federation had returned to their respective unions."[13]

Until 1935, industrial relations in construction were basically controlled by a newly formed employer group, the Industrial Association of San Francisco, operating on an open-shop basis. One of its rules was that about half of all building trades workers had to be nonunion. To police the system, all contractors were required to secure permits from the association before they could secure materials from manufacturers or wholesalers. Getting a permit was not only contingent on maintaining an open shop but also entailed paying the wage scale set by the Builders' Exchange, which continued in existence alongside the Industrial Association.

In 1925 the Brotherhood decided to challenge the system. The GEB visited California and determined on a program that included the restoration of the union shop and the control of negotiations by international officers. The employers refused to negotiate, and a strike was called on April 1, 1926:

> The strike was the most costly and hard fought industrial dispute that ever occurred among the building trades of San Francisco. Between four and five thousand workers were affected by the strike. There was a partial tie-up of all building work for eight months. The carpenters spent three-quarters of a million dollars, and the Industrial Association was supposed to have spent even a larger amount.[14]

Hutcheson told the 1928 convention that in order to supply the contractors willing to do business with the Brotherhood, the district council had to establish material yards.[15] The Industrial Association brought in strikebreakers, made sure of strict law enforcement by city authorities, and secured an injunction against any interference with nonunion contractors.

Finally, in December 1929, the Carpenters called off the strike. Hutcheson claimed that the union had won a partial victory in that the material permit system was abandoned and the Industrial Association agreed to see to it that all employers paid the scale of wages agreed upon.[16] In fact the union's main demand, elimination of the open shop, was not won; nor was it able to secure any change in the system whereby wages were essentially determined by the employers.

Frederick Ryan's summary of the outcome of the strike, based on careful research, seems justified:

It has been ackowledged by union leaders that the 1926 strike was a mistake. In the first place, the carpenters were poorly prepared for the conflict. At the beginning of 1926 they were in the process of recovering from the open shop strike of 1921-23, and the strike lost them the gains which they had made during 1924-25. In the second place, incompetent leadership failed to bring into the conflict the mass support of other unions that would have undoubtedly made the strike successful. The strike was called off just as it appeared that mass action might be taken.

Disastrous as the strike was to the union shop hopes of the carpenters, it was not without benefit to the members of that craft and to the workers as a whole. The agreement that ended the strike probably stated nothing that the Industrial Association was not willing to grant before the conflict, that is, that unions could make agreements with employers' organizations so long as the open shop principle of the Association was not violated. But the conflict had its greatest effect in putting an end to the complacency with which the Industrial Association viewed the results of the open shop campaign. The strike indicated that the "contented" workers . . . demanded that working conditions under the open shop should, at least, approximate those obtaining under union shop conditions.[17]

In sum, the open-shop drive proved a success in San Francisco. There were special circumstances: a strong determination by employers to rid themselves of twenty years of domination by the Building Trades Council under the leadership of P. H. McCarthy. To the credit of the Carpenters, they were the craft that offered the greatest resistance to the introduction of the open shop, and they were willing to devote substantial resources to the protection of the union shop, which they had enjoyed for many years. If the other trades had been as militant, the outcome might have been different. The effort was a costly one, and contributed to the failure of the Brotherhood to expand during the 1920's.

Chicago and San Francisco were not the only cities in which the Carpenters encountered and fought the American Plan. From 1920 to 1924, the Brotherhood paid out of its central fund a total of $1,026,000 in strike benefits, and from 1924 to 1928, the sum of $683,000. These amounts were in addition to expenditures by local unions on their own behalf. However, by 1928 Hutcheson was able to tell the members that "while there are still periodical attempts made by the employers to put into effect the open shop, or as they have been pleased to term it, the 'American Plan,' our membership (while they have been inconvenienced) have been able to combat those efforts so that the system has not become anything other than what might be termed 'Local,' only existing in a few localities."[18]

JURISDICTION (3)

The reentry of the Brotherhood into the Building Trades Department did not result in any permanent solutions to the Carpenters' jurisdictional problems. Old disputes continued, new ones cropped up. Even during the war, the Brotherhood was vigilant on the jurisdictional front. The agreement that had been reached in 1917 with the Slate and Tile Roofers was canceled the following year, the Brotherhood insisting that the Roofers cede jurisdiction over asbestos and asphalt shingles.[19] Further conferences with the Machinists on millwrights yielded no results. The 1917 AFL convention reaffirmed a decision that had been made in 1914 to give the Machinists jurisdiction over the millwrights, despite Hutcheson's contention that the 1916 convention had overruled the earlier decision. He put it this way: "As I view it, the Baltimore Convention set aside the action of the Philadelphia Convention, and they are two opposite actions . . . If, as stated, the Philadelphia decision is to be in full force and effect, then I would like to have explained to me what is the use of arranging for a conference?"[20]

The first major event on the postwar jurisdictional scene was the organization on August 11, 1919, of the National Board of Jurisdictional Awards in the Building Industry. It was the product of lengthy negotiations between the Building Trades Department and four employer groups. The board was bipartite in character, and its duties were to hear jurisdictional claims and to make awards. The employer was empowered to proceed "with such workmen as in his judgment he may see fit to employ pending a decision by the Board." The unions agreed not to engage in sympathetic strikes in jurisdictional disputes. If there was a tie vote, the case went to an umpire, whose decision was final.[21] Hutcheson was designated as vice-chairman. Nevertheless, the Brotherhood was less than enthusiastic about endorsing the plan, as Hutcheson explained to the 1920 convention:

> We discussed the matter of entering this board and becoming a party to it, and you will permit the language, I informed the [GEB] that more than likely we would be damned if we did and would be damned if we didn't, and the thing was to determine the lesser of the two evils. If we refused to become a part of this plan, and without us the board was created and decisions rendered that were detrimental to our organization and an infringement upon our jurisdiction; then we would be derelict in our duties judging from the results, and we would be looked upon as neglecting the best interests of our membership. On the other hand, if we entered this board and awards were made against us, we would be equally in a position of being censured by our membership. However, after looking at it from all phases, we decided that in order to do the best for our organization we would go along with the then proposed plan.[22]

As soon as the board opened for business, a substantial number of cases were submitted to it. The Brotherhood refrained from bringing any in, because, as Hutcheson stated, "I always felt that if we had to go into court, whether it be civil or otherwise, that we are in a stronger position as defendants than a complainants." The Building Trades Department sent in fourteen cases in which the Brotherhood was involved, but Hutcheson kept delaying action on procedural grounds. "Some of the cases that had come before that Board convinced me that there were some gentlemen on that Board who did not care as long as they got a decision; they did not give a continental as to the merits of a case as presented."[23] He said that before continuing to participate in the plan, the Brotherhood wanted to find out to what extent the contractor association members of the board were really representative of the nation's employers.

Without answering this question, the board met without the participation of the Carpenters and decided the dispute over metal trim in favor of the Sheet Metal Workers. It did not take the Carpenters long to decide that in endorsing the National Board, they had not chosen the lesser of two evils. Several weeks after the decision was rendered, the GEB ruled that "inasmuch as our General Constitution and Laws, which have been ratified by referendum vote of our membership for many years, distinctly specify our jurisdictional claims: the Board cannot accept the decisions as rendered nor will the United Brotherhood of Carpenters and Joiners of America participate further with the National Board."[24]

This gave rise to a heated debate at the 1921 convention of the Building Trades Department. President Donlin affirmed that the National Board would go on operating regardless of the Carpenters. "The general contractors, the architects and the engineers are parties to this and they do not feel that any organization is bigger than the rest of the building trades industry . . . We have to fight the united interests of capital, material men, bankers and everyone else, and we must have harmony." To which Hutcheson replied: "The Brotherhood of Carpenters is not looking for a fight, but if they have to fight they will fight all the way, and the sooner it is started the sooner it will be over."[25] The president of the Bricklayers accused Hutcheson of being evasive and of consistent refusal to comply with adverse decisions. The Sheet Metal Workers declared that they would persist even against the 300,000-strong Carpenters "because fortunately we have some work in our trade the carpenters and no one else can do, but it has to be done by an experienced sheet metal worker." A resolution requiring compliance was adopted by a vote of 35 to 25; the Brotherhood delegates, with 16 votes, abstained. Hutcheson asked whether the Carpenters' Union was suspended, and when he was told that it would be if it refused to comply, the Carpenter delegates walked out. A few weeks later, after conferring with a delegation from the Building Trades Department, the GEB resolved that the Carpenters were no longer members of the department, but that if the National Board award were set aside and new proceedings instituted, with an opportunity for all parties to be heard, the Carpenters would participate.[26]

The AFL executive council urged the Carpenters to rejoin the Building

Trades Department and ask for a rehearing of the case, but Hutcheson wrote back that this was not possible. He began to mount an offensive, which included softening up Gompers, whom he criticized for testifying on behalf of the Board of Jurisdictional Awards before the Lockwood Committee, which was investigating the corruption in New York involving Robert Brindell. Gompers replied: "I was there as a man and not as the President of the American Federation of Labor, as a man who was supposed to know something about the labor movement."[27] Nevertheless, Gompers appeared before the next Building Trades Department convention in his capacity as AFL president to ask that the metal trim decision be set aside and the case reopened. But the members of the department were in no mood to compromise. The president of the Bricklayers stated: "If I had my way I would strike every building in the United States by the loyal unions of this organization to force law and order and enforcement of that decision."[28] The convention voted to ask the employer groups involved to require observance of board decisions in their contracts, and instructed local building trades councils to unseat local unions that refused to do so.

When building trades councils in Cleveland, Detroit, and other cities encouraged strikes against the Brotherhood, a complaint was lodged with Gompers. He in turn took the matter up with the AFL executive council, telling it that the AFL had never approved compulsory arbitration and at no time encouraged strikes against any of its affiliates. The Building Trades Department was directed to inform its affiliates to discontinue such action.[29]

Gompers then went directly to the Jurisdictional Board and urged that the case be reopened. The board agreed to consider his request, but said that unless the Carpenters agreed in advance to abide by its decision, there was no point in going ahead. When Gompers reported this to the AFL executive council, Duffy declared that the Brotherhood would not agree to accept the board's decision, pointing out that the board had seldom reversed itself on a rehearing. Gompers tried to work out a compromise: "It is either expected or implied that both parties will agree to the decision, but neither will be put as a question precedent to the opening of the case if the Carpenters agree to present the case . . . My view of this is that the Carpenters have nothing to lose in the entrance into this rehearing." Finally, Hutcheson agreed that the Brotherhood would give evidence before the board if manufacturers of metal trim and the contractors involved were also invited to appear before it.[30] The board did reopen the case, and heard an number of contractors testify that metal trim work was customarily performed by carpenters. However, the board refused to issue a decision unless the Carpenters would agree in advance to honor it — and there was no chance of that.[31]

While these attempts at compromise were going on at the national level, the battle was also being fought out in various localities. The Carpenters formed rival building trades councils in Los Angeles and Cleveland to support them against the Sheet Metal Workers. The Building Trades Department, for its part, committed a cardinal sin: it sponsored local unions of carpenters dual to the Brotherhood. The Brotherhood protested to the AFL specifically about such action in Dayton and Cincinnati, and after a hearing, the AFL executive council issued the following directive to the department's chief officer:

> You, as President of the Building Trades Department of the American Federation of Labor, are hereby advised that it is the decision of the Executive Council of the American Federation of Labor that it is the duty of Department officers to insist upon the Cincinnati Building Trades Council to disassociate the dual carpenters' union or any other dual union which may be part of the Cincinnati Building Trades Council and that said Cincinnati Building Trades Council and the Dayton, Ohio Building Trades Council *be required to comply with this decision within a period of ten days from the date of this communication to you.* The Executive Council directs that every effort of this character be denounced and defeated as inimical to the interests of the working people in that industry as well as the rights, interests and welfare of the great mass of the working people of America.[32]

In his reply, President Donlin wrote that the department had no official knowledge of the alleged dual unions, and asked whether the Carpenters were being required "to put a stop to the pernicious practice . . . in obtaining injunctions either directly or indirectly for the purpose of thwarting the proper and rightful activities of the International Unions associated with the Building Trades Department." Then he added, with righteous indignation:

> It is most difficult to understand why the Executive Council should render so emphatic a decision on the grievance of the Carpenters while remaining profoundly silent on those of the Building Trades Department. From outward appearances it seems to be a case where the Building Trades Department is to be made the catspaw by the Executive Council in order that the chestnuts of the Brotherhood of Carpenters and Joiners might be pulled out of the fire.[33]

Hutcheson had turned the whole matter around very neatly. The Brotherhood was now the aggrieved party, defending itself from predatory attacks by other building unions.

The Building Trades Department attempted to fight back. A resolution adopted at its 1924 convention asked the AFL to suspend the Carpenters from membership: "We believe it to be the duty of the American Federation of Labor to use all its influence to maintain the reputation for integrity which has been injured by the unfortunate action of the Brotherhood of Carpenters."[34] The Carpenters refused to budge, and the dispute dragged on, causing a great deal of turmoil around the country.

The Brotherhood also continued to complain of rival unions organized against them. Duffy told the AFL executive council of a new International Union of Building Trades Carpenters established in Cincinnati that claimed 5,000 members. The Dayton Building Trades Council notified contractors that work performed by members of nonaffiliates of the council, including the Carpenters, would be considered unfair.[35] These complaints prompted William Green, who had succeeded Gompers in 1925, to warn the Building Trades Department again that "the American Federation of Labor has always been opposed to the formation of dual or rival unions. No independent organization of either alleged carpenters' unions or any other unions can be recognized by the A.F. of L."[36]

The logjam was finally broken when an agreement was reached with the Sheet Metal Workers. The Brotherhood was given undisputed jurisdiction over the erection and installation of all interior metal trim except toilet partitions and the setting of metal window frames. The Sheet Metal Workers were to hang and adjust metal sash, work on storefronts, erect metal column forms, and install metal lockers and shelving.[37] In presenting this agreement to the Brotherhood convention, Hutcheson commented:

> The consummating of this agreement will end a controversy of many years' standing, and while there have been times that our organization was criticized and censored by other building trades organizations, contractors and the building public, the consummating of the agreement clearly illustrated the justification of our claims and show what may be accomplished if our membership will insist upon the jurisdiction of our organization being recognized.[38]

This settlement did not automatically bring the Carpenters back into the Building Trades Department; there was still the problem of the Jurisdictional Board. Up to 1926 the board had been strongly supported by the unions affiliated with the department. The president of the department had expressed the view that "the present National Board for Jurisdictional Awards is the most effective method of adjusting jurisdictional disputes which has yet been presented to or conceived by the building trades organizations of this country."[39] But a year later, the department voted unanimously to withdraw from the board.

The reason for this about-face was the willingness of the Brotherhood of Carpenters to rejoin the department only on the condition that it renounce the Jurisdictional Board. The department was anxious to get its largest member back. Organizational work was lagging in a number of cities, and dual building trades councils in New York and Cleveland were impeding collective bargaining.

"It is inconceivable that the Building Trades Department should be able to operate successfully without this powerful [Carpenters'] union. Let us remember that this was not the first time the Carpenters had come back into the Department on their own terms. On two previous occasions the same little drama had been enacted. It was while he was vice-president of the American Federation of Labor that James Duncan of the Granite Workers Union said, "It has ever been my argument — I plead guilty to it everywhere, that I have argued in season and out of season that a Building Trades Department without the Carpenters is almost a joke."[40]

The Carpenters were suspicious of the Jurisdictional Board from the start, and their worst fears were confirmed by its decision in the Sheet Metal Workers' case. To have yielded to the dictates of the board would have opened up a jurisdictional Pandora's box, and would have been counter to a half century of tradition. A contemporary historian described the matter very well when he wrote:

The Carpenters were not organized to benefit society . . . nor to help labor as a whole, nor even to assist the other building trades unions. They were organized for the sole purpose of benefiting themselves. The history of the Carpenters is a stormy one, the story of a powerful organization protecting its own interests at almost any cost . . . The most powerful union in the building industry, and one of the largest and strongest in the whole trade union world, the Brotherhood of Carpenters has never had to bow its head in defeat in any important conflict . . . The Carpenters must and should be held responsible for the failure of the National Board of Jurisdictional Awards. But they cannot be condemned for the part they have played unless the trade union movement as a whole and the economic order itself are also to be condemned . . . The withdrawal of the Building Trades Department . . . marks another smashing victory for that belligerent union, the United Brotherhood of Carpenters and Joiners of America, in its ceaseless and sometimes ruthless fight to further the interests of its members.[41]

John Dunlop has called attention to another factor that was at issue in this dispute, as well as more generally in many others — competition between contractors:

The difference in types of contractors is a factor contributing to jurisdictional disputes . . . General contractors, specialty contractors, and sub-contractors are to some extent in competition with each other. Different contractors on a single project create boundary lines between the work which is to be performed by each. The fact that some unions work exclusively, as a matter of policy or custom, for particular contractors tends to convert competition among contractors also into jurisdictional disputes between unions. If a contractor is to hire such a craft he must also hire a sub-contractor.[42]

This dispute pitted the general contractors who employed carpenters against specialty contractors involved in sheet metal work.

While the dispute with the Sheet Metal Workers was the most important one in which the Brotherhood was engaged at this time, it was by no means the only one. Jurisdiction over millwrights remained an open issue. In 1920 the secretary of the Millwrights Protective Union, an independent organization centered in Buffalo, New York, that claimed control over 90 percent of the flour, cereal, and feedmill machinery installation work in the United States, asked the AFL for a national charter: "It appears to me that by granting this charter to a deserving craft, you would not only be doing justice to us but would also stop the turmoil, bloodshed and disgrace of having union men opposing each other, as has occurred in many places. Nothing but harm can come from such actions and no one but employers benefit." The applicant was informed that the Carpenters had jurisdiction over the trade, which was confirmed by Hutcheson when he wrote to Secretary Morrison of the AFL that "it very often happens that we have to use the power of our organization by informing the contractor that unless those men are put off of the job our members will not be permitted to work on said job." The reply from Buffalo was bitter: "We do not want a charter from them, that has been offered us many times in the past 20 years. If we wanted to destroy our trade, such a means would be a sure means of doing it."[43]

The Machinists were a more formidable adversary. In 1920 the Brotherhood informed Stone and Webster, one of the largest construction firms in the country, that it would strike all their work unless carpenters were allowed to install machinery.[44] The Machinists threatened to withhold their per capita payment from the AFL unless the previous AFL decision awarding them jurisdiction over millwright work was enforced, but they were simply advised by the executive council to continue negotiating with the Carpenters.[45] This they continued to do in a desultory manner; it was to be many years before the dispute was resolved.

There was also a controversy with the Maintenance of Way Employees over the construction of railway roundhouses and other structures. The Brotherhood claimed jurisdiction by virtue of an award of the 1919 AFL convention, but the Maintenance of Way Employees asserted that they had always built railroad stations and subsidiary structures. They were prepared to turn over to the Brotherhood all new construction but not repair work, and at the insistence of the Carpenters, they were suspended from the AFL. They were finally obliged to yield all the work except maintenance on the railway right of way proper as a condition for reinstatement.[46]

A controversy that was to erupt once again some years later was temporarily allayed; it involved the control of lucrative carpentry construction in the Hollywood motion picture studios. An agreement had been reached with the International Alliance of Theatrical and Stage Em-

ployees to the effect that all construction in the studios was to be done by
the Brotherhood, while IATSE was limited to the work necessary to fit
up the stage behind the proscenium, as well as work in theatrical shops
and storerooms.[47] The agreement proved to be too imprecise, and an investigating team had to be sent to Los Angeles to look into the matter. It
appeared that IATSE had not turned over studio construction work to
the building trades, and it was ordered to do so on pain of suspension
from the AFL, after the Carpenters had threatened to stay away from the
next AFL convention. IATSE finally obeyed this directive, keeping control of property men and those engaged in placing furniture, laying
carpets, and hanging pictures.[48]

An agreement with the International Longshoremen's Association
confirmed Brotherhood control over all dock carpentry. However, "The
Carpenters raise no objection to the Longshoremen nailing a board or
plank, or doing work where Carpenters cannot be had."[49] This seemed
straightforward enough, but trouble developed over the building of partitions, bins, stalls, and bulkheads in connection with the stowing of
cargo.[50] An adjustment satisfactory to the Carpenters appears to have
been made.

An excellent example of a dispute occasioned by the introduction of a
new material, which has been a prolific cause of jurisdictional controversy, involved Celotex. The Lathers Union claimed the right to install
this material, the Carpenters disputed it. An agreement was reached
whereby any material specifically marked "Celotex Lath" was to be installed by the Lathers, with all other Celotex reserved to the
Carpenters.[51]

This is by no means a complete catalogue of Brotherhood jurisdictional disputes and agreements during these years, but it does provide
some idea of the efforts that the organization made to protect and extend
the work available to its members. A contemporary analyst of this activity raised an interesting question: was all this worthwhile in terms of net
gains to the Brotherhood and to the labor movement? The conclusion
was that the Brotherhood paid a big price for its victories in the form of
the bitter hostility that other unions developed toward it, as well as the
heightened distrust of employers and the general public. Society lost in
terms of the extra costs occasioned by jurisdictional stoppages. Nevertheless, in the final analysis, "despite this great cost the Brotherhood is
stronger as it stands today than it would have been had it not striven for
the ends which were the goal of these controversies . . . The Brotherhood's policy has been in entire accord with trade-union principles."[52]

This seems a fair summary. If one were to undertake a cost-benefit
analysis of the question from the point of view of the Carpenters, there is
little doubt what the result would be. In addition to the costs enumerated
above, cash outlays for strike support, the time of union officials, and the

pay lost by members idled by disputes must be added to the cost side. But for the Carpenters, the costs would be overwhelmed by the benefits: the imposition of a standard rate over a broad area of work; the prevention of skill dilution through excessive specialization; and above all, the creation of a powerful organization capable of holding its own with building trades employers, who were not always amenable to collective bargaining relationships. This is what was behind Hutcheson's admonition to the membership in 1928: "The members of our Brotherhood should put forth every effort to see that our jurisdictional claims, as set forth in our Constitution, are closely and strictly adhered to, and to at all times remember that our jurisdictional claims are based upon, not the character and nature of material used, but upon the skill, knowledge and ability required to properly erect or install the material."[53]

ENCOUNTERS WITH THE COMMUNIST PARTY

The Carpenters' leaders who took over after Peter J. McGuire were prepared to tolerate socialists in their midst, although they were not friendly to them. The Industrial Workers of the World never had any appeal to skilled workers, and constituted no real threat to the Brotherhood. We can say with the advantage of hindsight that the communists were more a nuisance than a danger to the Brotherhood leadership, but they did not appear so at the time. Moreover, as subsequent events demonstrated, it was not out of the question for communists to gain positions of power in the American labor movement.

Under the leadership of William Z. Foster, an able organizer who ran a major steel strike at the behest of the AFL in 1919, the Trade Union Educational League (TUEL) was established in 1920. The league was controlled by the newly formed Communist Party and financed largely by the Soviet Union; it goal was to gain control of the American labor movement by "boring from within" the unions affiliated with the AFL. Before it was converted into a full-fledged rival federation in 1928 on orders from Moscow, the TUEL gained considerable influence within a number of unions, including the International Ladies' Garment Workers, the Fur Workers, the Machinists, the Painters, and the Mine Workers.[54]

There is no doubt about the attitude of the Communist Party toward the leadership of the Carpenters' Union. Foster, one of the leading functionaries of the party for many years, called Hutcheson "one of the very blackest reactionaries in the labor movement." He accused Hutcheson of breaking the 1916 strike in New York by helping to recruit strikebreakers, and claimed that Hutcheson had eventually surrendered to the Chicago Landis Award.[55] As far as the Brotherhood leaders were concerned, they had little difficulty in deciding how to handle communists. In his opening address to the 1924 convention, Hutcheson declared: "Our membership should remember there are only two 'isms' that should enter into our organization—that is unionism and Americanism, and that all other 'isms' or advocates of that sort of thing should be kicked out and kicked out quickly."[56]

It is interesting, in the light of subsequent historical developments, that Hutcheson was followed to the podium by Philip Murray, then the first vice-president of the United Mine Workers, who was a fraternal delegate in place of John L. Lewis. He recounted his experiences with communists inside the UMW, and told of attending an International Mining Congress in Prague, where unions in a number of countries were complaining about Soviet interference:

> Is it any wonder, my friends, in the face of those experiences that the trade unionists have undergone in Europe that the Mine Workers, the Carpenters, and all other great, big, strong, militant trade unions in America are fighting with all the energies they possess the Communistic movement in this great country of ours? I have told my union in International Convention, in the presence of more than 2,000 delegates, that so help me God, if I should lose my right arm in the fight to destroy this movement, I am going to use all the intelligence and all the energy I possess, in fighting, day in and day out, not only the large employing interests, but the Communists in America as well.[57]

If Murray—and Lewis—had remembered these words a decade later, the American labor movement would have been spared a great deal of agony.

The first open confrontation between the Brotherhood leadership and the communists came at the 1924 convention. The TUEL was publishing a newspaper, the *Progressive Building Trades Worker*, and had gained some support in a number of the larger cities. Several resolutions that were introduced by left-wing delegates were aimed at reducing the authority of the Brotherhood's general officers, but all were defeated. Two men were nominated to run against Hutcheson for the office of general president; one of them was Morris Rosen, a TUEL adherent in New York City Local 376. Hutcheson was elected in the subsequent referendum by an overwhelming majority, but a number of locals charged that the count was fraudulent. Among other things, Duffy was accused of holding back the votes of nearly half the locals. All these charges were rejected by the GEB, which confirmed the election results.[58]

While the 1924 convention was going on, the GEB met and took the first of a series of actions aimed at the elimination of communist influence in the Brotherhood. Several members of locals in Los Angeles had been suspended by Hutcheson because of membership in the TUEL. The GEB sustained this action, and ruled that they could be readmitted only if they submitted affidavits to the effect that they had severed all their ties with the TUEL. They were also barred from holding office in the union for five years.[59]

The next move was against the TUEL group in Detroit. An investigating committee was sent out to look into the activities of Local 2140, headed by William Reynolds, who was also vice-president of the Detroit district council. The committee found that in the district council, "business was transacted in bedlam fashion. Speeches and more speeches of a fiery nature were made accusing one another of deception, intrigue, dishonesty, obstinacy, trickery, playing politics and so forth. The General Officers were accused of running things in a high-handed fashion and not living up to the constitution and laws; of expelling members without a trial . . . of bossing the District Council and dictating how they shall run their business and many other things." The committee recommended that the district council be reorganized to eliminate the TUEL faction, which appeared to control a majority of the delegates, and that Local 2140 be directed to expel Reynolds immediately on pain of having its charter revoked. This was apparently done, for Detroit caused no more trouble. In Chicago, the TUEL had gained influence in Locals 81 and 62. Five of their members were expelled, but they were reinstated when they pledged that they had severed their relationships with the TUEL and formally apologized to their locals.[60]

The main bout involved Morris Rosen, who had joined Brooklyn Local 376 in 1923 and became its president a year later. He had run against Hutcheson in 1924 on a campaign platform that a GEB investigating committee said contained planks "in opposition to our constitution and laws and not in conformity with our obligation." Officers of the local replied that they were not aware of this or of the issuance of material of a defamatory nature. The Committee recommended that the local be reprimanded, and the GEB concurred.[61]

This was not the end of the matter. Communist agitation continued in New York and elsewhere, and on January 21, 1926, an official notice was sent to all locals alerting them to communist efforts to infiltrate the labor movement:

> While our obligation guarantees to every member no interference with his political opinions, this matter is not and cannot be considered a political matter or a party political matter, but on the contrary, an attempt on the part of the Communists and their agencies to hamper and cripple Labor Unions. The General Executive Board . . . warns all members not to join them or have anything whatever to do with them or similar or kindred organizations under penalty of forfeiture of membership in the United Brotherhood of Carpenters and Joiners of America.[62]

Several weeks later the treasurer of Local 376 was expelled by the local for an alleged shortage of $1,000 in his accounts. He appealed the expulsion to Hutcheson, who sent board member T. M. Guerin to New York with instructions to examine the records of the local, which the national office had the constitutional right to do. Rosen claimed later that the local was willing to let Guerin inspect the books on its premises, but that Guerin wanted to take them away with him. In any event, a GEB investigating committee was sent to New York, and Local 376 was directed to surrender its records or lose its charter.[63] It was suspended when it refused to comply. A number of the members of Local 376, including Rosen, joined other locals in which the TUEL faction was active.

The whole matter was aired at length at the 1928 convention, where Rosen was permitted to speak as a representative of the former members of Local 376. He asserted that the grounds on which the local was suspended were merely pretexts, and that the real reason was his challenge to the Hutcheson administration at the 1924 convention:

> Local 376 was the leading and most militant Local exposing many of the things which were not being properly done in our organization. I say in many sections of our country there was corruption existing to a great extent, and we were persistently exposing these things . . .
>
> I ask you to pass judgment on the actions of General President Hutcheson during the last four years. I ask you to support our appeal. Favorable action on our appeal may not only win the case for us, it will have the effect of putting a stop to President Hutcheson's irresponsible moves . . .
>
> If the convention approves the actions of Hutcheson it will set a precedent where Hutcheson can suspend any Local Union or disenfranchise any member that displeased him — not that Hutcheson needs such a precedent — he usually does whatever he pleases, but I ask you once and for all to show that the membership rules and not him.[64]

The counterattack came from Duffy, who was fully prepared. He delivered a long diatribe against the Communist Party in general and Rosen in particular. The latter, he asserted, wrote regularly for *The Daily Worker*, the party organ, under the name of Mike Ross. He accused Rosen and Local 376 of having carried on a long, defamatory campaign against Hutcheson, who was charged with establishing the open shop in construction and of "faithful service to the capitalistic class." Duffy continued:

> Now what is at the back of it? Why didn't they want to turn over the books? . . . They didn't want us to know that they paid money to the Workers' Communist Party of America and the Trade Union Educational League . . . He [Rosen] calls us fakers, misleaders — misleaders of what? He don't say. Reactionaries, stand-patters, persecutors, aristocrats, bureaucrats, the Czar [pointing to President Hutcheson] Czar Hutcheson. That's the pet name for him, and all of us are fakers, you included . . .
>
> There [pointing to Morris Rosen] is a Communist, a Trade Union Educational Leaguer, sitting on the platform here with us today, and as a member of the Credentials Committee I voted to allow other Communists coming from

their Local Unions in different parts of the country to be seated here as delegates, because I wanted to meet them face to face and tear the mask away.[65]

The convention upheld the actions taken by the GEB by a near unanimous vote. A motion was made to expel Rosen from Brotherhood membership. When it carried, Hutcheson told Rosen: "Please get out of the building and off the Brotherhood's property as quickly as possible." Duffy then named Robert Golden of Local 1164 as a communist; when he was also expelled, and attempted to challenge the vote, Hutcheson said: "Get out as quickly as possible and do not make it necessary for us to put you out." Thomas Schneider of Local 2090 and Joseph Lapidus of Local 1164 were given the same treatment. Others whose names were mentioned as communists asked for the privilege of making explanations to the Credentials Committee, and this was granted. Of eleven who appeared before the committee, three were expelled, two were cleared, and six were put on probation after claiming that their opposition to local leadership had nothing to do with communism.[66]

This marked the end of any communist penetration into the Brotherhood. By present-day standards, the method of dealing with the communists may seem despotic, and indeed, their expulsion without a trial might have violated the union's constitution. But it must be remembered that beginning with the formation of the TUEL in 1920, and right up to the expulsion of communist-dominated unions from the CIO in 1950, the communists were widely regarded as a serious threat to the stability of American trade unions. Among other things, charges of communism provided ammunition for anti-union employers. The communist tactic of infiltration into local and national unions, and the subsequent use of the unions by party functionaries for the political purposes of the Communist Party, had nothing to do with the collective bargaining functions of American unions. In the view of the communists, trade unions were merely tools of the party, "transmission belts" charged with the duty of conveying to the rank and file the decisions of the party. The Carpenters recognized the dangers early and were able to extirpate communist influence quickly.

Other unions were not as fortunate. The International Ladies' Garment Workers' Union, for example, was almost destroyed by an internecine struggle between communists and their opponents. A vice-president of that union, Julius Hochman, was apparently impressed by the manner in which the Brotherhood was dealing with communists, to judge by a letter he sent to Duffy: "In connection with our recent experience, we are very interested to learn the methods used by Communists in their attempt to undermine your organization. Will you please advise us in detail of some such experiences that you have had?"[67]

In 1928 the convention of the International Association of Machinists conferred authority on its president to expel communists, a power that was used on a number of occasions by President Wharton. Communists were barred from membership in the Mine Workers Union by constitution. The success of the major unions, including the Carpenters, in defeating infiltration led to a change of communist policy in 1928. "Boring from within" was abandoned, and the Trade Union Unity League was created for the purpose of setting up independent national unions. The TUUL was never successful in winning over any substantial number of workers, but it did serve as a training ground for organizers who later played significant roles in the CIO.[68]

It may also be well to mention that while Hutcheson and some of his AFL confreres were being branded as "misleaders of labor" and "labor czars"—standard communist terminology of the time—Stalin, the undisputed leader of the world communist movement, was systematically imprisoning and executing the leadership of the Soviet trade unions, some of whom had opposed Communist Party domination of their organizations. The German Communist Party and its allied trade unions were undermining the democratic labor movement in Germany and paving the way for the rise of Hitler. The carpenters who attended the 1928 convention and expelled the communists from their midst may have been unaware of these contemporary events, but their instincts were good.

THE TWENTY-FIRST GENERAL CONVENTION

Many of the issues debated at the 1924 convention have already been discussed—the employer offensive, jurisdiction—but there are several others worthy of note. A number of locals of the former Amalgamated Society of Carpenters, which had been semi-autonomous within the Brotherhood since 1914, decided to sever their affiliation and become independent again. The GEB thereupon abrogated the 1914 pact and offered Amalgamated members full membership in the Brotherhood with credit toward benefits for their years in the Amalgamated locals. All the U.S. locals came over, but for a time the locals in Toronto and Vancouver stayed out.[69]

Several constitutional amendments were adopted and subsequently approved. At the recommendation of Vice-President Cosgrove, a maximum initiation fee of $10.00 for apprentices was fixed; some locals had been charging them semibeneficial initiation fees and dues and thus keeping them out. Also, to facilitate readmission, ex-members were thenceforth to pay only the regular initiation fee, plus $3.00 to be forwarded to the local union in which they had formerly held membership. A minimum of $1.00 per month dues was fixed for all members, a measure designed to bolster the financial solvency of weak locals.

The virtues of a pension fund or a home for aged carpenters had been under debate for some years. Hutcheson had favored a home, but the idea had not met with favor. In the spring of 1923 the GEB set up a committee to look into the possibility, and on March 26 issued a circular calling attention to the plight of superannuated carpenters, and asking whether "after a lifetime of struggle as trade unionists, they are to be thrown on the 'scrap heap' in their old age, uncared for and unprotected, to die in poverty or the workhouse." A referendum was ordered to decide whether the board of trustees (which had been set up to hold the Carpenters' headquarters in Indianapolis) should be authorized to purchase land for a home, to be financed by an additional monthly per capita tax of 10 cents per month.[70] Both propositions received the necessary two-thirds affirmative vote. In December 1923 the board authorized the purchase of 1,684 acres at $375 an acre, for a total cost of $631,500, at Lakeland, Florida. Just prior to the 1924 convention, a special committee examined the site and began to make plans for construction. Much of the land was planted with citrus groves, and it was recommended that these be operated commercially with a view toward making the home economically self-sustaining.

Some opposition to the idea of a home surfaced during the convention. C. J. Mulcahey of Local 632, Providence, Rhode Island, led a move to force a referendum vote on the home issue alone, divorced from the proposed pension. The convention voted to table his motion, and gave the go-ahead signal for the home and a pension system. Pension payments were to start one year after the home opened, and were to be based on the cost of maintaining a member at the home.[71]

The opposition persisted, however. In March 1925 a proposition endorsed by the necessary twenty-five locals in twenty-five states was submitted to a referendum vote. It read: "Shall the Board of Trustees be instructed to sell the land owned and controlled by the United Brotherhood of Carpenters and Joiners of America, and known as the Home for the Aged Members, located at Lakeland, Florida? The proceeds derived from the sale of said premises to be placed in the Old Age Pension Fund and to be used for the payment of Pensions to the Aged Members." The GEB issued a statement promising that the prospective pension would not exceed $25 a month. With 2,500 pensioners on the rolls, the annual cost would have been $750,000, compared with the $360,000 currently going into the home and pension fund from the $10 monthly per capita payment. The board maintained that selling the Florida land would pay the pension for only a year, whereas the land might make a major contribution to revenue and thus obviate the necessity of raising the per capita tax. Perhaps impressed by this argument, the members defeated the proposition by a margin of better than 4 to 1.[72]

The Home was completed on March 1, 1928, at a total cost of $1,404,000, excluding the land. Local unions were asked to contribute to its furnishings, and they did so to the extent of almost $100,000. The Home was dedicated October 1, 1928, and it opened on January 1, 1929. The payment of pensions began April 1, 1929.

At the time the Home was completed, the liquid assets of the Brotherhood totaled $1,143,000. Thus, the investment of $2,150,000 in the Home was a major one for the organization. Moreover, the Brotherhood was obligating itself simultaneously to maintain the Home and to initiate a pension scheme for members who did not enter the Home. If the general officers could have foreseen the economic cataclysm that was to overtake the nation less than a year after the Home was dedicated, plus the subsequent establishment of a national old age pension system, they might have come to a different decision. But it must have been very gratifying for them to hear the words of the secretary general of the International Union of Woodworkers, an organization with which the Brotherhood had affiliated in 1925, that were spoken at the dedication ceremony:

> I will refrain from trying to put into words what this building has aroused and stirred in me. The biggest of all big words would seem mere trivialities compared with the richness and efficiency of this building. Only a lyrical poet could satisfactorily interpret the thoughts and emotions that assail us as we gaze up this building and think of the noble purpose it is going to serve . . . By erecting this Home you, the organized Carpenters and Joiners of America, have solved your part of the problem, for no longer will the old members of your organization, who for many long years have remained faithful to their trade union, be left to their fate and the chance of charity. In this Home, every stone of which has been due to the combined efforts of the entire membership, they will find a restful place in which to spend the remainder of their days, and that as a right to which their own efforts have entitled them. This is an achievment of which the many thousands of members of this great organization of Carpenters and Joiners of America may justly pride themselves.[73]

RELATIONS WITH THE AMERICAN FEDERATION OF LABOR

William Green succeeded Gompers to the presidency of the AFL in 1924. A coal miner by trade, he had been a member of the United Mine Workers and served as an official at every level of that organization, including the office of secretary-treasurer, which he held from 1913 to 1924. Although he generally favored industrial unionism because of his experience in the Mine Workers, his trade union philosophy was very much in line with that of the Carpenters' leaders. It was his fate to be buffeted between two strong men, Hutcheson and Lewis, and when the final choice had to be made between them, he came down squarely on the side of the former.

The year of Gompers' death, 1924, marked a change in the traditional political stance of the AFL. The Farmer-Labor Party, which had been formed several years earlier, nominated Senator Robert M. La Follette of Wisconsin for the U.S. presidency on a third-party ticket. John W. Davis, a conservative lawyer, was the Democratic nominee, and Calvin Coolidge, the Republican. La Follette was endorsed by the Socialist Party, though he repudiated communist support.

There were divided views in the AFL executive council on the proper course for the Federation to take. The major party candidates were considered anti-labor, and there was considerable grass-roots support for La Follette. A majority of the executive council favored the endorsement of La Follette, but a minority, including Duffy, were opposed. The council compromised by endorsing the record of the Farmer-Labor nominee as well as the party's program, and refrained only from endorsing him by name. Although La Follette received 5 million votes and carried his home state of Wisconsin, the hopes of those in the labor movement who favored independent political action were dashed.[74]

The Brotherhood remained strictly neutral during the campaign. A few months before the election, Duffy complained to Gompers that the action of the Wisconsin State Federation of Labor in endorsing the Farmer-Labor Party constituted interference with the rights of Brotherhood members:

> It is . . . specified in our Constitution that *party politics* must be excluded from our meetings. You can therefore realize the position our Local Unions in Wisconsin are in. I know they do not want to withdraw from the Federation of Labor of that state, and yet they want to comply with our laws , in accordance with the obligation they took to do so. If any of our Local Unions withdraw I would have communications from you or Secretary Morrison urging me to use my influence with these Unions to re-affiliate, but in cases of this kind I could not comply for I then would be in violation of our obligation and laws.[75]

There is no record of a reply from Gompers; he was ill at the time, and may not have had the opportunity to answer Duffy's letter. The letter was Duffy's way of warning Gompers that endorsement of La Follette might have adverse financial consequences for the AFL. As far as Brotherhood publications were concerned, one looks in vain for any reference to the fact that a presidential campaign was going on.

There were no scientific polls at the time, so we have no way of knowing whether the views of the AFL leadership were consistent with those of the membership in this first presidential election in which the AFL took a stong partisan position. What little evidence there is suggests that a majority of the members of the Brotherhood in 1924 intended to vote Democratic or Republican, and a minority for La Follette.[76] The rule of no politics at union meetings would have been a sound one if this were true.

Duffy watched AFL expenditures carefully, and was quick to protest anything he considered excessive. He noted that the AFL reported a cash balance of $236,000 on August 31, 1924, and asked why AFL organizers should not be used to assist international unions as well as local AFL bodies. Morrison replied with a long defense of AFL finances, pointing out that "in a prolonged strike or lockout of even a small part of the membership of our [federal] local unions, it would not take long to deplete the $201,131.56 in the defense fund."[77] Two years later Duffy opposed any AFL financial assistance to the British trade unions, which were conducting a general strike, on the ground that since the AFL could not call a general strike, it could hardly make a donation to support one in another country.[78]

In 1928 a letter from Green notifying executive council members that he was appointing W. C. Hushing to a temporary position as legislative lobbyist sparked off another protest by Duffy. Pointing out that the AFL already had two legislative representatives, Duffy wrote that the AFL's finances were shaky, partly because it was holding notes from some member unions that could not pay their per capita taxes. Green, who was not yet inured to the barbed missiles fired off by the Carpenters' general secretary, sent back a long letter, which ended rather plaintively: "I have no personal interest in the field workers employed by the American Federation of Labor. Not one working for the movement can be considered as my personal appointment . . . within the last four years the expenses of the American Federation of Labor have been reduced. The number of field workers is less than ever before . . . I am personally working to the extent of my physical limitations."

Duffy may have felt that he had gone too far, for he wrote to Green, "Old pal, you have got me wrong," and went on to say that in the Gompers period, "things were running wild with the A.F. of L. so far as finances were concerned." However, he ended up on a characteristic note:

I have the interest of the A.F. of L. as much at heart as any member or officer thereof, and as I am a member of the Executive Council I think I have the right to call attention to matters I feel should be looked after. Of course, if I have not that right there is no necessity of my being an officer of the A.F. of L. any longer. Mind you, I am not finding fault with you, nor do I propose to find fault with you. You are my choice for President of the American Federation of Labor; I stand solidly with you, but at that I claim the right to tell you how I feel about things. Surely you do not expect me to be a passive individual without stability or backbone. You know, Bill, I am not built that way.[79]

In his dealings with AFL officials, Duffy was quick to take offense at any slight or criticism, whether intended or not. For example, Matthew Woll was making a survey of unemployement relief services maintained by the national unions and asked that questionnaires eliciting relevant information be completed. When no answer was forthcoming from the Carpenters, he sent several letters to Duffy urging that the information be sent, and finally one to Hutcheson. Duffy replied to Green, not Woll, his letter beginning as follows:

> After serving more than a quarter of a century as General Secretary of the United Brotherhood of Carpenters and Joiners, I should be able to fill that office without instructions or directions from others. The laws of our organization specify my duties and I follow them out to the letter. I am elected by referendum vote of our entire membership. I am paid from the General Fund of the organization. I am not therefore subject to orders, instructions or directions from those outside our organization. As the General Secretary of the Carpenters I will not take orders from the American Federation of Labor or any of its officers.

He then went on to explain that he was very busy and the consensus in the Brotherhood office was that he should not answer. It was the complaint to Hutcheson that got him "sore." Green wrote back expressing his regret "that your feelings have been so badly hurt and that you are smarting under a deep sense of personal injury." He explained that the requests had been routine, that Vice-President Woll, "a prodigious worker," had delegated the work to others, and had not even signed the letters himself. Then he added:

> Your years of service have given you a standing in the labor movement acknowledged and conceded by all the officers of national and international unions and by all members of our movement who know of your work and your services. No man, either in or outside the organized labor movement, can question your honesty, your integrity or your willingness to sacrifice for and serve the labor movement. I would rather be the subject of criticism myself than to know that your feelings have been hurt by either warranted or unwarranted condemnation. I ask you therefore, not as President of the American Federation of Labor, but as your true friend and fellow trade-unionist, to forget the incident, dismiss it from your mind and regard it only as an incident in the discharge of your personal and official duties.

Duffy replied that he was as busy as Woll, but was willing to close the incident.[80] The Carpenters in general, and Duffy in particular, were not easy people to deal with, as Green was to discover over the years. But when the occasion arose, they were not ungenerous. In September 1929 Duffy wrote to Morrison at the behest of the GEB pointing out that a number of unions were not paying per capita taxes on their full membership to the AFL, but were nonetheless given the right to vote on total membership at AFL conventions. Specifically, he stated that the United Mine Workers had paid on 250,000 members but were nonetheless voting on 400,000. He tendered a check for five-eighths of the Brotherhood membership and demanded full voting rights, threatening to boycott the next AFL convention if this were not done.[81] At the next meeting of the executive council, Green explained that the Mine Workers had spent a great deal of money on a strike in Ohio and Pennsylvania that had been lost, and were in very poor financial shape. Duffy thereupon agreed that the Brotherhood would pay the balance of its per capita.[82] A letter that he had received from John L. Lewis several years earlier suggests that relationships between the Miners and the Carpenters were good:

> I am informed that on July 24th you will have completed twenty-five years of service as general secretary of the United Brotherhood of Carpenters and Joiners. This is indeed a record of which any man might be proud. It is also a special matter of pride to those of your friends and associates in the trade union movement, who, through contact and personal observation, have come to know of your sterling qualities as a man and your outstanding qualifications as a trade union officer.[83]

Before moving on to the events that occurred at the 1928 convention, an interesting sidelight is provided by a letter written to Duffy by Sumner Slichter, one of the most distinguished labor economists in the United States, who was then professor of economics at Cornell University and later moved to Harvard: "The copies of 'The Carpenter' in the Cornell Laboratory of Industrial Relations are used so frequently that I am having them bound to avoid loss and injury of copies. Assuring you that we appreciate the cooperation which your organization has extended to us the last several years and that we are finding your journal of real help in giving our courses on labor problems, I am . . . "[84]

THE TWENTY-SECOND GENERAL CONVENTION

The 1928 convention was held in Lakeland, Florida, at the site of the new Home for aged carpenters. The number of members in good standing was put at just over 295,000, down from 322,000 in 1924. However, if the number of members in arrears is included, the totals for the two years were only 4,000 apart.

The GEB reported that the wage settlements made during the previous four-year period, although they did not give locals all they asked for, were "honorable and show the best results ever attained in any four years of our existence, and at a low minimum cost to our membership in loss of time on account of strikes."[85] In fact, union wage rates for carpenters had risen by 14 percent since 1924, while the cost of living was almost unchanged, so there was cause for optimism.

The GEB had set rules for admission to the Home: 65 years of age, continuous membership for not less than 30 years, inability to earn a livelihood. To finance the Home and a pension of $15.00 a month for those not wishing to go to the Home, the convention voted to increase the per capita tax by 15 cents a month. This increase failed to receive the required two-thirds majority in a referendum vote, making it necessary to postpone the payment of pensions. However, another vote was held in September 1929, at the initiative of a Chicago local, and this time there was a bare two-thirds majority for a 25-cent increase.[86]

The only important constitutional change was a new stipulation that local demands were to require a two-thirds affirmative vote by the members affected before a strike could be called if GEB sanction was sought. Before this amendment, only a 55 percent majority was required. Apart from this provision, strike rules remained unchanged. District councils had control over all trade movements, but if a strike of all the locals in an area was contemplated, sanction of the national union was required, even if no strike benefits were sought. Nor could strikes take place until a three-member local committee met with the employers and attempted to adjust the controversy. The action of the 1928 convention continued a long-run trend in the direction of imposing conditions that reduced the likelihood of strikes. The Brotherhood was never a strike-prone organization where ordinary trade disputes were involved, and it became less so as time went on.

During the 1920's a number of unions had set up their own banks as a means of providing their members with cheaper services. By 1926 some thirty-six were in operation. There was mounting concern about the safety of these banks, and the AFL at its 1926 convention adopted a resolution urging that further extension of labor banking be stopped until the experience of those in existence had been evaluated. All of these banks, with the exception of that owned by the Amalgamated Clothing Workers (which is still in existence and prospering), were wiped out in the depression. The GEB reported to the 1928 convention that it "never looked upon . . . requests [to patronize the labor banks] with favor or enthusiasm, and therefore have refused to be influenced in any way by those soliciting the patronage of the United Brotherhood in behalf of institutions of this character . . . The General Executive Board has religiously and consistently adhered to the policy of directing the depositing of the funds of the Brotherhood in National Banks where the deposits are protected by guarantee."[87]

The incumbent general officers were reelected without opposition. In its closing hours, the convention was addressed by Secretary-General Woudenberg of the International Union of Woodworkers, who congratulated the delegates on their foresight in expelling communists from membership and pointed to the experience of France to illustrate "what baneful consequences are entailed, what damage can be done, if this propaganda is allowed to get a foothold in the trade union movement."[88] Based on observations made during his stay in the United States, Woudenberg wrote a few months later that organized woodworkers in New York could buy from one and a half to two times as much as their Amsterdam colleagues. However, he noted a big gap in organization: 15,000 woodworkers were employed in Grand Rapids, and the Brotherhood had not yet gotten a foothold. As for black workers: "The Brotherhood itself comprises negro workers in appreciable number, especially in New York, Chicago, and Philadelphia. In the southern states, too, there is quite a number of negro carpenters organized in the Brotherhood. At the Convention three negro delegates were present."[89]

The convention ended on an optimistic note. The beautiful surroundings of the Home for the Aged must have contributed to a feeling of euphoria. The worst appeared to be behind. The three years just past had been good ones for the construction industry, and employment was high. There was every expectation that the next convention, scheduled for 1932, would mark still another milestone in the Brotherhood's progress. As things turned out, the next convention was not held until 1936, and then under greatly altered circumstances.

SUMMARY

We can now return briefly to the question raised at the beginning of this chapter: why did the labor movement fail to grow in the relatively favorable economic climate of the 1920s? A definitive answer would require a much closer analysis than has been possible here. However, at least as far as the Brotherhood of Carpenters is concerned, the following hypotheses may be advanced.

1. The major adverse factor appears to have been the fierce onslaught mounted by the employers beginning in 1921 and continuing in full force for three or four years. During these years, the first union priority was to defend existing positions rather than to organize new territory. The effects of the concerted employer opposition were beginning to wear off as the decade came to an end, but even then it served to slow potential gains. The 1920s have sometimes been called the era of "welfare capitalism"; it was a period in which the prestige and power of American business management reached new heights. The Brotherhood was committed to accommodation with the existing economic order, but employers in many major centers wanted what has more recently been termed a "union-free environment."

2. Jurisdictional controversy was probably another cause of lagging union progress, contributing, among other things, to employer determination to operate on an open-shop basis. The Brotherhood emerged

victorious from its disputes with other building crafts, but there were substantial costs.

3. Internal political strife also took a toll. The Brotherhood did not suffer as much from this as many major unions; in the United Mine Workers, for example, it was one of the chief causes of severe membership losses and financial crisis. However, Carpenter locals in some major cities — New York, San Francisco, Detroit, and Los Angeles among them — experienced attacks from the left that did not do the union any good.

4. Christie attributed lack of Brotherhood progress to poor leadership:

> Partly because few new members entered the union and partly because of tight control from the top, no new blood entered the leadership . . . Like the carpenters they led, these men were fixed in their ways. New ideas about society at large, the regulation of the trade, or the conduct of the union were strictly forbidden . . . The new leaders . . . were staunch defenders of the status quo, conservative much in the same manner as the small midwestern businessmen of the 1920's.[90]

It is difficult to evaluate this proposition with any degree of precision. However, one might cite the fact that John L. Lewis, who was to turn out to be probably the best organizer the American labor movement ever had, would hardly have been accounted a success in 1928. He was a staunch defender of the status quo, a midwestern conservative. Nor does the subsequent record of the Hutcheson administration support the thesis of poor leadership. On any comparative basis, the Brotherhood leadership in the 1920s did a better job in keeping the Carpenters organizationally intact than was true of most other American labor unions.

5. Changing technology was not often cited in Brotherhood sources as a factor in job availability, but it may nevertheless have played a role. Between 1921 and 1928 productivity in contract construction rose by 25 percent. Thus, the near doubling in the value of contract construction between the two years was offset by a sharp increase in output per worker. Precisely what impact this had on carpenter employment cannot be determined in the absence of more detailed data than are available. However, if one considers five building trades in addition to the carpenters — electricians, painters, iron workers, bricklayers, and pipefitters — only the electricians experienced a smaller average wage increase than the carpenters between 1921 and 1928.[91] This suggests that changing technology may have had a greater impact on the demand for carpenters than for most other building tradesmen, and may have reduced the potential pool of carpenters eligible for Brotherhood membership relatively more than in the case of the other trades. If this were so, it would provide an additional explanation for the Carpenters' insistence on maintaining their jurisdictional rights.

These explanations are certainly not the whole story. Understanding the vicissitudes experienced by the American labor movement in the decade before the Great Depression will remain an unsolved problem until the histories of a number of unions other than the Carpenters are studied in depth, so that comparison are possible.

10

THE GREAT DEPRESSION
AND THE NEW DEAL

The depression that was inaugurated by the stock market crash in the fall of 1929 was the worst in American history. It is hardly necessary to dwell on its severity. The data on employment and wages presented in Table 10.1 bear eloquent testimony to the fate of the American economy during these difficult years. Unemployment reached the unbelievable figure of almost 25 percent of the labor force in 1933, the worst year of the depression. Union wage rates in general and for construction in particular held up relatively well, but this was in part illusory.[1] A combination of payment below the formal rates by union contractors and even lower payments by nonunion employers cut the average annual earnings of a full-time construction worker almost in half by 1933. The final col-

Table 10.1.   Unemployment and earnings in the United States, 1929–1935

Year	(1) Percent of civilian labor force unemployed	(2) Indexes of union wage rates in construction (1929 = 100)	(3) Indexes of Carpenters' Union wage rates (1929 = 100)	(4) Average annual earnings per full-time employee in construction (1929 = 100)	(5) Indexes of value of construction (1929 = 100)
1929	3.2	100.0	100.0	100.0	100.0
1930	8.7	104.5	104.1	91.2	84.1
1931	15.9	104.5	104.5	73.7	64.4
1932	23.6	89.4	85.4	54.2	40.8
1933	24.9	87.0	85.4	51.9	35.7
1934	21.7	87.8	86.6	56.3	44.4
1935	20.1	88.6	88.2	61.4	49.2

Sources: Col. 1: Bureau of the Census, *Historical Statistics of the United States*, 1957, p. 73; cols. 2 and 3: Bureau of Labor Statistics, *Handbook of Labor Statistics*, 1975, p. 228; col. 4: Bureau of the Census, *Historical Statistics of the United States*, 1957, p. 95; col. 5: ibid., p. 379.

umn in the table showing what happened to the value of construction reflects both the heavy unemployment and reduced wages.

There are no systematic figures for unemployment among carpenters during the depression. Duffy claimed that in January 1932, some 80 percent of all Brotherhood members were out of work.[2] This figure, representing the worst winter of the depression, may have been exaggerated, but fragmentary evidence from local sources fully confirms the fact that employment conditions worsened rapidly after 1929. In the February 1930 issue of *The Carpenter*, several letters urged that membership in the Brotherhood be limited and clearance cards denied because of the unemployment. Stay-away notices proliferated. The Chicago district council wrote in as follows:

> When it is remembered that 1929 was a very poor year and 1930 starting out with approximately one-third the amount of building construction as compared with the same period in 1929, and considering the fact that there are approximately thirty thousand carpenters in Chicago, this should be sufficiently convincing to induce members in other districts to stay away from Chicago at this time. But if these figures will not convince, we advise any carpenter coming here to bring a good size bank account and be prepared to take a vacation.[3]

Several years later, the Chicago district council submitted the following figures to underline its advice to stay away from Chicago:[4]

Year	Number of building permits issued in Chicago	Value of permits (thousands of dollars)
1928	9,594	315,800
1929	6,146	202,286
1930	2,434	79,613
1931	1,292	46,440
1932	467	3,824

The impact of the depression on Brotherhood membership can be seen from the following comparison:

Membership category	June 1928	June 1932
Members in good standing	300,086	134,059
Members in arrears	36,384	100,013
Honorary members	9,666	7,933
Total	346,136	242,005

What is notable is not only the drop in the total, but the large number of members who were in arrears with their dues — an increase from 10.5 percent of the total in 1928 to 41.3 percent in 1932. At the beginning of the depression, the rule was that a member owing three months' dues was not in good standing and was suspended from benefits; if six months in arrears, he lost his membership and could only be readmitted as a new member. On April 19, 1932, Hutcheson issued a dispensation granting locals the right to carry delinquent members for twelve months instead of six in order to maintain ties with unemployed members.[5]

Some Locals, in an effort to preserve work for their current members, established prohibitive initiation fees. The GEB decried this practice:

> The board does not consider this the right policy to follow. If all men eligible to admission were members of our organization we would not have to contend with the Open Shop proposition, wage reductions, uncalled for layoffs and other such grievances. Besides that, when we refuse to admit new members to our organization we are at a stand still, neither growing, advancing or developing. In fact we are on a decline, going backward instead of forward. To continue to exist and progress, new members must be admitted to the Brotherhood.[6]

The GEB directed that the maximum initiation fee be set at $100. This was still much too high, and in October 1933 locals were excused from payment of a $5 fee to headquarters for each new member on condition that initiation fees be reduced to a nominal amount.[7] The various dispensations were in effect until July 1, 1936, when employment conditions made possible a resumption of the previous dues levels.

One thing the national union refused to do was to compromise its long-standing policy of maintaining freedom of geographical mobility for members. When a local wrote in asking that a limit be set on the hiring of out-of-town workers, the GEB ruled that "members of our organization have the right to transfer by clearance card to any local and after doing so would then be a local member, and the GEB cannot sanction any movement or proposition curtailing or abridging the rights of members."[8]

The impact of the membership decline on national union revenues is clear from Table 10.2, which also shows the virtual halt of strike activity during the worst years. Stringent economies had to be imposed to keep the organization solvent. In April 1932 the GEB reduced the travel allowances of union staff by 40 percent. General officers and national representatives donated on week of free service each month beginning September 1, 1931. As of January 1933, the salaries of all national officers were reduced by $50 a week.[9]

Table 10.2.   United Brotherhood of Carpenters and Joiners, revenue and strikes
1929–1935

Year	National union revenue from per capita tax (thousands of dollars)	Trade movements considered by GEB	Strike sanctions granted
1929	2,423	95	36
1930	2,494	132	55
1931	1,956	44	19
1932	1,211	13	7
1933	1,340	7	6
1934	1,304	n.a.	58
1935	1,550	45	45

Source: United Brotherhood of Carpenters, *Proceedings*, Twenty-Third General Convention, 1936, pp. 75, 173.

Drastic cuts in benefits had to be made. Funeral benefits for wives and disability benefits were discontinued in 1932. A pension system had been inaugurated on January 1, 1930, but the amounts were progressively reduced until they reached a low of one dollar per month from July 1933 to December 1935. Hutcheson told the 1936 convention that this experience "convinces me that we should not have called it a pension but that it should have been called 'Relief' and should be disbursed in a manner that would be most helpful to those who need assistance."[10] However, the Home for the Aged continued to operate; from 1929 to 1936 some 476 men had been admitted, but with deaths and departures, the number in residence on June 30, 1936, was 299. The few who gained admission were the fortunate ones. The cost of maintaining the Home from 1929 to 1936 was $1,906,000. Although this was offset somewhat by receipts of $500,000, mainly from the sale of fruit, the cost of about $2,000 a year to maintain each resident stands in sharp contrast to the pensions received by those who remained at their homes.[11] At best, the disparity between Home and pension costs would have been great; the depression served to exacerbate the situation.

Another economy measure was postponement of the convention scheduled for 1932. Duffy explained this decision as follows:

Some of our so-called friends and would-be enemies are accusing General President Hutcheson of postponing that convention. Let me say to you that General President Hutcheson did nothing of the kind . . . In January, 1932, [Local Union No. 11] sent out a circular letter to all our Local Unions asking that the convention be postponed until 1936. The reasons advanced were that on account of the depression many of our men were out of work and walking the streets . . . because of that and because of the cost of the Local Unions of sending delegates to a convention in 1932, Local Union No. 11 proposed that it be postponed. That proposal was endorsed by Unions from thirty-five states. It was sent out to referendum vote—democracy in the United Brotherhood of Carpenters and Joiners of America—and it carried.[12]

Duffy neglected to say that the vote was 33,280 for postponement and 22,598 against, well below the two-thirds margin that had been required for measures of such importance in the past. This discrepancy was handled by an interpretation by Hutcheson that "inasmuch as the proposal was not an amendment to our General Constitution, but a proposition to postpone until a definite period [1936], it did not require a two-thirds majority."[13] It should be pointed out, however, that the Brotherhood was not the only union to postpone its convention during the depression. The Machinists, for example, postponed their 1932 convention "until industrial conditions improve," which turned out to be 1936.[14] Nor would a 1932 convention have been representative, since many locals would not have had sufficient funds to send delegates.

UNEMPLOYMENT RELIEF

At its 1931 convention, the AFL rejected endorsement of a national system of unemployment insurance. As the depression deepened, however, pressure for legislative action increased. As the following interchange of correspondence suggests, the Brotherhood was a bit ahead of the AFL officialdom in assessing the mood of union members. Duffy wrote to Green in June 1932 "on a matter of vital importance to the organized wage workers of our country and to the American Federation of Labor in particular. To be plain; I mean the question of unemployment insurance or relief of some kind for those out of work, willing to work but cannot get it." He warned Green that unless the AFL acted, the workers would take the bit in their teeth and say: "Good Bye, American Federation of Labor."[15]

Green's reply was cautious. He stated that the AFL was asking Congress to appropriate funds for immediate relief, but added: "It is difficult to understand how men arrive at the conclusion that the enactment of unemployment insurance legislation would solve the economic problem which prevails."[16] Nevertheless, the AFL executive council changed its position a month after this letter was written, and instructed Green to prepare a legislative draft. The 1932 AFL convention came out in favor of national unemployment insurance.[17]

It was with some hesitation that the Carpenters, as well as the AFL,

reversed their position on unemployment insurance. A circular issued by
the GEB in May 1932 informed the locals that a committee calling itself
the New York A.F. of L. Trade Union Committee on Unemployment Insurance and Relief had no connections with the AFL, and was in fact led
by communists. The circular went on to say that the remedy for
unemployment was work, not insurance, and warned that government
unemployment insurance was not without its dangers to the labor movement:

> A worker receiving unemployment insurance must accept work when offered
> to him, no matter what the nature of the work may be, even though it be in
> some other line or trade or in a non-union shop, and if he refuses to accept
> work, or to go to work in a non-union shop, his unemployment insurance im-
> mediately ceases and he is then worse off than ever. Do we want to place our
> members in that position? We cannot have unemployment insurance without
> unemployment exchanges. We must report; we must register; we must subject
> ourselves in every way to the control of the laws or our insurance stops. Are we
> prepared now to accept monetary relief without regard to what may happen?[18]

In reply to this circular, Local 2717 of Brooklyn, New York, issued an
"open statement" to the Brotherhood members charging that the
"Hutchesons and the Duffies" were disregarding "the sufferings of our
carpenter membership of whom 75% are totally unemployed and the
others working part time at bootleg scale wages," and denying that the
Committee on Unemployment was communist-dominated. Duffy
responded with a ten-page letter, asserting that the matter should have
been taken up with the New York District Council of Carpenters rather
than with an outside ad hoc committee, explaining in detail the AFL
position, and concluding: "Let it now be understood once and for all that
the A.F. of L. is not opposed to unemployment insurance or relief, provided the rights of organized labor are safe-guarded and the rights of our
members protected."[19] Shortly thereafter, locals were told that affiliation
with the New York committee might lead to their suspension.

There is no record of any action against the New York locals, but
several others were disciplined for communist activity. Local 143 of Can-
ton, Ohio, was put in receivership, and Hutcheson assumed control of
the Seattle district council. Local 1051 of Philadelphia put out a circular
letter containing statements that were labeled as untrue by the GEB, and
submitted it for publication to the Communist newspaper, *The Daily
Worker*, where it appeared on August 11, 1933. For this act, the local
was disbanded and its members transferred to other locals.[20]

Considering the success of communist organizations among the
unemployed during these bleak years, the Brotherhood got off lightly. In
any event, by 1934 the Brotherhood was solidly behind unemployment
insurance; an editorial in *The Carpenter* considered the various plans

that were then under discussion and came out for some form of government benefit system.[21] The fear of government intervention in the labor market could not stand up against unemployment rates of 20 percent and more.

POLITICS

James J. Davis resigned from the office of Secretary of Labor to run for the U.S. Senate in 1930. Months before his resignation, Duffy wrote to Green suggesting that the AFL intercede with President Hoover on behalf of Hutcheson:

> After the election of President Hoover, several labor men were looking for the appointment of Secretary of Labor. When you had a conference with him you told him, after mentioning several names, that any of them would be satisfactory to the American Federation of Labor. In this case I do not want you to do that. I want you to tell him straight from the shoulder that Brother William L. Hutcheson of the Carpenters is our choice.[22]

Green's reply was characteristically cautious; he noted that Davis intended to serve until the election in November, but promised to watch the situation closely. When Davis won the primaries in May, Duffy renewed his request, but Green threw cold water on the idea: "I was not favorably impressed with the reaction of the President or his representative with whom I conferred to the suggestion that President Hutcheson be appointed or that some one associated with the American Federation of Labor should be selected to serve as Secretary of Labor."[23] In fact, Green had submitted a list of candidates acceptable to the AFL, including, in addition to Hutcheson, John L. Lewis, Matthew Woll, John Frey, and John R. Alpine. But Hoover chose William Doak, a member of the Brotherhood of Railroad Trainmen, which was then unaffiliated with the AFL.[24]

Despite this rebuff, Hutcheson came out for Hoover's reelection, and accepted appointment as director of the Republican Labor Bureau. In a letter addressed to the AFL executive council urging it to examine Roosevelt's labor record, he charged that Roosevelt, when governor of New York, had failed to prevent contractors on public works from paying low wages and had made no effort to get relief for 40,000 transit workers compelled to work seven days a week. He asked also whether the Democratic candidate favored "fiat money and abandonment of the gold standard, thereby automatically cutting American wages."[25]

It is generally believed that Roosevelt had the strong support of organized labor in 1932. In fact, the executive council of the AFL declared that "the organization was maintaining its traditional neutrality; that the interests of labor have been protected and conserved through a strict adherence to a non-partisan political policy . . . The wisdom of such action is clearly apparent when it is considered that the American Federation of Labor is composed of men and women who entertain different political opinions."[26] This was certainly correct as far as the top leadership of the AFL was concerned. Hutcheson and John L. Lewis were staunch Republicans, while Daniel J. Tobin of the Teamsters and Joseph A. Franklin of the Boilermakers worked hard for the Democratic Party. However, there was no doubt where American workers stood: they voted overwhelmingly for Roosevelt, who continued to have their allegiance for the rest of his life.

Hutcheson came in for some criticism from within the Brotherhood for accepting the position of director of the Republican Committee's Labor Bureau. Local 58 of Chicago sent a letter to the GEB protesting this action, but the GEB replied that he was acting as an individual and not as an officer of the Brotherhood. "This is the privilege of any member of the organization. The Board is convinced that the General President in accepting the position . . . is not neglecting the business of the Brotherhood."[27]

After the election, Green lobbied for the appointment of Tobin as Secretary of Labor; when Roosevelt appointed Frances Perkins, he issued a statement expressing keen disappointment over the choice:

> Labor has consistently contended that the Department of Labor should be what its name implies and that the Secretary of Labor should be representative of Labor, one who understands Labor, labor's problems, labor's psychology, collective bargaining, industrial relations, and one who enjoys the confidence of labor. In the opinion of Labor the newly appointed Secretary of Labor does not meet these qualifications. Labor can never become reconciled to the selection made.

Green sent this statement to Duffy for comment; he replied that he had expected better from Roosevelt, who had always professed friendliness toward the AFL. "I consider your statement too tame. I am afraid if I were issuing it I would make it much stronger. That is why I have refrained from writing you sooner on the matter. I have always been a Democrat, but I am through now." A few days later, he wrote Green again to say that the Labor Department had become "an old woman's department, supervised, controlled and operated by Mother Perkins, Aunty Abbott and Grandmother Anderson. To my way of thinking it has outlived its usefulness; so the sooner it is put out of business the better."[28] Despite all these forebodings on the part of Green and Duffy, Frances Perkins went on to a distinguished career as Secretary of Labor, and proved to be a staunch defender of the interests of organized labor.

It is interesting to note that Duffy was able to work closely with Hutcheson despite the fact that the two men were on opposite sides of the fence politically. Not long after the election, Duffy delivered a talk at the University of Notre Dame in which he indicated his concern over the possibility that Hoover might have been reelected: "Before the inauguration of President Roosevelt, rebellion and revolution were staring us in the face. I give the people of the United States, the workers, credit for holding back. Why did they hold back? Because a new president was coming in. I am afraid that if they had elected the old president — and I have nothing in the world against him [Hoover] — that we would have had a revolution."[29] That Duffy would have been harboring these dismal thoughts while his colleague was working for Hoover provides a good illustration of what political neutrality meant for American trade unions. A partisan policy could have torn the organizations apart.

Another event indicating that the Brotherhood may have been even more committed to neutrality than other unions was a resolution adopted by the GEB in 1935, at a time when there was talk of amending the U.S. Constitution in order to curb the power of the Supreme Court, which had consistently declared New Deal legislation unconstitutional. Brotherhood delegates to a forthcoming AFL convention were directed to try to table any action along these lines, since "a positive or a negative action by the convention at this time would be used by politicians as committing organized labor to either one party, or to the other, during the coming presidential campaign."[30] This resolution undoubtedly owed a great deal to Hutcheson, who remained a staunch defender of the Supreme Court, regardless of its decisions in individual cases.

THE NATIONAL INDUSTRIAL RECOVERY ACT

One of the first acts of the Roosevelt Administration was to secure the passage of the National Industrial Recovery Act, popularly known as the NRA. The purpose of the law was to create "codes of fair competition" for individual industries in order to secure price stabilization and increased employment. As far as labor was concerned, the most important provision of the act was Section 7(a), which declared it to be the public policy of the United States that employees should have the right to organize and bargain collectively without interference, restraint, or coercion by employers. This clause was mandated as a provision of every industrial code.

The Carpenters, like the rest of the labor movement, supported the NRA. They saw in it a potential for furthering new organization for the first time since the onset of the depression. "The National Industrial Recovery Act gives a golden opportunity to employees in the mills in the

Northwest to organize themselves into bona fide trade unions. For this purpose there is one and only one organization, namely, the United Brotherhood of Carpenters and Joiners of America."[31] Half a year after the enactment of the law, Duffy stated: "We were satisfied with it. It is doing the work." He quoted with approval a letter from Green, which asked: "If this plan fails, to what can we turn, or what other remedy is available?"[32]

Section 7(a) was accompanied by a burst of organizational effort. The Carpenters, for example, sponsored a conference of mill men in Chicago, at which forty delegates representing nineteen locals in seven states were present. The purpose of the conference, in addition to furthering organization, was to raise wages. The mill men's contract then in effect based the scale within a radius of 500 miles of Chicago on the average wage of all union shops, plus 15 percent for Chicago itself. But better roads and larger trucks were resulting in stiffer competition from outlying areas, and the conference urged all locals to eliminate initiation fees and to work for a thirty-hour week.[33]

The code of most concern to the Carpenters was that for the construction industry. A code drawn up by the Construction League, an employer organization, was accepted temporarily by the NRA administrators in August 1933. Among other things, it stipulated a wage level below that of 1929, as well as a forty-eight-hour week. The Building Trades Department fought these provisions and succeeded in having Roosevelt order that a new code be prepared with the cooperation of the department. The revised code provided for a forty-hour week, a 40-cent minimum wage, and a tripartite National Construction, Planning, and Adjustment Board, a partial victory for the labor point of view.[34]

Since the Carpenters' Union was not a member of the Building Trades Department at the time, its officers did not participate in negotiations over the code. However, the National Board was directed by the President to set up some mechanism for settling jurisdictional disputes, and the final code contained a provision outlawing jurisdictional strikes. A three-member National Jurisdictional Awards Board, consisting of neutrals, was created and empowered to make binding awards. The president of the Building Trades Department was given a key role in administering the plan.

The prospect of being excluded from the jurisdictional award process galvanized the Carpenters into action. They sought to affiliate with the Building Trades Department, giving rise to a new crisis in the affairs of that organization. The presidents of several unions that had remained in the department stated openly that the only reason the Carpenters wanted to come back in was to stop adverse awards. Not surprisingly, one of those who objected was the head of the Sheet Metal Workers.[35] As it happened, the Jurisdictional Awards Board had scarcely begun to function when the NRA was declared unconstitutional by the Supreme Court in the Schechter case, so it was never put to the test of dealing with the Carpenters.

RELATIONS WITH THE BUILDING TRADES DEPARTMENT

The Carpenters had rejoined the Building Trades Department in 1927, but the honeymoon was short. At the first convention attended by the Carpenters after their return, Hutcheson moved that "in the printing of the proceeding all decisions affecting the Brotherhood of Carpenters, as rendered by the Board of Jurisdictional Awards, be eliminated." This motion carried unanimously.[36] However, at the 1929 convention the Carpenters asked that the per capita tax paid to the department be reduced, and this was turned down. The Carpenters raised complaints that local building trades councils were enforcing prior jurisdictional decisions against the Brotherhood. The department undertook to set up a new agency to decide jurisdictional disputes, the Board of Trade Claims, which functioned from 1930 to 1934. This board, which operated jointly with the National Association of Building Trades Employers, was supposed to consider questions submitted by unions, and eventually to approve decisions made by an umpire jointly chosen by the parties.[37]

The Carpenters refused to have anything to do with the new board, which they felt was simply a reinstitution of the old Board of Jurisdictional Awards, to which they had objected. In 1929 they withdrew once more from the Building Trades Department. The department retaliated by introducing a resolution at the 1930 AFL convention condemning the Carpenters for repeated violations of other unions' jurisdictional rights; it was referred to the AFL executive council.[38] At a subsequent council meeting, Duffy declared that "not the Executive Council and not even the convention in session can force down the throats of the Carpenters any decisions they might render." He was assured by Green that the council would take no such action.[39]

Relationships between the Brotherhood and the Building Trades Department continued to deteriorate. In 1930 James P. Noonan, president of the International Brotherhood of Electrical Workers and a member of the AFL executive council, died. Duffy, fearing that he would be replaced on the council by a building tradesman affiliated with the department, urged Morrison to prevent the nomination of someone who "would not work in harmony with us." He continued:

> General President Hutcheson and myself have gone very carefully and seriously over the entire matter. We looked at it from all angles and all viewpoints. You will not be surprised when I tell you that we have no confidence

whatever in the Building Trades Department, its representatives, officers, or Executive Council. You realize to put one of them on the Executive Council of the American Federation of Labor would not please the carpenters, in fact it would raise their antagonism. You know also we have been the target of the Building Trades Department for years and we are still; whenever they get a chance they hammer us right and left.[40]

Duffy recommended the appointment of Daniel Tobin, president of the Teamsters Union, another Indianapolis-centered union at the time, which had affiliated with the Building Trades Department but had opposed the creation of a new jurisdictional board. He wrote a similar letter to Martin Ryan, president of the Railway Carmen, asking for his support and promising to return the favor in the future. "If we allow the matter to go to the convention we don't know who we may get."[41]

Tobin owed his election to the AFL executive council largely to Brotherhood support. He had always been a strong proponent of the Democratic Party, and was as close to Roosevelt as any trade union leader. Many years later he had this to say of Hutcheson, one of his main political opponents in the AFL:

Jack Lewis and Bill Hutcheson are two of the greatest men in the labor movement . . . of all the men I know in the labor movement I respect and admire [Hutcheson] for his courage and his guts. He will disagree with you if you are the Pope or the King of England when he thinks he is right. Then after the meeting is over he will call you down to the bar and have a glass of beer and all the past is forgotten . . . I am glad to be able to say one word of praise for the man that belted and flayed me.[42]

The Bricklayers' Union, then second in size only to the Carpenters in the building trades, had left the Building Trades Department in 1927 because of dissatisfaction with jurisdictional rulings. In 1931 the Carpenters and the Bricklayers entered into an alliance for mutual aid and protection after reaching an agreement on the installation of cork. Local disputes between the two organizations were to be referred to the national presidents for adjustment. They were joined a few months later by the Electrical Workers, the third largest building union, and also a fugitive from the department's jurisdictional machinery.[43] Almost coincidentally, the Carpenters renounced their 1920 agreement with the Ironworkers on the ground that the latter were violating it.[44]

This so-called Tri-Party agreement was greeted with a mixture of dismay and indignation by the smaller organizations that remained loyal to the Building Trades Department. It was branded by the department's president, M. J. McDonough, as "nothing less than a dual organization of building tradesmen." He wrote to the department's affiliates:

In view of the high-handed, flagrant and ruthless manner in which the Carpenters are attempting to destroy the concord and harmony that we have succeeded in establishing in the building industry, as well as the intention of that organization to create turmoil and industrial discontent at a time when we should all be engaged in an effort to rehabilitate the building industry and bring about a revival in building investment, we feel that all local unions in affiliation with our Associated Internationals should be promptly advised of the destructive methods it is evident the Carpenters intend to pursue.

The president of the Hod Carriers declared that while the withdrawing unions had been called pirates, "that word isn't strong enough. That Alliance was not formed to defend themselves against people who would come in and take their work, it was formed to take work away from someone else." The Sheet Metal Workers' president urged that all locals of the Carpenters, Bricklayers, and Electrical Workers be expelled from local building trades councils. To this suggestion, which would have amounted to an open declaration of war, a department official replied that an attempt had been made to do this, but that the local councils had refused to comply.[45]

The two groups remained apart during the next three years, and only the doldrums in which the construction industry found itself, combined with the reluctance of local building trades councils to kindle the flames, prevented the outbreak of large-scale stoppages. In 1934, when the Building Trades Department became the principal labor negotiator in establishing the NRA construction code, and with the prospect of a government-backed jurisdictional board facing them, the Tri-Party group moved toward reaffiliation. On June 14 the three organizations notified the department that they were coming back in. Hutcheson sent a circular to Brotherhood locals notifying them that this did not affect the Tri-Party agreement, and that local building trades councils were to remain neutral in jurisdictional disputes.[46] The executive council of the department unanimously approved acceptance of the application for reaffiliation.

But when the Building Trades convention opened in September 1934, the credentials committee refused to seat the delegates of the three unions on the theory that the convention needed first to approve their applications for admission. The following colloquy took place:

> Delegate Hutcheson: Excuse me for laughing, Mr. Chairman, if I have to laugh.
> President McDonough: You are out of order.
> Delegate Hutcheson: I am still in order.
> President McDonough: You are not in order until you are affiliated.[47]

The committee on the report of the executive council recommended against acceptance of the three unions, saying that the solidarity forged by the loyal trades during the depression "will be disrupted by the three aforementioned crafts now seeking affiliation." Hutcheson and Bates

(Bricklayers) asked for the floor but were denied it. The convention finally rejected the applications by a voice vote.[48]

William Green appeared before the convention the next day and told the delegates that he had promised the three unions that they would be readmitted, and that the convention had no choice but to do so under AFL rules. "The reasons assigned by your committee for excluding these organizations from affiliation or excluding them from representation are not valid. There is no question of law. It is an expression of their opinion. You cannot exclude people merely on the ground that something is going to happen."[49] However, the bitterness of the delegates against the three applicants was so great that even Green could not prevail; all they would do was to give their executive council the authority to negotiate the conditions under which the three would be admitted.

During the next half year there took place a succession of plots and counterplots. The three unions appealed to the AFL executive council for redress. The Building Trades Department countered that Hutcheson's circular letter constituted evidence of bad faith, and that the three unions were deliberately overstating their membership in order to gain control of the department. The executive council recommended to the AFL convention, which was then taking place, that the three unions had a legal right to admission, and that if no agreement could be reached within forty-five days, Green should call a special department convention to accomplish this purpose. In the debate on the recommendation, Duffy denied that the three unions would have enough votes to control the department. Hutcheson declared that the Carpenters were doing very well under the Tri-Party agreement, and that the Carpenters had applied for affiliation only on the urging of William Green. Coefield (Plumbers) stated that the three unions had the support of several others, including the Teamsters, and would thus dominate the department. Notwithstanding the protests of department spokesmen, the AFL convention approved the executive council recommendation by a vote of 19,399 to 3,826.[50]

The AFL convention also decided, among other things, to increase the size of its executive council, and Hutcheson was elected tenth vice-president. The Carpenters thus had two members on the council, Duffy having served in that capacity for many years. There was no question where the AFL as a whole stood on the building trades controversy.

Despite the pressure brought to bear by the AFL, the Building Trades Department stood its ground. Green thereupon called a convention for November 26, 1934, with seven unions in attendance—the Tri-Party group plus the Operating Engineers, the Laborers, the Marble Polishers, and the Teamsters, all former members of the old department. J. W. Williams, a member of the Carpenters' GEB, was elected president, and a constitution was adopted. The membership represented by the new department was roughly twice as great as that of the old one. The new body entered a legal suit to obtain the books and records of the old one, but the court held that the new officers had not been elected according to the department's constitution. On the other hand, the judge ruled that the terms of the officers of the old department had expired, so that in fact there was no department official who could function legally.[51]

Hutcheson demanded that the AFL recognize the new body, which was done by the executive council. The old department vacated the offices supplied it by the AFL and moved to quarters provided by the Machinists. It continued to operate, claiming that it was still recognized as the official construction body by government and industry.

The deadlock continued. In June 1935 the AFL executive council authorized Green to call still another convention if a three-man mediation committee failed to bring about a settlement.[52] The old department declared that the chairman of the mediation committee was biased against it, and no progress was made.[53] The new group reconvened on August 1, 1935, and reelected the same officers, which the AFL hoped would give it full legal status.

The whole matter flared up again at the 1935 AFL convention. The seating of J. W. Williams as president of the department was protested; the president of the original group asserted that the AFL had never revoked his charter. Philip Murray urged that the matter be deferred on the ground that the AFL should not intrude into the internal affairs of its affiliated departments, and with the support of the Miners, this motion carried. There appeared to be no way out.

Fortunately, through the intermediation of George Harrison of the Railway Clerks, a compromise was worked out during the convention. A committee of six persons from each side, plus a neutral in the event of a deadlock, was appointed to work out a merger of the two groups. The old constitution of the department was to be taken over, but the committee could recommend changes. Until agreement was reached, neither group was to be recognized as the official Building Trades Department of the AFL.[54]

Finally, a special convention was held beginning March 25, 1936. The negotiating committee had agreed that the officer positions of the combined body were to be divided equally among the two contending groups, but no agreement could be reached on the presidency. Harrison, as a neutral, broke the deadlock by assigning the position to J. W. Williams and giving the secretary-treasurership to the old group, a clear victory for the Carpenters. However, as part of the reorganization the Carpenters had to accept a new scheme for the settlement of jurisdictional disputes, according to which the president of the department was to make an initial decision on disputes that was final and binding only for the individual sites at which the disputes arose. The national unions could then appeal the decisions to a referee, who could formulate a national rule. John A. Lapp was selected as referee, and he began to hear cases on January 1, 1937.[55]

The Carpenters apparently felt that they could live with this arrangement, but over their objection, the executive council of the new department set up a plan for *local* bipartite arbitration boards, which were required to meet within 48 hours of the occurrence of a dispute and render a written decision within the next 48 hours. If this decision were not obeyed within 24 hours, the employer could give the work to the craftsmen whom he believed were best qualified, and these men would have the right to the work pending review of the referee.[56]

This went far beyond what the Carpenters were willing to accept. At the 1937 convention of the Building Trades Department they protested that this plan was not contemplated by the 1936 merger agreement. In July 1937 the GEB resolved to withhold its per capita tax to the department until the plan was rescinded.[57] A split was averted when at the 1937 department convention the Carpenters' group captured the two top offices plus five of the eight vice-presidencies, giving the old Tri-Party bloc effective control. The executive council of the department set the local plan aside, and in 1938 John Lapp resigned as referee, after hearing about a dozen cases, when he was not able to secure compliance with several of his awards.[58] However, the national referee system continued.

It is unlikely that the Carpenters would have been willing to accept even the fairly ineffectual plan adopted in 1936 had it not been for the rise of a greater evil—the CIO—which posed a more serious threat to its jurisdiction than did the other building crafts. Faced with this new challenge, the AFL unions drew together, and it proved possible finally to achieve a unified Building Trades Department.

JURISDICTION (4)

While all this was going on, the Brotherhood was fighting a parallel battle with an old adversary, the Machinists. The issue was the same as it always had been, jurisdiction over certain work performed by both millwrights and machinists. At the initiative of the Machinists, the presidents of the two unions agreed in 1930 to establish a committee, three from each side, to determine whose members were actually doing the work. The committee found that more than 50 percent of all the machinery being installed was being done by nonmembers of the two unions; that more than 75 percent of the maintenance work was also being performed by non-members; that there was no clear line of demarcation between the Carpenters' millwrights and the Machinists' erectors; and that it was impractical for the work to be done by members of both unions together. It recommended that one organization be given full jurisdiction over the work, with an appropriate transfer of members. The only trouble was that no agreement could be reached on which union should have the work.[59]

Arthur Wharton, president of the Machinists, suggested to Hutcheson that in view of the relatively small number of carpenters involved the Brotherhood should cede jurisdiction to the Machinists. The answer was predictable: no![60] However, a tentative agreement was reached providing that all disputes between the two organizations were to be referred to the presidents without any work stoppages.[61] This did not last long; Wharton was soon back at the AFL executive council, complaining that some local building trades councils were threatening strikes if Machinist millwrights were employed. He demanded that the AFL send out a statement to the effect that its previous decisions gave jurisdiction to the Machinists, and he issued a circular to his own locals carrying this message. The council refused to comply after Green said that to do so would have made the situation worse: "Understanding the movement as we do, and the Building Trades Councils as they are and the influences that move them, do you think for one single moment that would change the situation . . . It may take some time and the exercise of a lot of patience, but we do not want to see open warfare running through this movement."[62]

After more negotiation, the Carpenters and the Machinists agreed to divide up the jurisdiction on the basis of the specific work done. The Machinists were given the work of building, assembling, erecting, dismantling, and repairing machinery in machine shops, buildings, and factories, while the Carpenters were to install line shafting, pulleys and hangers, spouting and chutes, conveyers, lifts, and hoists, except for conveyers that were integral parts of a machine.[63] It looked as though this dispute of twenty-five years' standing had finally been resolved.

At this point the Building Trades Department intervened and declared that some of the work in question belonged to the Elevator Constructors, the Iron Workers, and the Sheet Metal Workers, and that their interests had to be protected. When this new obstacle was raised, Green expressed his unhappiness: "Now I confess I am innocent as a lamb. I never knew there was another building trades organization interested in this jurisdictional dispute between the Carpenters and the Machinists over millwright work, and it has taken all the joy out of my life to find out here that there are other organizations apparently interested, because I thought we had accomplished something wonderful this year."[64]

The whole matter was referred back to the AFL executive council to determine whether in fact the interests of other unions were involved. This was too much for the Carpenters, who apparently felt that they had already been too generous and they simply canceled the agreement.

When Wharton objected that the Building Trades Department had no right to intervene, and that the agreement made no provision for cancellation, Duffy replied that forty Brotherhood locals had maintained that the agreement entailed a change in the constitution and required a referendum vote. Wharton thereupon asked the AFL to set forth publicly the historical Machinists' jurisdiction, adding: "We have exercised patience in this dispute with the Carpenters. Never have I seen such dilatory tactics. The Carpenter without due notice sends out notice throughout the country setting aside the agreement and notifies firms to that effect and we find ourselves confronted with threats of strikes by representatives of the Building Trades councils."[65] The AFL executive council, under threat of withdrawal by the Machinists, agreed to publish the definition of their jurisdiction as decided by the 1913 AFL convention, but with an addendum insisted upon by Duffy detailing all subsequent convention action on the controversy.[66] This was done, and there the matter rested for another five years, when it flared up once again.

THE TWENTY-THIRD GENERAL CONVENTION

Much of the time of the 1936 Brotherhood convention was devoted to problems raised by the formation of the CIO, particularly the organization of the lumber workers. These issues will be dealt with in the next chapter. Here we are concerned with some of the actions and discussions that throw light on the eight years that had elapsed since the previous convention.

The slowly improving economy enabled pensions to be raised from the almost meaningless $3.00 per quarter at the depth of the depression to $9.00 on October 1, 1936, and to $12.00 three months later. Officers' salaries were restored to the levels stipulated in the constitution. The Brotherhood was back on its feet financially.[67] Membership, however, had not yet reached the predepression level of 346,000. Duffy reported that as of June 30, 1936, there were 301,875 members in all, which represented a gain from the 242,000 in 1932. But of the 1936 total, some 131,000 were nonbeneficial members, who were largely newly organized lumber workers and woodworkers; the building carpenters numbered only 161,000. Unlike the situation in 1932, however, the number of members in arrears was not great.[68]

The apprenticeship program had been one of the victims of the depression. Many of those employed as apprentices were simply kept on at the apprentice wage rather than being promoted to journeyman status. The system was just beginning to pick up once more in 1936.

Mill work was another casualty. The minimum wage at which the union label could be secured had to be reduced to 50 cents an hour. "We had considerable of a battle to hold even to that low minimum, and in several cases were obliged to withdraw the label from the mills that did go below this rate."[69] The mill rate issue actually generated a good deal of convention debate. The district councils of the large cities approved a constitutional amendment that would have barred the use of the union label where less than 60 cents an hour was being paid to mill men. They raised the same complaint that had been made many time before. A representative of the Chicago council stated: "Every time we go to negotiate an agreement with our manufacturers in the city of Chicago, they tell us to go out and get the wages up for the outlying towns. We have had shops out one hundred miles from Chicago and they were given an agreement for 30 or 35 cents an hour less than the city of Chicago is paying." A delegate from New York added:

> I fully realize the hardships that are met with in the various localities where the wage scale is considerably lower, but in such localities they should not attempt to deliver into a district or city where the wage scale is considerably higher. I don't believe it is fair to those members who have lived all their lives and have held membership in an organization for twenty or thirty years in the same place. They are entitled to protection. We don't want to build a fence around the city of New York, but we do ask for fair competition among our own membership . . . We have in the city of New York some forty mill manufacturers who are under agreement with the District Council . . . Contract after contract went out of the city of New York to low wage communities for the reason that the label had been granted to those communities.[70]

Hutcheson, with the interests of the national union in mind, warned that the imposition of a mandatory 60-cent minimum would make it very difficult to organize the large nonunion mill sector. His advice was to drop the idea of a minimum and operate on an ad hoc basis:

> When these [mill] agreements come up for renewal questions could be asked when they make application for the label. Where are your products used, what percentage of your products do you ship out of the city in which your mill is located? We could ascertain where they are shipping, and if they go into a locality where the wage is higher than that provided in the agreement for that particular place, then let your Council officers say no, you can't have the label if you are going to ship it there in competition with the men who are getting a higher scale.[71]

This advice was accepted. The convention voted down the proposed minimum.

No important constitutional amendments were adopted. However, the admission ritual was changed to include a promise by the candidate that he would forfeit his membership if he joined the Communist Party. When a delegate arose, identified himself as a member of the Socialist

Party, and asked whether the change applied to him as well, he was told that the Socialist Party was recognized as a legitimate one.

One interesting footnote can be added. Duffy reported that there were locals of blacks in thirteen cities. "We have always tried to organize the negro carpenter. Years ago we put a negro organizer on the road. General President Huber, who is now dead, put a colored organizer on the road, thinking we would do better in that way, and whenever we have a chance we organize them because we know that otherwise they would be a detriment to us."[72]

When it came time to elect officers, some opposition to Hutcheson manifested itself; James M. Gauld of Boston and J. W. Williams, who was president of the Building Trades Department, were nominated to run against him. A delegate from Kirkwood, Missouri, seconding Williams' nomination, said that he had been sent to the convention with instructions to clean house. Duffy arose to plead with Williams to withdraw, saying that he was needed in the Building Trades Department to protect the interests of the Carpenters there. Williams declared that when he arrived at the convention, he had no intention of running for the general presidency. He was approached by a "large number" of delegates and was asked to stand. He attended several caucuses, and agreed to run provided there was no other opposition candidate. "Large numbers of the delegates coming to this convention impressed me with the fact, as they put it, that there was a demand for a change." Williams added that this did not apply to Duffy or Neale, the treasurer, with whom he had worked closely for many years. He claimed that Gauld had committed himself to run for vice-president, but had accepted the nomination for the presidency "after his promise on his word of honor to me that he would not be a candidate." Williams thereupon withdrew his name.[73] Shortly thereafter, Gauld's name was also withdrawn.

The opposition had collapsed. It is not clear from the record on what issues the movement to unseat Hutcheson was based. The motives may have been political, for Hutcheson had worked against the reelection of Roosevelt in the presidential elections that had been held a month before the convention opened. In any event, all the officers were reelected unanimously, and Hutcheson never again faced an opposition nominee for the general presidency.

The impact of the Great Depression on the Brotherhood was eloquently summarized by the GEB in its report to the 1936 convention:

Our Local Unions and our members as individuals were innocent victims of this depressed condition. Millions of idle men sought any means of livelihood and it was only because of the courage and fortitude of our members that we were held together . . . Our members refused to believe that low wages would create work. They have maintained their unions and recovered their wage scales. While sheer poverty made it impossible for many thousands of our members to pay dues they never lost or surrendered their union principles.[74]

With the depression behind it and its organization intact, the Carpenters' Union was able to look forward to renewed growth. Now it faced a new and in some ways more dangerous challenge — the rise of the most important dual union movement in the history of American labor.

11

THE CHALLENGE OF INDUSTRIAL UNIONISM

The United Brotherhood had ceased being a pure craft union, catering only to the skilled worker, when it began admitting millmen. The degree to which it had progressed on the road to industrial unionism is suggested by the fact that by 1936, 43 percent of its members had a nonbeneficial status. These were the men working in the mills and the woods, whose wages were lower than those of construction carpenters.

The jurisdictional claims of the Brotherhood expanded steadily until they embraced an entire industry — woodworking. However, little effort had been made to organize the men in the woods and sawmills. The Carpenters had enough trouble protecting their existing locals on construction sites and in mills. Duffy asserted that "although these [lumber] workers rightfully belong to us, we did not claim them in the past for the reason the work they performed was in the rough."[1] In 1903 the AFL had chartered a national union of Shingle Weavers, which, after several mergers with timber workers' unions, had gone out of existence in 1923. A company union formed during World War I, the Loyal Legion of Loggers and Lumbermen, was the only labor organization of lumber workers from 1923 to 1933. As a consequence of the stimulus given to unionism by the enactment of the National Industrial Recovery Act, the AFL itself began an organizing campaign, and by the close of 1934 it had chartered 118 federal locals of lumbermen, loggers, and shingle weavers.[2] It was the attempt of these locals to band together in a new international union that aroused Brotherhood interest in the possibility of taking them over.

THE DEBATE ON INDUSTRIAL UNIONISM

The debate within the AFL that eventually led to the formation of the CIO began in 1933, when efforts to organize the mass production industries, which had come to naught in the post–World War I years, were resumed. AFL policy was to issue its own federal charters to newly formed locals of factory workers rather than to charter new national unions. One of the reasons for this policy was the unwillingness of the craft unions, particularly those in the metal trades, to cede jurisdiction over skilled craftsmen in automobile plants, steel mills, and other industrial establishments. The federal local device was considered a means of holding workers together until they could be apportioned among the jurisdictional claimants, but the craft unions feared that this might eventually jeopardize their ability to take over the men who "belonged" to them. Wharton, president of the Machinists' Union, put it this way:

> It is true that we were informed that eventually those members of Federal Unions who properly belong to our International Union will be transferred to us, but that transfer may be postponed for a long time, for mechanics properly belonging to us who begin their trade-union career on a basis of low initiation fees and low dues will not be enthusiastic over being transferred, and, in addition . . . it will be necessary for us to accept these members without their paying an additional initiation fee.[3]

When the 1934 AFL convention opened, the rising tide of organization in the factories made it impossible to postpone the issue any longer. After a lengthy discussion, a compromise resolution was adopted that authorized the AFL executive council to issue national union charters in the automobile, cement, aluminum, and other mass production industries, with the proviso that the executive committee "fully protect the jurisdictional rights of all trade unions organized upon craft lines and afford every opportunity for development and accession of those workers engaged upon work over which these organizations exercise jurisdiction."[4]

The interpretation of this proviso was left to the executive council. The proponents of industrial unionism, led by John L. Lewis, argued that it would be impossible to divide workers in a single factory among various craft unions. Lewis urged the executive council to grant a broad charter to the Automobile Workers; he was supported by Green, who, it will be recalled, had once been an official of the United Mine Workers. Green declared:

> It is impossible for us to attempt to organize along our old lines in the automobile industry . . . I must confess to you that I am come, you will come, and all of us will always come face to face with the fact, not a theory but a situation actually existing, that if organization is to be established in the automobile industry it will be upon a basis that the workers employed in this mass production industry must join an organization en masse. We cannot separate them.

Hutcheson, who had been made a member of the executive council in 1934 when the council's membership had been enlarged from eight to seventeen, took the lead in opposing the grant of an industrial charter. He maintained that the "American Federation of Labor has been brought to the point it has through following the lines of craftsmanship. I think we should adhere to that with the thought in mind of protecting these craftsmen and at the same time giving these men an opportunity to organize." By a vote of 12 to 2, the council issued a charter reserving craftsmen to the skilled unions.[5]

Why were the Carpenters, who were in the process of asserting jurisdiction over the entire woodworking industry, opposed to industrial unionism in other industries? Part of the answer was given by Hutcheson when he spoke against chartering a new union of steel workers against the opposition of the Amalgamated Association of Iron, Steel, and Tin Workers, which held an AFL charter but was virtually defunct:

> It seems to me it behooves us not to give too much consideration to dissatisfaction expressed by groups in this or any other organization. It has been necessary for my International to compel observance of its laws. We have had expressions of dissatisfaction. It became necessary for me to suspend 65 locals. If that group had received recognition from the organized labor movement of the country, you can imagine what a chaotic condition our organization would have been in.[6]

Hutcheson was referring to various opposition movements within the Brotherhood during the previous two decades, and was concerned lest chartering new unions might afford an opportunity for radical political groups to gain a foothold in the labor movement. He and other craft union leaders preferred the expansion to take place within the traditional framework of the jurisdictional lines that had been mapped out by the AFL, affording a stronger possibility of controlling the character of the new unionism. He could point to the long-established principle of exclusive jurisdiction that had been embodied in the constitution and practices of the AFL. There was also the important additional fact that he did not want to give up his claim to the millwrights and the maintenance carpenters who worked in automobile plants, steel mills, and other industrial establishments. When Lewis, in asking for an industrial charter for the Rubber Workers, pointed out that the United Mine Workers enjoyed an industrial charter, Hutcheson replied: "There is a vast difference between the Mine Workers in the manner they are employed and the rubber workers. The rubber workers and automobile workers work in factories and buildings and those buildings have to be kept in condition. There is a big difference between these buildings and the mine buildings and I think there is a vast difference between the two groups of workers."[7]

The climax of the debate came in the 1935 AFL convention. Twenty-one resolutions on industrial unionism were submitted to the committee on resolutions, which came in with a majority and minority report. The majority supported the craft union position, while the minority held that only through industrial unions could the mass-production industries be organized. The majority report was adopted by a margin of almost 2 to 1.

The discussion was quite heated at times: at one point the Mine, Mill and Smelter Workers Union, which had been chartered by the AFL in 1911, protested that the craft unions were impinging on its jurisdiction, and Lewis came to its support. Hutcheson remarked that "listening to the previous speaker [Lewis] you would imagine that all the craft organizations affiliated with the American Federation of Labor were a set of pirates." He asserted that the Mine Mill union had never acquired jurisdiction over skilled workers in smelters, and added: "Are we who represent craft organizations going to allow this abridgement of our rights? I, for one, am not going to do it."[8]

The most dramatic moment of the convention came when the resolutions committee reported negatively on a proposal to grant an industrial charter to workers in the rubber industry. Hutcheson raised a point of order against further debate on the ground that the matter had already been decided in the earlier vote on the industrial union resolutions. Lewis arose to argue that the issue was a new one, and added: "This thing of raising points of order all the time on minor delegates is rather small potatoes." Hutcheson retorted:

> Mr. Chairman, I was raised on small potatoes. That's why I am so small. Had the delegate who has just spoken about raising points of order given more consideration to the questions before this convention and not to attempting, in a dramatic way, to impress the delegates with his sincerity, we would not have had to raise the point of order at this late date, we would have had more time to devote to the questions before the convention.[9]

There are different versions of what happened next. According to one, Lewis walked over to Hutcheson, and after an exchange of oaths, "Lewis caught Hutcheson flush on the jaw and took a weaker right in return. Then, both men wildly clutching each other, they crashed through a table and down to the floor."[10] Hutcheson's biographer upgraded his response: "Lewis unleashed a right to Hutcheson's jaw, and took a stinging right in return."[11] A reporter who was at the convention told me that he saw no blow struck, and that what appeared to have happened was a shove by Lewis that caused Hutcheson to lose his balance. In any event, the effect was the same. "This was more than a clash between physically and politically powerful labor leaders; it was also a sign that the differences had reached a point where they would not be easily conciliated."[12] Even before the convention was over, the proponents of industrial unionism met to consider the next steps, and within a month, the Committee for Industrial Organization (CIO) was established.

From the start, Hutcheson and the Brotherhood were implacable foes of the CIO. In July 1936 Hutcheson demanded that the AFL act against the unions affiliated with the CIO.[13] In his report to the 1936 Brotherhood convention, he claimed that victory for the CIO would mean the end of the Carpenters' Union, since factory buildings would be constructed by industrial workers "and take the labor movement back to the days of the Knights of Labor. History will show that in attempting to organize the workers under the banner of the Knights of Labor the worker soon came to realize that that was the wrong form of organization, and from that failure came the American Federation of Labor."[14] A resolution was adopted to the effect that craft unionism was the only sound form of organization for the skilled mechanic, and locals were instructed to give no aid and comfort to the proponents of industrial unionism. "The efforts of Lewis and his wild dreams of industrial unionism were condemned."[15]

When the Rubber and Automobile Workers engaged in sitdown strikes in 1936 and 1937, the GEB declared that although the Carpenters had always defended the right to strike and picket,

> We do not believe in the stifling of industry by workers taking possession of properties that do not belong to them. We have always conducted ourselves as law abiding citizens and will continue to do so, and we do not intend to be influenced by any hysteria or new methods that are being adopted, and herewith declare that we, the General Executive Board, will not countenance any action on the part of our members in departing from the well tried methods of organizing that have been successful in years gone by, nor will we permit members of the Brotherhood to take possession of property that does not belong to them, through the method of sit down strikes.[16]

The position of the Carpenters was affirmed by the U.S. Supreme Court, but not before the United Automobile Workers had employed the sitdown strike with great effectiveness in bringing General Motors and Chrysler to the bargaining table.

At the 1937 AFL convention, the delegates of the United Brotherhood objected to the seating of Charles P. Howard, president of the Typographical Union, who was acting as secretary of the CIO. Duffy delivered a long attack on the CIO, and delegate William Kelly declared that "we ought [not] to tolerate among our membership anybody who has become a traitor to the movement, and nobody can serve as Secretary of the C.I.O. and sign charters for international unions and deny the fact that he is a traitor." Howard was excluded by an overwhelming margin.[17] A year later, Hutcheson wrote Green of rumors that the AFL-CIO dispute was going to be submitted to arbitration, and he expressed his strong opposition on the ground that there were no issues to arbitrate. "I might inform you that our Board definitely took action that if in endeavoring to settle the controversy between the two groups arbitration was agreed to it would mean the beginning of the disintegration of the A.F. of L. as our organization could no longer, under those circumstances, remain affiliated with the American Federation of Labor." Green replied that while Secretary of Labor Perkins had proposed a plan of arbitration, the AFL had not considered it and would not approve it.[18]

THE LUMBER WORKERS

The major area of confrontation between the Brotherhood and the CIO was in the Pacific Northwest. Although many of those who worked as loggers had special occupational skills, most who were employed in sawmills were semiskilled. The Brotherhood had been able to deal with semiskilled mill men successfully, but the sheer number of the men in the Northwest, and their traditions, created new and difficult problems.[19]

The AFL began chartering federal locals in the Pacific lumber industry in 1933 with the assistance of the Oregon and Washington State Federations of Labor. In July of that year, the Northwest Council of Sawmill and Timber Workers was formed at a conference in Enumclaw, Washington. The Carpenters were still out of the picture. An AFL organizer, Harry Call, urged Green to induce them to assume jurisdiction. The president of the defunct International Union of Shingle Weavers had made that suggestion earlier, but the Brotherhood had not been receptive. "At the time this proposal was made our organization was torn by strikes and dissention, and about all we had to offer the Carpenters was much trouble and little organization." Now, however, there were well-established locals capable of exercising discipline. "The Carpenters already have jurisdiction over some parts of this industry, box factories and planing and finishing mills, a situtation which can easily lead to jurisdictional battle some day."[20]

It was eight months before the Carpenters reacted. On August 29, 1934, Duffy wrote Morrison claiming Brotherhood jurisdiction over all branches of the lumber industry, and asked that all federal AFL locals be turned over to it.[21] This was followed several months later by another communication pointing out that the sawmills were now turning out not rough work but finished products—floors, door and window sills, trim and so on—that were sold in competition with the products of finishing mills:

> The unions chartered by the American Federation of Labor want this material to be labeled to show it is union-made. That would mean two labels on the market as far as woodwork is concerned and would cause confusion, annoyance, trouble and possibly strikes in some instances, as our members would refuse to handle, erect and install woodwork not bearing the union label of the United Brotherhood of Carpenters and Joiners of America.[22]

Green answered by saying that the matter would have to be referred to the AFL executive council, to which Duffy replied testily: "It is your duty in accordance with the law of the American Federation of Labor to see that the transfer of these workers from the American Federation of Labor is made to the United Brotherhood of Carpenters and Joiners of America. We cannot see where the Executive Council has anything to do in matters of this kind." Green turned the other cheek, and after giving a history of AFL attempts to organize lumber workers, going back to 1903, he explained to the council that the newly formed Northwest Council wanted its own charter based on the jurisdiction of the former International Shingle Weavers Union, and that this potential conflict in jurisdiction was properly within the domain of the AFL executive council.[23]

The prospect of a potential rival being chartered galvanized the Brotherhood into action. At the next meeting of the AFL executive council, both Duffy and Hutcheson argued strongly for transfer of the federal locals to the Brotherhood. The latter asserted that "today in the Northwest they take these logs in one end of the mill and when they come out at the other end it is flooring and finished material for building." On the motion of Vice-President Weber, *seconded by John L. Lewis*, the executive council ordered the locals turned over to the Carpenters, thus confirming its jurisdiction.[24] Some 7,000 workers were involved in the transfer.

On February 20, 1935, the AFL formally notified the lumber locals about the terms of the transfer. They were given the option of beneficial membership, which meant at the time minimum dues of $1.00 per month and per capita payment to the national office of 75 cents per month, or nonbeneficial status, which would enable them to fix their own dues level provided they paid a per capita tax of 25 cents per month. Duffy insisted that each local pay a charter fee of $15.00, which the AFL agreed to contribute.[25]

There was a great deal of dissatisfaction with the transfer to the Carpenters among the locals affected, particularly in Oregon. Leo Flynn, an AFL organizer, reported to Morrison that locals were being advised to refuse to join the Brotherhood by the opponents of the Carpenters in the Building Trades Department as well as by would-be leaders of a new national union.[26] Abe Muir, a member of the GEB who was to play a major role in the lumber controversy, was dispatched to attend a meeting of the Northwest Council on March 23, 1935, and he convinced the locals affiliated with the council to join the Brotherhood.

This harmony proved to be short-lived. A major strike began in mid-April, and a month later about 90 percent of all the operations in the Douglas Fir region were shut down. Muir eventually negotiated an agreement that was resisted by a number of locals. On June 5, a rump convention of insurgent unions met in order to wrest control of the strike from him. Participating actively in this movement were former members of a communist union that had voted to dissolve and to "bore from within" the AFL. The communists were still a small group at this time, however; the opposition came mainly from lumbermen who were unhappy with Muir's leadership. Some local central labor councils gave support to the anti-Muir faction.

Clashes between the rival groups occurred at various centers, but the strike declined slowly and the men went back to work. Muir revoked the charters of a number of local unions and replaced the insurgents with men loyal to the Brotherhood, but the seeds of dissension had been sown. Jensen summarized the significance of the 1935 strike as follows:

> By holding out against any settlement with the union the employers opened the way for left-wing elements to enter the industry labor scene. The militant opposition of most employers and the resort to state police and the militia, in some cases, in opening the mills was also shortsighted and contributed greatly to ill will in the industry. The intense dislike of outside leadership among the workers contributed to insurgency, but the problem of leadership could have been solved peacably in time, once stable relations were established. The course of events, and the bitterness which developed, seriously jeopardized worker unity for many years, and gave an unhappy direction to labor-employer relations in the industry.[27]

Once the dust of the strike had settled, Muir called a convention of all the sawmill and timber locals. The Northwest Council was abolished and a number of district councils established; they were kept deliberately small in an effort to splinter the opposition. The Brotherhood claimed to have organized 70,000 workers in 200 locals.[28] As far as the Brotherhood was concerned, jurisdiction was no longer an issue. Hutcheson dismissed the idea that while the Carpenters might have jurisdiction over the sawmills, this did not necessarily extend to the lumberjacks. In his report to the 1936 convention, he stated:

> To most people there would seem to be little, if any, connection between the men who go into the woods to cut down the trees and the men who erect finished products in a building, but it should be called to mind that while today there is a large number of employees performing the various operations involved in preparing the materials for use, nevertheless they are closely allied and therefore should be together for mutual benefit and it should also be remembered that years ago it was the custom of the workmen when erecting buildings to go into the woods, select the trees out of which they would construct the building, and they would not only select them but would shape and fashion and put the material in place.[29]

Opposition to the Brotherhood leadership persisted, however, and on September 18, 1936, representatives of all the district councils met in

Portland and established the Federation of Woodworkers. Harold Pritchett, president of the British Columbia Council, a faithful follower of the communist line if not an actual Communist Party member, was elected president. The federation, despite its support for the newly-established CIO, nonetheless decided to send delegates to the Carpenters' convention that assembled at Lakeland, Florida, on December 7, 1936.

The convention was a stormy one. The GEB had ruled that representatives of nonbeneficial locals were not entitled to be seated as full delegates with voting rights. At the start, when a delegate from one of the lumber locals moved that all the representatives present be treated as regular delegates, Duffy took the floor to denounce the Federation of Woodworkers. He accused it of planning to join the CIO; of publishing its own newpaper, which had attacked the Brotherhood; and of having set up a skeleton rival union. He warned them of what would happen if they persisted in this course:

> Let me tell you this — and this is no threat. Go on out of the Brotherhood and we will give you the sweetest fight you ever had in your lives. First we will notify all local unions not to take any notice of you, not to give you any support, moral, financial or otherwise. The next step will be that we will notify all the big firms with which we have contracts covering hours, wages, and working conditions for the timber workers that if they want to continue employing you outside of the Brotherhood we will put them on the unfair list and your manufactured stuff won't be handled elsewhere.[30]

Delegate Campbell of the Kelso (Longview) district council, asserting that he had two of the largest lumber mills in the world in his town, asked that the lumber delegates at least be given a vote on matters that pertained directly to their industry. When they joined, he added, they had been told that they would have full rights within the Brotherhood except for benefits. During the past year they had contributed $150,000 in per capita tax to the Brotherhood, and represented a third of its total membership.[31]

A subcommittee of the GEB was appointed to confer with the lumber and sawmill delegates, and at the conclusion of some discussion, it agreed to have a survey made of the lumber industry as the basis for future action; to design a label for the group; to place enterprises that had no contracts with the lumber workers on the unfair list; to conduct an organizing campaign; and to permit lumber workers to address the convention. This report, which was approved by the convention, fell short of the lumber delegates' demands. Dennis Nichols of Aberdeen took the floor to deny that their representatives were communists, while O. M. Orton, also from Aberdeen, stated: "Our place is in the United Brotherhood. They are the ones who can give the most help and assistance."[32] Orton was later to become president of the communist-dominated CIO woodworkers' union.

Perhaps the most sincere plea for delegate rights was made by Don Helmick, a leader of the Columbia River district and one of the most outspoken opponents of the communists. He repeated that the lumber workers had been told they would have all Brotherhood rights except benefits, and concluded: "We belong with you. We realize very definitely and appreciate most sincerely the value of being members of this great Brotherhood."[33] But the Brotherhood leadership would not yield. The fear of having 100,000 additional voters in the organization, many of them under left-wing influence, must have been a factor in their decision to take a hard line. Not only was the influence of communists involved; there was also a residue of IWW ideology among the loggers. For example, Worth Lowery, who was to become the first anticommunist president of the International Woodworkers' Association, had once been an IWW member.

On their way back from the Lakeland convention, the lumber delegates stopped in Washington to confer with John L. Lewis, but he had his hands full with the steel and automobile organizing drives, and gave them little encouragement. Another Woodworker Federation convention was held on February 20, 1937, where the delegates to the Brotherhood convention, despite their rebuff, advised that the federation remain within the Carpenters' Union, citing the advantages of the union label and protection against raiding by other AFL unions.[34] Their recommendation was adopted, but with the mounting success of the CIO organizing drives, sentiment for an independent union grew. A conference of district officials was called for June 7, 1937, to consider the possibility of joining the CIO. William Hutcheson, who was in the Northwest at the time, and John Brophy, a top CIO official, were invited to attend. Hutcheson refused on the ground that the Federation was not part of the Brotherhood, and sent Muir, who got a cool reception. Brophy appeared together with Harry Bridges, head of the West Coast longshore union, and promised the lumber workers a CIO charter plus $50,000 for organizing expenses. A resolution calling for a referendum on the issue was adopted, although five district councils refused to go along. These councils were suspicious of the left-wingers, but on the other hand, they were alienated by the emphasis placed by the Brotherhood on craft unionism. While the referendum was in progress, the pro-CIO group issued a statement that said in part:

Hutcheson and his agent Muir have used every possible means of selling us out by boss collaboration. In the case of the Longview workers during the '35 strike who were courageously fighting the largest lumber concern in the world, Muir was meeting in hotel rooms with the company stooges and formulating the infamous McCormick agreement and when the workers refused to accept this agreement in all of its silliness, Muir yanked their charters and destroyed their unity.[35]

The outcome of the referendum was a vote of better than 3 to 1 in favor of accepting the CIO offer. After five days of discussion at the next convention, a majority of the delegates resolved to join the CIO. Those loyal to the Brotherhood left; the Columbia River district council remained in the Federation of Woodworkers because of its commitment to industrial unionism. On July 20, the International Woodworkers of America emerged as a fullfledged CIO union. The Carpenters now faced their first rival in many years.

THE STRUGGLE WITH THE INTERNATIONAL WOODWORKERS' ASSOCIATION

The Carpenters' strength was concentrated mainly in the Puget Sound region; the International Woodworkers' Association (IWA) dominated the remainder of the Northwest and commanded the allegiance of a majority of the lumber workers. Hutcheson set out to show that the most powerful union in the AFL could not be affronted with impunity. On August 11, 1937, a circular letter was sent to all Carpenter locals with the following message:

> The [G.E.B.] found that there were Communists and adverse influences boring from within for the purpose of trying to destroy the activities of the United Brotherhood, and the building up of a dual International Union of Wood-workers . . . to combat this dual movement it becomes necessary to notify all our Local Unions, District, State and Provincial Councils of the Brotherhood that our members must not handle any lumber or mill work manufactured by any operator who employs C.I.O. or those who hold membership in any organization dual to our Brotherhood. The C.I.O. has challenged us, and we must meet that challenge without hesitation . . . Let your watchword be, "No C.I.O. lumber or millwork in your district," and let others know you mean it.[36]

A list of mills organized by the Woodworkers was circulated to retail lumber dealers, many of whom canceled their orders. A meeting of the locals that remained loyal to the Carpenters was held, and the Oregon-Washington Council of Lumber and Sawmill Workers was formed.

The first major battle between the two unions was fought at Portland, Oregon, where six large mills were involved. The employees of these mills had formed a single local, which switched from the AFL to the CIO. The Carpenters began to picket the mills, and they were closed down. The employers decided to open them pending a National Labor Relations Board election, but when the mills reopened the Carpenters resumed picketing, aided by the Teamsters, who refused to deliver lumber and fuel and overturned trucks driven by IWA men. Men on both sides were beaten. The Carpenters picketed the rivers in small boats, and logs could not be brought in because the sailors refused to go through the picket lines. As a result, the mills were closed once more.[37]

The NLRB certified the IWA as bargaining representative for the workers involved on the basis of a check of membership cards, but the Carpenters refused to withdraw. The controversy continued until 1939, when NLRB elections finally gave three mills to the Carpenters and three to the IWA.

The IWA, in the meantime, held its first regular convention in December 1937, with a dues-paying membership of 45,000. The political controversy between communists and anti-communists that was to plague the organization until 1941 broke out almost immediately. The communist faction was in control of the union, although the opposition, with Don Helmick in the lead, constituted a substantial minority. The convention debated an offer from the Carpenters' Washington-Oregon Council for unity under the banner of the Brotherhood, with a guarantee of autonomy for the lumber workers, the right to use the Carpenters' label, the commitment that all organizers would be from the lumber industry, and a reasonable return for local organizing of per capita taxes paid to the Carpenters from the Northwest. This might have worked a year earlier, but with the IWA a going concern, it was a futile gesture.

A depression hit the lumber industry in the summer of 1937, hindering organization on both sides. When construction turned up in July 1938, conflict broke out again. A majority of the Carpenters' local in Bellingham, Washington, had gone over to the IWA, but a minority remained loyal to the Brotherhood. During 1938 the IWA leadership was split on the advisability of accepting a wage cut, and the group favoring the reduction split off and formed an independent local. The IWA called a strike, whereupon the Carpenters granted a charter to the independents. An NLRB election held soon thereafter resulted in victory for this group, which was granted a union shop, eliminating all IWA influence in the area.

Continued internal dissension in the IWA, together with the Carpenters' offensive, contributed to IWA decline. By 1939 its paid membership was down to 20,000. Don Helmick made accusations against the IWA leadership at the union's 1939 convention that went as far as any that had been made by the Carpenters:

Yes, I am opposed to the Communist Party, and I'll tell you why, because I as a logger believe that I have the free right to oppose the Communist Party or any other political group, and that when [President] Harold J. Pritchett says that I'll either get in the Party or they will run me out of the industry, then I say they have asked for a fight . . . Somebody says this is red-baiting, we hear this cry. Every time that the question of whether Communists are actually and bonafidely controlling or attempting to control a given organization, if anyone mentions red—ah, ah, communism, they are red-baiting . . . When people like Jim Murphy, the Director of the Communist Party in Portland, who as far as I know wouldn't know a chocker from a frying pan is going to participate in laying down policy for unions then I object, and if Rapport, one of the leading officers of the Communist Party of Seattle is going to meet with woodworkers in regard to political affiliation of their persons OK let them meet. More power to them. But if they are going to meet and lay down policy for the IWA there are many of us that object.[38]

Ed Benedict, who was to become secretary-treasurer of the IWA a few years later, was even more specific in his charges:

. . . as we know very well International officers, members of the office staff, former district officers have over a long period of time been under the influence, some of whom are actually members of the communist party . . . That they consult from time to time with Morris Rapport, Section Organizer of the Communist Party, with headquarters in the Smith Tower Building in Seattle. We know they consult with Lou Sass, Organizational Secretary of the Northwest District, that's his official title.[39]

The Brotherhood was not without its own problems in the Northwest. Early in 1940 the GEB called for a special assessment of 50 cents per member per month for a six-month period to finance the defense of an antitrust prosecution suit that had been brought against the organization.[40] The assessment was levied on the individual lumber locals without going through the Oregon-Washington Council. When the Grays Harbor district council refused to pay the tax, a subcommittee of the GEB went out to the Northwest and told the council that it would have to pay or local charters would be revoked. This considerably reduced enthusiasm for the Brotherhood in the Grays Harbor area.[41]

The CIO sent Adolph Germer, an experienced organizer and its director of organization, to the Pacific Coast in late 1939. He concentrated on the largest companies. In October 1940 the IWA signed an agreement with the giant Weyerhaeuser Timber Company for its Everett mills, and it won an important NLRB election in Longview at plants of both Weyerhaeuser and Long-Bell in March 1941.

The IWA negotiated agreements in 1940 providing for wage increases of 4 to 5 percent. The Carpenters made demands on behalf of their members in the Douglas Fir area exceeding the IWA settlement, and called a strike when the employers refused. During the strike they had the cooperation of the anti-communist forces in the IWA, but eventually they had to accept the same terms as the IWA had secured.

The IWA continued to be hampered by internal strife. At its 1940 convention, a constitutional amendment that would have barred communists from membership was defeated by a margin of only 134 to 124. Finally, a strike in 1941 proved to be the undoing of the communists. The Carpenters' locals had settled for a 7½ percent wage increase and a maintenance of membership clause. Essentially the same offer was made to the IWA, but its communist leaders refused and called a strike. This occurred early in 1941, just prior to the Nazi attack on the Soviet Union, when the Communist Party was advocating resistance to U.S. defense preparations as part of Stalin's commitment in his pact with Hitler. Philip Murray, who had succeeded John L. Lewis as president of the CIO, publicly urged the IWA to accept the agreement, which became a recommendation of the National Defense Mediation Board, and the strike was called off temporarily pending further negotiations. The German attack on the Soviet Union changed everything, and a contract was reached along the lines recommended by the Mediation Board.[42] Partly as a result of these events, the communists were defeated at the 1941 IWA convention and a nonpolitical leadership installed.

The IWA now claimed about 50,000 members. The Brotherhood claimed 35,000 members in lumber locals working under closed-shop agreements, and "several thousand" more who were not under contract.[43] A study conducted by the Bureau of Labor Statistics in 1944 yielded the conclusion that of 130,000 workers then employed in the basic lumber industry of the Far West, four-fifths were working in unionized operations, slightly more than half under IWA contract and the rest under Carpenter contract.[44] The IWA had the preponderance of the workers in the Douglas Fir region of Oregon and Washington, while the Carpenters controlled almost all the California pine and redwood areas. Neither union had made any headway in the pine regions of the Southeast.

The Brotherhood committed some tactical errors in its initial handling of the West Coast lumber workers. A more conciliatory stance toward their delegates at the 1936 convention might have prevented the split that occurred. The communists may well have sought to create a rival union, although Lewis might have been cautious about giving them a CIO charter. The Carpenters might have been able to hold the noncommunist elements.

Having committed an error, however, the Carpenters showed their strength by fighting their way back into the industry in the face of a strongly flowing CIO tide. At least three factors made this possible. The first, and probably the most important, was the intense distrust with which many lumber workers regarded the communist policies of the IWA. As between craft unionism and communism, the first was the lesser evil for them, particularly when a modified form of industrial union structure was installed through the regional council device. Second, there was the boycott. "The Brotherhood boycott and the support which the Carpenters received from other AFL unions during the jurisdictional fracas were probably the most effective tactics used in recouping the losses suffered by the Carpenters in the initial switch."[45] A final factor was the generally pro-AFL attitude of the employers when confronted with the choice between the AFL and the CIO.

POLITICS AND THE AFL

With the success of the New Deal in moving the country toward economic recovery and in securing the enactment of pro-labor legislation, particularly the National Labor Relations Act, the labor movement moved closer to the Democratic camp. This was bound to cause some problems for the Carpenters, since Hutcheson remained an unreconstructed Republican. Among other things, he still distrusted government intervention in the labor market. For example, at the 1936 AFL convention, he spoke in opposition to a resolution that would have the AFL ask Congress for legislation mandating the six-hour day—a popular solution for unemployment at the time. Hutcheson was not opposed to the idea for federal employees, "but when it comes to private employers, I say, establish your hours by negotiation and not by law. If the Federal Government can establish a six-hour day there is no reason in the world why some future Congress cannot establish a ten-hour day. What they can give us they can take from us."[46]

The AFL maintained its traditional policy of neutrality during the 1936 presidential campaign. However, a group of labor leaders, headed by George Berry, president of the Printing Pressmen's Union, formed Labor's Non-Partisan League to work for the reelection of Roosevelt. Although the AFL executive council did not endorse the League, many of its members were favorably inclined toward its objectives. This prompted Hutcheson to resign from the council in October 1936. The formal excuse, as presented by Duffy, was that since Lewis's resignation, the Carpenters were the only union with two representatives on the board, and Hutcheson felt this to be wrong. But, Duffy added, "President Hutcheson further said the Council and President Green in the [forthcoming] report would laud acts of Congress and the New Deal, the majority of which or a number of them had been declared unconstitutional by the Supreme Court of the United States . . . President Hutcheson did not want his name attached to the report."[47]

Hutcheson aligned himself with the Republican National Committee to work for Alfred Landon, the Republican nominee. Tulsa Local 943 criticized him for this action in a circular letter. Hutcheson replied that he was supporting Landon "because of his concepts of principles of government as set forth in the Constitution of the United States." He criticized Roosevelt for showing disrespect for the Constitution: "In no time in the history of our country have we had a President who attempted to set himself up as a dictator in reference to the conditions under which people should transact business, as the present President." He reasserted his personal right to back any candidate he preferred, and said that he was not involving the Brotherhood in any way. This statement was approved by the GEB.[48]

Roosevelt's sweeping victory did not change Hutcheson's mind. In a letter to Green early in 1937, he asked for the AFL executive council's poisition on several proposed pieces of legislation: (1) to increase the number of justices on the U.S. Supreme Court; (2) to give the Secretary of Labor authority to subpoena employers and employees; (3) to establish minimum wages and maximum hours. "For your information, I desire to say I am opposed to the above referral of proposed legislation, and as a representative of an organization affiliated with the American Federation of Labor desire to enter protest against the endorsement by the American Federation of Labor referred to, or the legislative agents of the American Federation of Labor working for the enactment into laws of the proposals." Green replied that the AFL had opposed the second proposal successfully. However, it did endorse the "court-packing" scheme (which never got out of Congress after the Supreme Court changed its mind and held some key New Deal legislation to be constitutional), and backed *state* rather than *federal* minimum wage legislation.[49] Green met shortly thereafter with the Carpenters' GEB, which expressed disapproval of the AFL position on the court scheme, but no threats of withdrawal from the AFL were made.

In 1937, after Berry had left his union to accept appointment as U.S. Senator from Tennessee, John L. Lewis became chairman of Labor's Non-Partisan League, while Sidney Hillman of the Amalgamated Clothing Workers was made vice-chairman. The AFL thereupon denounced the League and directed all its subordinate bodies to separate themselves from it. Strong reaffirmation by the executive council of its political neutrality was undoubtedly one of the factors that induced Hutcheson to rejoin it. At the 1939 AFL convention, Duffy declined nomination as first vice-president, a position he had held for some years, and nominated Hutcheson in his stead. According to AFL protocol, Hutcheson should have come in as a junior vice-president and worked his way up the ladder of seniority. In this case, however, tradition gave way to reality, and Hutcheson became first vice-president of the American Federation of Labor.

The AFL remained neutral once more in the 1940 presidential campaign, but Hutcheson did not. He rejoined the Republican National Committee to work for Wendell Willkie. With the Brotherhood under criminal indictment for violation of the antitrust laws (see later in the chapter) neither he nor the Brotherhood pulled its punches. An unsigned article in *The Carpenter*, entitled "FDR, the CP and the CIO," accused the administration of placing communists in government positions.

"There are competent and unbiased observers who are convinced that Frances Perkins' Department of Labor is heavily rouged with red sympathy." Roosevelt, it claimed, had shown "unconcealed disgust" for the Dies Committee, a House of Representatives body set up to explore anti-American activities. "Those who believe Dies has gone to extremes are far outnumbered by those who declare he hasn't gone far enough."[50]

Six months later, *The Carpenter* noted that Roosevelt had finally seen the light and approved the work of the Dies Committee. However, the attack on Perkins was renewed. "Since Mr. Dies had been given the President's blessing and the FBI to carry on his work, Labor would like Mr. Dies to give some of his attention to the government Labor department and the background of some of its personnel . . . President Roosevelt has kept Madame Perkins in his cabinet despite the opposition of Labor and her ineptitude."[51]

On the eve of the presidential election, Hutcheson used the columns of *The Carpenter* for an open letter to all district councils and locals. Starting out with the premise that every member of the union retained full freedom of political action, he pointed out that the Attorney General of the United States, "who is appointed by the President of the United States," was engaged in persecuting members of the Brotherhood through antitrust indictments:

> It seems strange indeed that although the law referred to has been on the Statute Books for fifty years no administration prior to the New Deal administration has ever attempted to say, or even inferred that men exercising their rights as Trade Unionists and as free Americans would be violating the provisions of said law. The acts of representatives of the New Deal administration clearly show that they are not friends of our Brotherhood. Therefore the members of our organization should remember and follow the long practice and custom of the American Federation of Labor namely, "Assist your friends and defeat your enemies."[52]

We do not know how many carpenters followed this advice, but the fact that two of the most prominent labor leaders of the nation were opposed to Roosevelt (John L. Lewis had gone on the radio with a dramatic speech in which he came out for Willkie and promised to resign as CIO president if Roosevelt were elected) did not appear to have affected the outcome. Roosevelt was reelected for an unprecedented third term. Lewis carried out his promise to resign. Hutcheson remained undaunted. When a delegate from Local 105 in Cleveland arose at the 1940 convention to criticize him for his role in the campaign, saying that the New Deal had accomplished a lot for the worker, he was called on a point of order that Hutcheson sustained. A local wrote in to object to his "using his office for the purpose of promoting some candidate for office. This activity has caused the United Brotherhood to lose all influence in Government affairs . . . The membership of Local 448 goes on record condemning the partisanship of President Hutcheson."[53] None of this led to any change of views on the part of the general president.

JURISDICTION (5)

The conflict with the CIO muted but did not eliminate jurisdictional strife within the AFL. As we have seen, the Carpenters returned to the Building Trades Department in 1936. Their concern over the new system of adjudicating jurisdictional disputes turned out to be unfounded. The resignation of John Lapp as referee and the illness of his successor, William P. Carroll, executive manager of the Cleveland Building Trades Employers Association, resulted in Carroll's failure to make any decisions before his death in 1941. His successor was Peter W. Eller, a construction executive, who issued only one decision prior to his resignation in 1944. His successor was William L. Hutcheson, who heard appeals from decisions of the president of the Building Trades Department and rendered five decisions until he retired from the position when the Taft-Hartley Act became law in 1947.

The dispute with the Machinists was renewed when President Wharton of that organization asked the AFL executive council for a statement confirming its jurisdiction over millwrights. The council agreed to the preparation of a statement, with Green remarking that "the convention decisions leave the impression that the convention has decided in favor of the Machinists in this millwright work."[54] The issuance of the statement was held up when Hutcheson agreed to negotiate, but this proved fruitless. The Machinists came to the council once again, a year after their initial request, and produced letters from local building trades councils warning contractors to employ carpenters on millwright work. In a statement submitted to the council, the Machinists declared:

> The time has arrived when the A.F. of L. must, if it hopes to continue as the parent body of all legitimate labor organizations in this country, take some definite steps to eliminate these jurisdictional questions . . . it is not that menace from the outside of the A.F. of L. which has retarded our progress mostly in recent years but it is just such questions as we are discussing here today that have disheartened the members of various organizations and have destroyed the confidence of employers and the public in the ability of the A.F. of L. to deal with these troublesome questions.[55]

Nothing happened, and the Machinists were back a year later with a request for action. The AFL had begun sending out the jurisdictional information requested by the Machinists, but stopped when the Carpenters threatened to withhold their per capita payments. Green suggested that the Machinists send out an earlier AFL pamphlet in which all decisions relating to the Carpenter-IAM controversy were set forth in detail, but the Machinists demanded a clear, concise statement of their jurisdictional rights and threatened in turn to withhold their per capita taxes unless this were done. The executive council refused to act because Hutcheson had convinced it that public statements or negotiations of any kind might affect adversely the outcome of the pending antitrust action against the Brotherhood.[56]

The Machinists then asked for some assurance that the Carpenters would not maintain an offensive during the suspension of negotiations. "President Hutcheson and his associate officers are less worried than is the Executive Council about their interests being placed in jeopardy—otherwise they would not use the support given them at your last Council meeting to continue in the most flagrant manner their raiding of the Machinists' work jurisdiction to the extent of using their economic power to compel employers to violate contracts with our Association."[57]

When the Carpenters were out from under the federal indictment, the Machinists again asked the council to act. Hutcheson said he had no objection to having the original AFL pamphlet sent out, but Harvey Brown, the new IAM president, said that it was too long and equivocal, and that contractors would not read it. The executive council directed Green to appoint a new mediation commission.[58]

Failing to get redress from the executive council, the Machinists turned to the AFL convention. Although a resolution they introduced was recommended by the resolutions committee, it failed to carry. One of the reasons why the convention did not vote against the Brotherhood is suggested by the following comment made by Hutcheson:

> Not one of the representatives of the Brotherhood took the floor to oppose the report on that resolution which favored and recommended concurrence with the [Machinists'] Resolution. Our minds were made up then and we have not changed them one iota. If the Convention had seen fit to agree in that Resolution we would have been out and gone. Our delegates were instructed to neither vote one way nor the other. We made up our minds that if the Federation did not want our affiliation we would go out.[59]

William Green's observation on this fracas was very much in point: "This is one of the most aggravating disputes in the history of the Federation."

The Machinists tried again at the 1941 AFL convention. This time the resolutions committee recommended that the matter be referred to the executive council, which is what the convention did. Harvey Brown was bitter. He told of meeting with Hutcheson, who was willing to cede to the Machinists about 15 percent of all the machinery installation in the country. "My answer to General President Hutcheson was 'Your generosity is only exceeded by your good looks.' The Carpenters' policy was might makes right. Our membership throughout the United States and Canada for the last three and a half years have been asking this question: Are the Carpenters' per capita tax dollars brighter or bigger than the per capita tax dollars of the Machinists?"[60]

Not even U.S. entrance into the war brought about a truce. The management of an ordinance plant in La Porte, Indiana, refused to negotiate with either union until jurisdiction was clarified. Hutcheson offered the following alternatives: (1) divide the work; (2) leave it to the management; (3) leave it to the men doing the work to decide whether they want to be machinists or millwrights. The IAM, fearful of clouding its claims, refused to accept any of them.[61] Controversy erupted over some aircraft plants that were using woodworking machines for plywood fabrication. Hutcheson offered to enter into a joint organizing campaign with the Machinists, the Carpenters to retain only woodworkers, but the Machinists refused on the ground that this would redound to the advantage of the CIO, which could argue that all the workers should be in a single union. Daniel Tobin, a member of the executive council, stated that "some of us at least are getting to the point where we feel we must make some concessions where there are a few men within an industry; that is perhaps the question that brought about the CIO." Hutcheson insisted, however, that he could never instruct any woodworker to join the Machinists.[62]

Finally, on May 27, 1943, the IAM, after a membership referendum, withdrew from the AFL. Negotiations looking toward their return were begun immediately, and they were persuaded to come back on October 6, without having gained anything except the promise of more conferences. It was to take another decade, and the retirement of William L. Hutcheson, before even a temporary solution to this controversy could be found.

There were a few other relatively minor jurisdictional skirmishes during this period. The Brotherhood had claimed jurisdiction over furniture workers ever since its merger with the Amalgamated Wood Workers, and had chartered a number of furniture locals. In 1933, under the impact of the NRA, there were efforts to organize a national union of furniture workers. Evansville, Indiana, was the focal point of the effort. As might be expected, the Carpenters were strongly opposed to the granting of a federal AFL charter, let alone one for a new national union. Duffy claimed that the Brotherhood had spent $130,000 trying to organize the Evansville workers, but that they had been influenced by radicals who insisted on having an industrial union charter.[63] The AFL refused the request for a national charter, but the CIO established a national union of furniture workers in 1937. The Carpenters refused to set up internally a separate furniture workers' department on the ground that this would tend to fragment the trade, and some locals went over to the CIO.[64] The president of the CIO Furniture Workers, Morris Muster, resigned from that organization in 1946 because of communist control, at a time when the union claimed a membership of 42,000.[65]

The Carpenters were also involved in a dispute with the Upholsterers International Union. That organization had changed its title to "Upholsterers, Furniture, Carpet, Linoleum and Awning Workers International Union," to which the Carpenters objected. Summoned before the AFL executive council to explain this action, President Sal Hoffman stated that from 8,000 to 10,000 upholsterers had gone over to the CIO, leaving him with 10,000 members, and that he was trying to recoup the lost ground. He was directed to stop using the new name.[66]

The upholsterers then ceded jurisdiction over linoleum mechanics to the Brotherhood, but in the process of settling a dispute with the Painters' Union, recognized their jurisdiction over rubber and cork tiling work, which the Brotherhood also claimed. Outraged, Hutcheson accused the Upholsterers of taking advantage of the Painters' ignorance, and asked that the Upholsterers-Painters agreement not be recognized by the executive council.[67]

A meeting between the Brotherhood and the Upholsterers was held at the request of the AFL to try to settle a number of local disputes involving various aspects of upholstering. The Carpenters suggested that the Upholsterers affiliate with the Brotherhood as a way out of the difficulties, but "this did not appeal to Hoffman."[68] The Brotherhood alleged that it had taken in only furniture workers, of whom it currently had between 30,000 and 35,000. As had been the case with so many other unions, the Upholsterers made no headway against the Carpenters, who simply refused to budge from the jurisdiction they claimed.

UNITED STATES v. HUTCHESON

In late 1939 and early 1940, the Anti-Trust Division of the U.S. Department of Justice launched a major offensive, mainly against the building trades, for the stated purpose of encouraging competition through the elimination of jurisdictional strikes, featherbedding, extortion, union refusal to work on new materials, and collusive price fixing. Some thirty-five AFL unions were involved.

The AFL responded with indignation. Joseph Padway, its general counsel, charged that the indictments were "the most reactionary, vicious, outrageous attempt in the last dozen years on the part of any department of the Government to bring labor unions under the provision of the Anti-trust laws. Labor stands aghast and horrified at this bold attempt."[69]

The motives that led to this attack on the building trades are still not clear. Thurman Arnold, who headed the Anti-Trust Division, had become convinced, and may have convinced Roosevelt, that the only way to restore competition and stimulate building was by a wholesale campaign against the unions. On the other hand, reviving the threat of antitrust prosecutions, which had presumably been laid to rest by the Clayton Act, might have been expected to result in violent AFL reaction against the administration on the eve of the national election campaign.

The CIO, which was just beginning an attempt to gain a foothold in construction, greeted the indictments with undisguised satisfaction. Some of the smaller AFL unions, having confronted the imperialistic tendencies of the larger organizations, may have welcomed them as well, though they could not say so openly. The administration, a bit apprehensive over Roosevelt's bid for a third term, could appear before the country as a champion of free enterprise and an opponent of monopoly. That there was no love lost between Roosevelt and Hutcheson may also have been a factor.

The Carpenters were the main target of the government attack. Seven grand jury indictments were brought against them. The charges included refusal of their members in some cities to work on low-wage out-of-town materials, even though the producing mills were operating under Carpenter contract; and the boycott of materials produced by mills in which the NLRB had certified another union, the IWA in particular, as bargaining agent for its employees. A number of the indicted national unions pleaded *nolo contendere* and paid small fines. The Carpenters were determined to resist; this was a heaven-sent opportunity for Hutcheson to show that his opposition to the Roosevelt administration was justified. Charles H. Tuttle, a corporation attorney and one of the leading members of the New York City Bar, was retained as chief defense counsel.

The government elected to try as its strongest case an indictment brought in St. Louis that arose out of the Carpenter-Machinists jurisdictional dispute. Involved was Anhaeuser-Busch, the large brewing company, which had employed members of the Machinists to perform millwright work. The Carpenters demanded that they be given the work, and called a strike in June 1939 to emphasize their position. A national boycott of Anhaeuser-Busch beer was instituted. The indictment charged:

> Beginning so many years ago that the Grand Jurors are unable to fix the date, the United Brotherhood of Carpenters and Joiners of America has been engaged and it is still engaged, in a so-called jurisdictional dispute with International Association of Machinists . . . the said combination and conspiracy were formed . . . not to obtain higher wages, shorter hours of labor, or any other legitimate objective of a labor union, but only with the unlawful and wrongful object and purpose of inducing and coercing an employer to violate

a contract with one group of employees and to replace them with another group.[70]

One can only wonder at the assertion that protecting its jurisdiction was not a "legitimate objective" of an American labor union. The lawyer who wrote the above-quoted lines badly needed an education in labor history. The government nevertheless advanced this theory, and after the indictment was dismissed by a lower court, took direct appeal to the U.S. Supreme Court. That tribunal, in May 1941, handed down a decision in favor of the Brotherhood, written by Justice Felix Frankfurter. The Court took direct issue with the government's basic theory when it declared: "So long as a Union acts in its self-interest and does not combine with non-labor groups, the licit and the illicit under Section 20 [of the Clayton Act] are not to be distinguished by any judgment on the part of the Attorney-General regarding the wisdom or unwisdom, the rightness or wrongness, the selfishness or unselfishness of the end of which the particular Union activities are the means."[71]

It was indeed a glorious victory. A few days after the decision had been handed down, Roosevelt suggested to Green that the AFL general counsel meet with Attorney General Jackson to agree on a plan to bring about the termination of all the outstanding antitrust proceedings. The government campaign against the unions had collapsed.

In an address to the 1946 Carpenters' convention, Charles Tuttle made a comment on this episode that is worth quoting:

I won't attempt to try and understand why a national Administration which professed to be the friend of organized labor should allow so distinguished a representative as the head of the Anti-Trust Division to bring [the] principles of organized labor into jeopardy through Anti-Trust prosecution, and I won't attempt to do otherwise than to leave it to you to guess why your organization, the United Brotherhood, and your distinguished General President, were . . . singled out as the laboratory test tube or as the guinea pig.[72]

The Carpenters thought they knew the answer, but perhaps some research will be done in the future that will establish the motivation beyond doubt.

THE TWENTY-FOURTH GENERAL CONVENTION

Before considering the developments that took place at the 1940 convention, a few remarks on some internal events during the preceding four years are in order. Local 72 of Rochester, New York, challenged the right of the general president to attempt to bring about the settlement of a dispute on the ground that it unduly impaired the authority of locals and district councils. The rule had been adopted by the GEB many years earlier. The board pointed out that this rule applied only to a general trade movement for an increase in wages or a reduction in working hours and not to strikes for other causes affecting individual shops or jobs.[73] This ruling made it clear that the national office retained the right to monitor negotiations involving area-wide agreements.

Early action against the communists spared the Carpenters the internal turmoil that many unions experienced during the 1930s. However, there were a few cases in which the GEB found it necessary to take action on political grounds. Local 34 of San Francisco was suspended because of its refusal to sever relations with the communist-controlled Maritime Federation of the Pacific. Local 2090 of New York City had its charter revoked for refusing to use the standard ritual, which contained an anti-communist pledge, in initiating new members, as well as for sending out material defaming national union officers.[74]

Finally, the following statement by Hutcheson appears in the GEB minutes for April 13, 1938:

> Due to the passing on to the great beyond of Vice-President Lakey a vacancy was created in the office of First Vice-President, and according to the constitution the Second Vice-President automatically becomes First Vice President; however Vice President Meadows requested that he be permitted to remain as Second Vice President, therefore the vacancy to be filled is that of First Vice President, and in conformity with the constitution of our Brotherhood, and authority invested in me as General President, I have appointed Maurice A. Hutcheson to that vacancy.[75]

William Hutcheson's elevation of his son to the second highest post in the Brotherhood was bound to cause a great deal of eyebrow raising. Duffy, in a famous speech to the 1940 convention, took upon himself the responsibility for the appointment. He claimed that after Lakey's death, he went to Hutcheson and recommended that he appoint his son to the post:

> And, in the language of the street, Bill Hutcheson hit the ceiling. He wouldn't listen to me. He said, "Nothing doing, that is out of the question." And I said, "Wait a minute, Bill." He said, "If I do that, and it is approved by the General Executive Board, then they will say it is a Hutcheson family affair." I said, "Well, they say that anyhow. They say we have a Hutcheson-Duffy machine." Of course, there is nothing to that, but some of them say it just the same.
>
> Maurice A. Hutcheson is my protege. I am breaking him in as a General officer. I think I have served my apprenticeship as General officer over and over again in this organization.[76]

William Leiserson, in his classic book on democracy in American trade unions, made the following observation on the appointment:

There is no reason to believe that the son was less qualified for president than any one of the older vice-presidents who would normally be designated by the officialdom of the union to succeed to the presidency, and then be unanimously elected by the convention. The significant fact is the ease with which an extra-constitutional arrangement could be made to bring the young staff employee into the official family as first vice-president and in line to succeed to the presidency.[77]

Leiserson is correct in saying that the normal procedure would have been for Meadows to move up to the first vice-presidency. However, when Meadows declined to do so, Hutcheson had full authority to appoint his son to the post. There was nothing "extraconstitutional" about his action, contrary to Leiserson's assertion. Moreover, Maurice Hutcheson, who had served the union in various posts for twenty-six years, was not exactly "young" at the age of 43 when he became first vice-president. William Hutcheson had been 41 years of age when he became general president of the Carpenters' Union, while John L. Lewis was 39 at the time of his ascendancy to the top position in the United Mine Workers.

There was an incipient revolt at the 1940 convention against the election of Maurice Hutcheson to the post to which he had been appointed, but it was quickly scotched. After a long apology for doing so, a delegate nominated Harry Schwarzer, a district organizer, to run against Hutcheson. When another delegate rose to hint darkly that the opposition had some connection with a "Black Legion," Schwarzer declined the nomination and Hutcheson remained the sole candidate for the first vice-presidency.[78]

The 1940 convention found the Brotherhood with 319,848 members, of whom 227,742 were beneficial and 74,346 nonbeneficial, while the rest were in honorary and other categories of membership. Total membership had risen by 18,000 since 1936, but the gain was greater than the totals indicate. The number of beneficial members increased by almost 67,000, offset by a nonbeneficial loss of 57,000 men. As we have seen, that portion of the nonbeneficials who were in the West Coast lumber industry had only a weak attachment to the Carpenters' Union. The gain in construction tradesmen far outweighed the loss in semiskilled lumber workers. However, the representatives of the 35,000 nonbeneficial members in lumbering who had remained loyal to the Carpenters were seated as delegates with full voting rights.

The convention adopted a number of constitutional amendments, some of which were defeated in the ensuing referendum vote. Surviving the vote was a change in the category of the lumber workers from nonbeneficial to semibeneficial status, which integrated them more firmly into the union structure. However, the members defeated a suggestion made by Hutcheson that the pension plan be eliminated because of the recently enacted Social Security Act; they also voted down salary increases for the general officers and an increase in the initiation fee. Perhaps the most important action of the convention was the adoption of a detailed statement setting forth national and local standards for carpentry apprenticeship. This was necessary because of the revival of the apprenticeship system after its virtual demise during the depression.

World War II had already begun when the twenty-fourth convention was held, but the United States was not yet involved. The six years that followed until the next convention — the convention scheduled for 1944 had to be canceled because of the war — were to see great changes in the economy, which were reflected in the structure and operation of the Carpenters. By 1940 the Brotherhood had recovered from the Great Depression and was prepared to take full advantage of the opportunities for growth afforded by the full employment and the system of industrial relations that characterized the wartime labor market.

12

WAR AND RECONVERSION

Recovery from the Great Depression was slow. After a drop from the high level of unemployment in 1932, the rate of unemployment rose once again to 19 percent in 1938, and was still 14.6 percent in 1940. Construction activity picked up in 1939, but the value of construction in 1940 did not yet reach the 1929 level. Total employment in contract construction was 1.3 million in 1940 compared with levels of 1.6 million in the late 1920s. Against this background, the record of Brotherhood membership progress during the last half of the 1930s in the traditional construction area was by no means a poor one.

All this changed abruptly with the inception of the national defense program in the latter part of 1940. The gross national product rose by 9 percent from 1939 to 1940, and by 15 percent the following year. Construction activity, particularly in the public sector, increased by leaps and bounds. The number of employees in contract construction went from 1.3 million in 1940 to 2.2 million in 1942. The depression was finally over, but it took a war to do it.

The Carpenters' Union committed itself to full cooperation with the national defense effort. This was at a time, incidentally, when the communist-controlled unions in the CIO were calling for strict neutrality in what they termed a capitalist war with which workers should have no concern. The GEB issued a declaration accepting for its membership "a full share of the burden and such sacrifices as may become necessary to the end that there may be retained for the citizens . . . rights under our democratic way of life." It continued: "The Brotherhood declares its full support of the National program for financial aid to Great Britain, consistent with the thought and purpose that the defense of America must come first, and further declares itself willing to aid all nations who are holding democracy's battle line against tyranny or various forms of isms such as Communism, Nazism or Fascism."

A message of "ardent admiration and fraternal sympathy" was sent to British and Commonwealth trade unions. The GEB promised the government that the Brotherhood would do its best to supply competent labor where needed, and added a paragraph that looked back to the controversial labor policy of the U.S. government in World War I: "The Brotherhood fully recognizes that where a defense project has been undertaken upon proper terms as to wage rates, the project should be carried through, and that the right to work as a union man upon such projects should not be burdened with exorbitant or unusual fees and dues or be impaired by the introduction of men not of the household of our Organization."[1]

This was in line with the AFL position on what was then still a European war, and on national defense. William Hutcheson, however, was personally more isolationist than the generality of AFL leaders. In September 1941 he joined the national committee of America First.[2] This organization, which included among its members the famous aviator Charles A. Lindbergh, had as its purpose the prevention of any American intervention in the war. Hutcheson's personal views do not appear to have influenced the attitude of the Brotherhood toward defense or assistance to the embattled British empire.

In May 1940 Roosevelt appointed an advisory committee to the Council of National Defense, which had been set up to coordinate the defense program. He selected Sidney Hillman, president of the CIO Amalgamated Clothing Workers, to be head of the Labor Division, an appointment that was not popular with the AFL. Among other things, John L. Lewis had launched the Construction Workers' Organizing Committee under the leadership of his brother, A. D. Lewis, and the building trades feared that Hillman might favor that organization. The AFL executive council refused to appoint anyone to work with him, but Hillman secured the cooperation of the Building Trades Department by assuring it that he would deal exclusively with AFL construction unions.[3]

The attitude of the AFL executive council was undoubtedly influenced strongly by the attitude of its first vice-president. When Hillman was invited to address the 1941 convention of the Building Trades Department, "Hutcheson voiced vigorous objection to permitting Hillman to speak 'not because I have any objection to the gentleman personally or any criticism of his official conduct, but because he is the head of the CIO union.'" Hillman was defended by the department's president, John Coyne, who said that he had been most helpful to the AFL unions by recognizing them on all government construction work.[4]

Hillman had in fact taken up the cudgels for the AFL unions. An agreement was reached between the Building Trades Department and defense officials setting forth conditions of employment on government work. Commencing August 1, 1941, overtime was to be paid at time and a half rather than double time; the unions agreed to three shifts without shift premiums; they promised that there would be no work stoppages over jurisdiction; and wages were to be determined by union rates in each locality. A government-labor Board of Review was established to adjust disputes, including jurisdictional questions, arising under the agreement.[5]

A contract for 300 prefabricated homes for defense workers had been let to the low bidder, the Currier Lumber Company of Detroit, which had a closed-shop agreement with the CIO Construction Workers Union. Hillman ordered cancellation of the contract, and despite protests by John L. Lewis and threats of anti-trust action by Thurman Arnold, he stuck to his guns. The work was eventually completed by AFL workers. Philip Murray, president of the CIO, tried to get Roosevelt to set aside the stabilization agreement with the building trades, but he was unsuccessful.[6]

THE WAR YEARS

The attack on Pearl Harbor led to an immediate promise of cooperation from the Carpenters. Hutcheson wrote to Roosevelt:

> Heretofore I have opposed sending an expeditionary force overseas. Now that our country has been attacked it is the duty of every American to help in every way he can to supply and produce the necessary munitions of war so that we can speedily overcome our enemies and show them that we cannot be ruthlessly attacked without retaliating. There should not now be any need for Congress to give consideration to anti-strike legislation as I am sure that members of the various labor organizations regardless as to their affiliation will show their patriotism as real Americans by refraining from committing any act or taking any action that would in any way handicap or slow up the program in preparing and manufacturing the necessary munitions of war and to that end I desire to offer for myself and the members of the Brotherhood our cooperation and service in any way that it may be needed.[7]

The Building Trades Department, in the first flush of patriotism after the declaration of war, promised that there would be no strikes on defense work for the duration, and committed itself to permitting flexibility in the hiring of workers where labor shortages prevailed. But Hutcheson, in telling the Brotherhood membership that the union agreed to this proposal, added a proviso going back to his experience in World War I:

> It naturally follows that it would be with the understanding that work coming under the jurisdiction of our organization should be done by our members. In other words it is not the intent of this understanding that we stand passively by and have others do work that should be done by our members, and every member of the Brotherhood should be alert to see that there is no infringement on the work that we have been in the habit and custom of doing.[8]

Jurisdictional stoppages were infrequent and of brief duration during the war. The Board of Review, now under the Office of Production Management, helped settle the disputes that did arise. The Carpenters did not relax their customary vigilance over their job area, however. On the contrary, every effort was made to ensure that the Brotherhood got its fair share of the available work. For example, when projects were located in areas where there were no locals, or where the existing locals were too small to police large projects (the plants built in Tennessee and Washington to produce atomic weapons were examples), special representatives were placed in charge of the projects and were required to report weekly to the Brotherhood general office on the conditions in effect, the number of members employed, and other relevant information. Some of these projects employed as many as 6,500 carpenters at one time.[9]

The Carpenter carried an editorial favoring assistance to the Soviet Union. "If, with Stalin's aid, Hitler is whipped, Russia will be no menace. Between the two there is but one choice, It is aid to Russia, to the limit."[10] It was made clear, however, that support extended only as far as governmental aid. The GEB, noting that plans were afoot to set up collaborative relations between British, American, and Soviet trade unions, declared that since "the so-called Trade Union movement of Russia is dominated, controlled, and directed by the Soviet Communist Government, and is, therefore, not a free Trade Union Movement, we cannot endorse or participate in any such procedure . . . The United Brotherhood of Carpenters and Joiners of America can in no way cooperate or collaborate with the Communist Trade Union Movement of Russia."[11] The CIO, with Hillman in the lead, worked with the British and Russians to create the World Federation of Trade Unions, a step that it was later to regret.

A tripartite National War Labor Board was created by the government to determine wages and other conditions of labor during the war years.[12] The building trades succeeded in securing a separate body to handle their cases, titled the Wage Adjustment Board, which was also tripartite. Actually, this body administered the Building and Construction Trades Wage Stabilization Agreement of May 22, 1942, which preceded by four months the conferring of general wage control authority on the War Labor Board.[13] While it operated within the general framework of the regulations established by the War Labor Board, the Wage Adjustment Board had a good deal of independent authority to handle the complex affairs of the construction industry. Maurice Hutcheson was an alternate labor member during the first year, and then was replaced by William Blaier, a member of the GEB. The board functioned smoothly and helped prevent a repetition of the unfortunate events of World War I.[14]

The Brotherhood favored an anti-inflationary economic policy; the GEB called for the stabilization of wages and farm prices for the duration and for the compulsory purchase of war bonds, and even went so far as to advocate the enactment of a national sales tax to reduce consumption.[15] However, Hutcheson was uneasy about a complete wage freeze, and felt that if union conditions were not recognized, workers should retain the right to strike.[16]

A proud moment in the history of the Brotherhood occurred in October 1942 when the Kaiser Shipyards launched a ship named the "Peter J. McGuire." Hutcheson spoke at the ceremony, referring to McGuire as the father of the Brotherhood.[17]

JOHN L. LEWIS AND THE AFL

The president of the Carpenters served during many of the war years as a member of a three-man committee appointed by the AFL to negotiate with the CIO for unity in the labor movement, but nothing came of the effort. However, an interesting development occurred in 1943, when the United Mine Workers applied for affiliation with the AFL. To the argument that this might impede possible merger with the CIO, which the Mine Workers had left in 1940, Hutcheson declared that "there is no sincerity there and they do not even intend to come into the Federation for the simple reason that the activities of some of them are what we would call subversive activities, not speaking about the head of the CIO, but the back of it." On a motion to defer action on the Miners' request, Hutcheson cast the sole negative vote; he favored their immediate admission.

The discussion was renewed at Hutcheson's insistence. There was concern about District 50, Lewis's attempt to build a catchall general union on the British model. Hutcheson said that Lewis had promised him that District 50 would cause no problems; the Miners would come back with their original jurisdiction. A committee was appointed to confer with Lewis on the specifics of jurisdiction.[18]

Lewis was not an easy man to pin down. It was his position that "the United Mine Workers under present conditions has no interest in questions of hypothetical jurisdiction. After the fact of reaffiliation any and all questions of jurisdiction having factual or realistic premises can be considered procedurally by the American Federation of Labor." He promised not to interfere with the activities of the Progressive Mine Workers, a coal miners' union centered in the Illinois fields that had been chartered by the AFL when the United Mine Workers left. The negotiating committee urged that the matter be held over to the next AFL convention, but Hutcheson moved in the executive council that the Miners be admitted on the basis of their original jurisdiction. This motion failed to carry on a tie vote when Green warned that Lewis had put $1.6 million into District 50 and was not likely to abandon it easily.[19]

The 1943 AFL convention authorized further negotiation, but at the same time called on other unions to file any complaints about UMW invasions of their jurisdictions. When confronted with this list, Lewis said, "Take us in as we were. You will have no trouble." However, he insisted that through the activities of District 50, he had acquired jurisdiction over chemical workers, and refused the proposition that his chemical workers be merged with several AFL federal locals to form a new national union. In a subsequent letter to George Meany, at the time secretary-treasurer of the AFL, who was handling the negotiations, Lewis relinquished any jurisdiction over construction and railroad workers (of whom he had very few in any event) but said nothing about chemical workers. Hutcheson urged that the UMW be readmitted on the basis of this letter, but Green's motion that admission be predicated on acceptance of the pre-1936 jurisdiction prevailed.[20]

These marathon negotiations continued almost entirely because of Hutcheson's persistence. When further talks between Lewis and Green could not resolve the remaining differences, Lewis sent Green a letter that was typical of the man:

Press reports reveal that the Executive Council of the American Federation of Labor has again with characteristic servility to the Roosevelt Administration failed to take affirmative action with respect to the pending application for reaffiliation by the United Mine Workers of America . . . Throughout this period of a year, the majority of the members of your Executive Council have lacked the courage to either vote "Yes" or vote "No" on the question of acceptance. Instead they have constantly muttered and mumbled and indulged in fearsome incantations over the fallacious and hoary question of jurisdictional rights. It is an amazing exhibition of base hypocrisy approximating moral turpitude.

He asserted that the AFL had become a puppet of the Democratic Party, "in fact, [had] achieved the status of a political company union," and he asked that his application fee be returned.[21]

Despite this vitriolic communication, Green, who had once been secretary-treasurer of the United Mine Workers and would undoubtedly have liked to see his old union come back, continued to meet with Lewis. A year later, Lewis proposed that the Miners return as they were, with immediate talks on jurisdiction to begin upon readmission, and with the executive council to make a final decision on jurisdiction if no agreement could be reached. On Hutcheson's motion this offer was accepted, but one final obstacle arose: Lewis wanted a seat on the executive council, and the council would not agree.[22] It took almost another year to resolve this obstacle, but Lewis was finally appointed thirteenth vice-president in place of Harvey Brown, president of the Machinists' Union, when that organization left the AFL. Hutcheson had achieved a neat ploy: he was able to replace a bitter foe of the Carpenters with a man who had become a staunch ally. Lewis announced when he appeared at the council in January 1946, "We have always recognized the high principles of this Council, and the leaders of the American Federation of Labor have followed their convictions, and they have performed their duty with a high sense of responsibility to their membership."[23] No more base hypocrisy or moral turpitude.

But the honeymoon proved to be brief. The building and metal trades unions complained to the 1947 AFL convention that Lewis had refused to negotiate jurisdictional conflicts involving District 50 of the Miners. A majority of the resolutions committee wanted to condemn the Miners, despite Lewis's warning that "you can't pass resolutions here for the benefit of the newspapers to force the United Mine Workers of America to emasculate a great, growing concern like District 50." Hutcheson pleaded with the convention to defer action. "No one on the Executive Council did more to endeavor to get the United Mine Workers of America back into the American Federation of Labor than I. There are many delegates on this floor who know that in the past I have taken a determined stand on many issues, but I say to you here today, forget any little differences, any little feelings that may exist. If we haven't followed the proper procedure, let's start doing it now." The convention followed his advice.[24]

Refusing to yield on jurisdiction, despite his implied commitment to do so, Lewis seized upon another issue to criticize his opponents in the AFL. The Taft-Hartley Act required all members of the executive council, who were also vice-presidents of the AFL, to sign noncommunist affidavits. Lewis was adamant in his refusal to do so, saying that the requirement made second-class citizens out of labor leaders. The presidents of the Building and Metal Trades Departments appeared before the council to urge that the affidavits be signed, since their affiliates could not participate in NLRB elections until this was done. When it was then proposed that the AFL constitution be amended to change the title of vice-president to executive board member, thus obviating the need for the affidavits, Lewis's indignation knew no bounds. He regarded it as a cowardly retreat "that would strip the members of this Council of whatever honorary cognomen they may attach to their positions." Lewis cast the sole negative vote on this proposition, and on December 12, 1947, he sent the following message: "Green—AF of L. We disaffiliate. Lewis."[25] It can well be imagined that Hutcheson was not at all happy at the prospect of having to declare himself formally not to be a communist, but he did so because he felt that the interests of his members required it.

There were many parallels in the careers of Hutcheson and Lewis, for many years the two most influential men in the American labor movement. Despite their clash in 1935, the close identity of their political and economic philosophies made them natural allies, and by 1940 they were reconciled. They were powerful personalities who were able to dominate a group of men with no shrinking violets among their number. They were not lovable characters, but they were supremely effective in protecting and advancing the interests of the men they represented. In the end, it was Hutcheson, despite Lewis's formidable organizing ability, who proved to be the more flexible and successful. Lewis left behind him a leadership vacuum from which the United Mine Workers has still not recovered, while Hutcheson's place was ably filled by a succession of able administrators who kept the Carpenters on an even keel.

Carpenters and Miners are different types of men—in their manner of life, their attitudes toward the outside world, their personalities. But they had some things in common: they were proud of their skills and built strong unions. Both made a major contribution to the growth of the American labor movement, the Miners by leading and financing the struggle for organization of industrial workers, the Carpenters by initiating the American Federation of Labor and helping it survive through lean years, by keeping its ranks intact, and by contributing toward establishing the building trades as the hard core of American unionism.

SOME INTERNAL DEVELOPMENTS

Little occurred during the war years to disturb the smooth operation of the Brotherhood. However, there were a few events worth mentioning. The membership had voted down an increase in officers' salaries recommended by the 1940 convention. With the onset of inflation, Hutcheson raised all general salaries but his own, claiming that this action was within his constitutional authority. A member of Local 116, Bay City, Michigan, filed a charge alleging that in fact he had no such authority, and that the officers were thus receiving what amounted to stolen funds. The GEB ruled that this accusation was defamatory, and directed the local union to prefer charges against the individual concerned. He was tried, found guilty, and fined $200, which he paid.[26] Whether Hutcheson's action was constitutional is a complicated question; it certainly was not customary.

While the war was still in progress, members and officers of the Brotherhood began to think of what the postwar world would need. The New York and Chicago district councils sponsored a resolution asking the GEB to consult with the relevant authorities on planning a reconstruction program to avoid unemployment and bring about a speedy conversion to civilian production.[27] In an article written for *The Carpenter,* Hutcheson emphasized that the intervention of government in industry must be temporary, to be relinquished once the war was over. He set forth six basic principles that he believed should govern the postwar economy:[28]

1. The self-administration of industry by cooperating functional groups.
2. Reasonable profits; "the profit motive is the mainspring of business initiative."
3. Labor's right to bargain collectively.
4. The safeguarding of human rights.
5. The maintenance of full employment.
6. Preservation of the free market.

A subsequent editorial in the *The Carpenter* entitled "Let's Keep Realistic," probably written by Duffy, carried a similar message:

> If any sort of postwar planning is going to be successful, it must be realistic. The plain facts in the case are that we live in a capitalistic system. Under the system we have gained more of the good things of life than any other peoples of the world . . . If there are abuses in industry which must be corrected, let us look toward the correction of them as rapidly as possible. But let's never lose sight of the fact that our real hope for postwar jobs for all lies in industry . . . If industry is going to have to be the backbone of our postwar world, blindly damning business is solving none of our problems.[29]

THE ELECTION OF 1944

The 1944 presidential election was repetitious of the three preceding ones, as far as the Carpenters were concerned. A long article by Hutcheson, appearing in *The Carpenter,* stated the following premise: "It is my conviction that the New Deal Administration has shown itself incapable; that its methods and policies have, themselves, created new threats to our national economy more disquieting than those which we have been attempting to escape; that wage earners have been victims of a cruel political deception in that our economy has been entrusted to visionaries." Labor, Hutcheson averred, wanted the following:

1. The preservation of free enterprise. The New Deal has "waged a continuous war of attrition against the psychological and economic support of our free enterprise economy."
2. The abatement of bureaucracy. "For labor, bureaucracy creates a truly poisonous atmosphere. In the war between conflicting boards and authorities, labor's problems and interests become the football of political careerists."
3. A halt to paternalism. The Roosevelt administration, in his view, had enfeebled the labor movement, had prolonged divisions in labor, and had visited reprisals on union leaders who refused to toe the line.

4. The creation of postwar jobs through private industry. "The channels of enterprise must not be choked by government controls or punitive tax policies on the one hand, or by self-defeating monopolistic or cartel policies by industry on the other."

5. Maintenance of labor's social gains.

6. Protection of the national interest. Too much should not be promised to other nations.[30]

The AFL maintained official neutrality in the election, although many of its national union leaders served as Democratic delegates and took some credit for the replacement of Henry Wallace by Harry Truman in the vice-presidential spot. It was the CIO's Political Action Committee, however, that was most active in the elections and helped provide Roosevelt with what turned out to be a fairly narrow margin of victory. When the election was over, *The Carpenter* once more warned, in an editorial, that the labor movement should eschew any dependence on government; that organized labor "cannot play the part of stooge to any political party or group of politicians and still retain its integrity and independence."[31] But the Brotherhood's journal did eulogize President Roosevelt when he died a few months later: "It has not always been possible to agree with his decisions, but whether we have considered them right or wrong it has never been possible to question the quality of his courage or the sincerity of his ideals. Yes — the world has lost a great man, a man whose personal courage will shine like a guiding beacon over a troubled world for generations to come."[32]

THE TWENTY-FIFTH GENERAL CONVENTION

The 1946 convention opened on April 22, some nine months after the cessation of hostilities. It was a very tumultuous nine months. Millions of men released from military service had to be reintegrated into civilian jobs. There was an enormous demand for the civilian goods that had not been produced during the war, and housing in particular was in short supply. The volume of new residential construction more than tripled from 1945 to 1946, while that of industrial and highway construction doubled. All of this meant more jobs for carpenters.

It also meant a large jump in Brotherhood membership. Table 12.1 gives the comparative membership figures for 1940 and 1945. The unavailability of data for the number of members in arrears in 1940 makes it difficult to compare membership for these two years precisely. However, even if we are to assume that all members in 1940 were in good standing, membership in this category rose by 65 percent between the two years. For the semibeneficial category (all nonbeneficial members

Table 12.1. Membership in the United Brotherhood of Carpenters, 1940 and 1945

Membership category	June 30, 1940	December 31, 1945
Beneficial members in good standing	227,742	377,281
Beneficial members in arrears	a	115,611
Semibeneficial members in good standing	74,346	59,793
Semibeneficial members in arrears	a	74,593
Members in military service	a	42,125
Honorary members	13,981	44,202
Others	3,779	8,787
Total	319,848	722,392

Sources: United Brotherhood of Carpenters, Proceedings, 1940 and 1946 conventions.
a. Members in arrears not shown separately from those in good standing.

had been thus reclassified after 1940) the conclusions are more ambiguous, because of the large number of such members in arrears in 1945.

Looking only at the totals, membership more than doubled. During the same period total AFL membership rose by 63 percent, so that even if one were to omit all members in arrears (many of whom would in fact subsequently regain good standing once they secured jobs and paid up their back dues), the Brotherhood still did better than the average AFL union.

Two additional facts can be discerned from the data that were presented to the 1946 convention. One is that not only the Brotherhood but its members as well were beginning to age. This is indicated by the large increase in the "honorary" category; most of these members were in retirement or semi-retirement. The second fact concerns Canada; in 1945 there were 17,881 members in beneficial and 5,708 in semibeneficial locals, a total of 23,589 Canadian members in all. Although this was a relatively small percentage of total membership, it had become sufficiently large in absolute numbers to support growing separatist tendencies.

No important constitutional amendments were adopted by the 1946 convention. A resolution that would have given the convention authority to fix the salaries of general officers and organizers without the requirement of a ratifying referendum vote was defeated. Another resolution sponsored by a group of black locals to bar all racial discrimination in the union was introduced: "We earnestly request that white members of our union join with their colored brethren and work on the same project." It was urged that all locals immediately investigate charges of discrimination, and that the national union do so as well. Delegate Reed of Knoxville defended the necessity of the resolution in the following terms: "The Brotherhood should come to our assistance. It should come to our assistance to see that we get fair play. That is not all, but that is much. We have striven hard in the past for many years to adjust our problems locally, but to no avail." He cited a number of specific cases, including the atomic installation at Hanford, where blacks were not permitted to work with whites, even where they were needed.

The resolutions committee did not concur on the ground that the problem was local, rather than national, and was better handled locally. Those opposed to the resolution argued that it would make organization more difficult in the South. The committee report was eventually adopted by a large majority.[33] The Brotherhood had admitted black members from its very beginings, but it was not yet ready to enforce non-discrimination on a uniform national basis. That took a few more years.

The process of nominating officers, generally routine, was accompanied by a great deal of acerbity this time. With Vice-President Maurice Hutcheson in the chair, Duffy renominated William Hutcheson for re-election. After a long nominating speech, nominations were closed, and Hutcheson returned to the chair. The following exchange then occurred:

> Delegate J. O. Mack, Local 61, Kansas City: Is it the prerogative of the chair to ask for other nominations in any convention? I wonder if the President is afraid of opposition in this convention.
> President Hutcheson: H — —, no.
> Delegate Mack: Mr. Chairman, there were delegates on this floor for nomination and they were not allowed the floor, and there was a man standing right at this microphone, Hanson, from New York, that I wish it to be recorded in the minutes of the proceedings of this convention that there was a man standing on the floor and refused —
> President Hutcheson: You are out of order. Please take your seat. The presiding officer who was presiding will make an explanation.
> Vice-President Hutcheson: I asked if there were further nominations and there were none.
> A Delegate: I would like to ask a question. One man got up six or seven times and he was called out of order.
> President Hutcheson: You will take your seat. The chair wants to state that from his observation no one arose at the time to raise a point of order and in answer to the question of the delegate again I will say, h — — no, I am not afraid of any opposition. I have had so much opposition while I have been serving the Brotherhood from those within who would tear down the Brotherhood and those from without who would try to steal from our Brotherhood and I don't fear opposition, no matter where it comes from.[34]

The record does not show that any call for other nominations was made. Perhaps having thought the matter over, Hutcheson decided, after nominations for all other offices had been made, to reopen the nominations for president. J. O. Mack, president of the district council of Kansas City, who earlier had protested the closing of nominations, was promptly nominated. There then occurred an event that was without precedent in the history of the Carpenters. Vice-President Stevenson arose to say that Hutcheson did not have the right to reopen the nominations, and proceeded to criticize both the second nominee and the man who nominated him: "[The nominator] met [the nominee] about eight weeks ago out in Des Moines, Iowa, and it requires no further recommendation of this type of individual, except to meet in a foreign city with others who are disgruntled and dissatisfied . . . This organization is too important to suffer that type of obnoxious disruption."[35]

Instead of calling Stevenson out of order, a motion not to reopen the nominations was made and carried. This episode illustrates a point often made by students of trade union democracy: there is a tendency to equate opposition to incumbent officers with treason to the union. It would have been interesting to see how many votes Mack could have secured against Hutcheson. His chances of winning were remote, but it strengthens, not weakens, an organization to test minority support.

That opposition to the Hutcheson administration was substantial is indicated by the fact that the incumbent vice-president, Maurice Hutcheson, was opposed for election by W. A. Meyer of Oklahoma City. Hutcheson won by a majority of 93,904 to 51,195, but the fact that 35 percent of those voting cast their ballots against the incumbent suggests that the opposition was more than a small clique out to disrupt the Brotherhood.

THE HOLLYWOOD AFFAIR

The jurisdictional lines between the Carpenters and the International Alliance of Theatrical and Stage Employees (IATSE) over the construction of sets for Hollywood movies, which had presumably been settled by a 1925 agreement, were the occasion of a controversy that broke out at the conclusion of the war. The politics of this dispute, in which the Carpenters were only one of the players, could have provided a good scenario for a film thriller.

This dispute was the culmination of a long struggle between IATSE, which had been plagued with a corrupt leadership, and a group of craft unions for control of the lucrative work in the Hollywood studios. In 1939 IATSE had won what appeared to be a decisive NLRB election, but the crafts, united in a loose organization, the Conference of Studio Unions, reorganized and stepped up their organizational efforts. The leader of the Conference was Herbert Sorrell, business agent of the

Painters' Union in the studios, who issued charters to a wide range of occupational groups, including cartoonists, publicists, story analysts, office employees, and set decorators.[36]

Organization of the latter group threatened IATSE, for set decorators sometimes acted as property men, who were members of an IATSE local. After negotiations and a brief strike, followed by War Labor Board intervention, had failed to resolve the dispute, the Painters called another strike on March 12, 1945, in which they were supported by five crafts, including the Carpenters. The conflict was broadened from seventy-seven set decorators to several thousand men.

IATSE responded by chartering locals of painters and carpenters and reaching an agreement with the Producers' Association in which it promised to supply all the necessary personnel. To complicate the situation further, President Walsh of IATSE charged that Sorrell was a communist, allied with Harry Bridges, the West Coast Longshoremen's head. This allegation was supported by President Weber of the Musicians' Union, who told the AFL executive council that "the organization formed by Mr. Sorrell out there was for the purpose of breaking down the international organizations in the studios which is what Sorrell is out for. He stated the communists are trying to use the communists in the organization to break the power of our international unions in the studios." The executive council ordered IATSE to revoke the craft charters it had issued and to negotiate a settlement of the controversy.[37]

An NLRB election was held in May, but the many challenges that were made held up a decision. The strike continued, marked by mass picketing, violence, and nationwide publicity of the kind that the labor movement did not need. On October 12 the NLRB found that the decorators had voted for the Painters. IATSE agreed to revoke the charters that had been given to carpenters and painters, but asked that the AFL appoint a committee to investigate the entire situation, including alleged communist and CIO activity in the studios.[38]

Eric Johnston, representing the Motion Picture Producers and Distributors of America, appeared before the AFL executive council to urge that immediate steps be taken to bring the strike to an end. After a long discussion, the council concluded that work should be resumed immediately; that all employees should be returned to the positions they held prior to the strike; that meetings should be held between IATSE and the striking crafts in an attempt to reach an agreement; and that if none was reached within thirty days, a committee appointed by the council would arbitrate the dispute, its decision to be binding on all parties. There was general agreement to this solution.[39] The strike was called off on October 31, although some of the strikers refused to return as long as workers who had crossed their picket lines remained on the job.

Direct negotiation between the parties proved fruitless, so the AFL committee, consisting of three neutral AFL council members, Felix Knight (Railway Carmen), W. C. Birthright (Barbers), and William Doherty (Letter Carriers), went to Los Angeles to investigate. Their decision was satisfactory to the Painters but not to the Carpenters, who lost about 350 jobs when IATSE was assigned jurisdiction over miniature sets, nonpermanent property building, and the construction of platforms for lamp operators, among other things. Hutcheson vigorously protested the decision to the AFL executive council on the ground that he had not been given sufficient opportunity to confer with the committee before it rendered its report. A Carpenters' general representative had been present at the hearings, but Hutcheson contended that he had been assured of a chance to present his evidence personally. He reviewed the history of the Carpenter-IATSE disputes, and contended that the 1925 agreement between the two organizations had never been approved by the Carpenters' general office, but only by a local whose charter was subsequently revoked.

The arbitration committee members maintained that they had made no commitment to the carpenters beyond assuring them that their representative would have an opportunity to present evidence; they felt that they had fulfilled their mission, and were not prepared to make any change in their decision. Hutcheson then charged that the committee had failed to give due weight to convention resolutions adopted in 1920, 1921, and 1922, and demanded that the committee report be amended to include these earlier decisions: "Vice-President Hutcheson stated that surely no one on this Council today would imagine a committee of the Council saying that they were delegated authority to set aside an action of this Council, or one taken 25 years ago, or set aside an action of three Conventions taken by the Federation 25 years ago."

Eric Johnston appeared before the council once again and told it that the producers were keeping IATSE "replacements" for the craft unions on their payrolls pending final settlement of the dispute, at a cost to them of several million dollars. The committee reiterated its determination against making any changes in its award, and stated that all the Hollywood jurisdictional disputes had been settled with the exception of the erection of sets. Hutcheson then presented a statement to the council in which he declared that the Brotherhood national office refused to recognize the award; he said that he could not order the local to accept the award, but that he had "sent word to his Local out there that it is up to them entirely; if they want to accept this decision it is all right, if they do not want to accept it, it is all right; in other words it is in their hands."[40]

A few months later the Carpenters' convention condemned the refusal of the AFL executive council to amend the award, but the Hollywood local decided against resuming the strike. This did not mean, however, that the Brotherhood had thrown in the towel; far from it. Hutcheson told the next AFL council meeting that "the membership were deeply moved and he did not think that they would ever acquiesce in the decision, because they claimed the decision had taken work away from them that they had performed for a long period of time."[41]

The matter was laid over to the next council session, which began with a report from an AFL representative who had been sent to Los Angeles to see what was happening. His report was not encouraging. There had been a great many brief stoppages, and another general eruption was threatened. Carpenters' Local 946, which had jurisdiction over the studios, had 1,500 members in all. If it won set erection work, there would be additional employment for 500 men, with another 250 jobs depending on interpretations of prop work and the term "miniature sets." Many lucrative jobs were at stake.

Hutcheson proposed that the investigating committee issue a "clarification" of its intent in the award. After conferring, the committee obliged by stating that "the word erection is construed to mean assemblage of . . . sets on stages or locations. It is to be clearly understood that the Committee recognizes the jurisdiction over construction work on such sets as coming within the purview of the United Brotherhood of Carpenters and Joiners of America's jurisdiction." What this statement did essentially was to give the Carpenters all *new* construction of sets, reserving to IATSE the assembly of sets that had already been built and were taken from storage. This in effect reversed the original award, and Hutcheson was happy to accept it.[42]

An interesting sidelight on this "clarification" was provided by Ronald Reagan, then a member of the Screen Actors' Guild committee that attempted to mediate the controversy. According to Reagan, "It was a curious document. It came purportedly from the arbitration trio, mimeod and unsigned. It came from the HQ of the Hutcheson's carpenters' union. It repeated what had been said before—with the addition of some confusing phraseology which indicated there was a carpenter in the woodpile."[43]

Representatives of the Screen Actors' Guild, including Reagan, went to see William Green at the AFL convention in Chicago and told him that the Guild "was prepared to fly stars to every big city in the United States to make personal appearances and show films of the [picket line] violence outside the studio gates, and to tell the people that one man—the first vice-president of the AFL, Bill Hutcheson—was responsible."[44] When Green said he was powerless to act, the group went to see Hutcheson, who refused to budge. The Guild did not make good on its publicity threat, but thereafter it refused to honor the picket lines maintained by the Conference of Studio Unions, strengthening the IATSE position.

IATSE, of course, refused to accept the "clarification," and threatened to shut down the studios if the producers assigned the disputed work to the Carpenters. The producers decided to go along with the original award despite a Carpenter threat that the "sets will be declared hot and we won't work on them." On September 12 all carpenters who refused to work under the terms of the nonclarified award were laid off, and they were followed by fellow craftsmen. Replacements were provided by IATSE, leading once again to mass picketing and violence. With the actors willing to work, the studios continued to operate.

The 1946 AFL convention called for the establishment of some mechanism for solving the Hollywood crisis. In a letter to Green, Hutcheson indicated that he would have no objection to an AFL survey, but "we cannot agree to anything as to what the result will be for the simple reason that at the present time we have no members of our International organizations employed in the major studios." He followed this up by asking the executive council to recommend to the next convention the revocation of IATSE's charter, and in the meantime to publicize the "clarification," and this was agreed upon.[45]

Nothing much happened until September 1947, when a meeting was held at the request of Congressman Carroll D. Kearns, chairman of a House of Representatives' subcommittee that had been looking into the situation, but this was also unproductive. Hutcheson then went back to the council and "expressed the opinion that unless there is something done by this Council the [Carpenters'] Board will consider handing in their resignation to the A.F. of L."[46] The council agreed only to report to the next convention what had been done to try to solve the controversy, and to ask for future guidance. This motion was carried over two negative votes, those of Hutcheson and John L. Lewis.[47] The Carpenters tried to get a ruling in their favor at the 1947 convention, but IATSE managed to prevent it. The strike gradually collapsed, though picketing went on until the end of 1948. IATSE won the NLRB election in 1949: "This was the end of the CSU [Conference of Studio Unions] and of many of the studio locals affiliated with it, including the Carpenters, Painters, and Machinists."[48]

This was not a good time for waging a fight to the finish on jurisdiction. The Taft-Hartley Act had just been enacted over a presidential veto, and jurisdictional strikes were made subject to government adjudication. The traditional Brotherhood tactic of perseverance, accompanied by job action where necessary, no longer worked.

SOME UNION PREOCCUPATIONS

There was still an occasional skirmish on the communist issue. A member of Local 563 of Glendale, California, was removed from the post of business agent by the general president. He appealed his removal to the GEB, which ruled that the 1946 convention had directed the removal from office of anyone who "is known or recognized as a Communist or who has ever been registered as such on a political register," and upheld the action.[49] Delegates from Local 634, Los Angeles, which had its charter suspended, appeared before the board to ask for its restoration. The board had set a number of conditions, including the filing of noncommunist affidavits by all members, which the local had refused to do. In answer to the petition, the board decided to grant clearance cards to some of the members so they could work in other areas, and fixed a deadline for the rest, but refused to restore the charter.[50]

On July 1, 1948, Frank Duffy resigned the post of general secretary, which he had held for forty-seven years. He was designated Secretary Emeritus, and was to continue receiving his salary. He had attended every Carpenter convention since 1896 and every AFL convention since 1902, which must be a record unparalleled in the labor movement. His impact on the Carpenters' Union was rivaled only by that of McGuire and William Hutcheson. Duffy was an able administrator and a tireless writer and correspondent; not among his meanest achievements was *The Carpenter,* which was published monthly without interruption under his editorship. His successor as general secretary was Albert E. Fischer, a member of Local 1602 of Cincinnati, who for some years had been Duffy's assistant.

While the Machinists' jurisdictional controversy was temporarily in abeyance, the Carpenters were involved, during the latter half of the 1940s, in a dispute with the Upholsterers' Union. The Brotherhood complained that the Upholsterers were chartering locals of furniture workers and cabinet makers, and prevailed upon the AFL executive council to issue a cease and desist order even before the Upholsterers had a hearing. The Upholsterers replied, with some indignation, that workers involved were former members of the CIO United Furniture Workers who switched to them after an internal rebellion by a number of locals in protest against communist domination of the CIO union. The Upholsterers had financed the dissident group to the tune of $150,000, and doubted that they would have joined any other union. The Brotherhood insisted that the men in question nonetheless be transferred to its jurisdiction. The Upholsterers asked that negotiations take place, but Hutcheson was unwilling to meet with them until they had complied with the council's order.[51]

In a letter to William Green, President Hoffman of the Upholsterers repeated a litany that had been uttered often before by unions in conflict with the Carpenters:

They simply take the position that, having exacted a favorable decision from the Executive Council without first complying with the traditional rules and time-honored procedures of the AF of L, they now have nothing to talk about . . . A question arises in the minds of our General Executive Board—do the rules and traditional procedures of the A.F. of L. dealing with matters of jurisdictional controversy between affiliated unions apply equally to all affiliated unions? I know that such a thought is furthest from your mind. Nevertheless the facts of this case have raised this question in the minds of our General Executive Board. If it ever should appear that there is one law for big unions and another for small, the patient pioneering of Sam Gompers for forty years would be undone in an instant.[52]

The Upholsterers learned something that both Gompers and Green well knew—that the AFL could not provide equal protection to all its affiliates. Large unions were much more likely to prevail in jurisdictional matters than small ones.

THE TAFT-HARTLEY ACT

The growth of anti-labor sentiment in the United States consequent upon the wave of postwar strikes that engulfed the country led to strong congressional sentiment for the enactment of legislation restricting union activities. A year before Congress finally acted, Hutcheson expressed the attitude of the Carpenters to such legislation in no uncertain terms:

The United Brotherhood of Carpenters and Joiners of America is a firm believer in free enterprise and free unions . . . The half-dozen anti-union measures now before Congress will be opposed to the last ditch by the United Brotherhood. The right to strike, the right to work or not work alongside a non-union man, the right to use or refuse to use non-union materials are all sacred rights of labor secured after much struggle and sacrifice.[53]

These sentiments were echoed by the GEB, which reaffirmed its commitment to free enterprise and the benefits it had brought to the Carpenters, but added: "We must again point out that a free labor movement must always be an integral part of the free enterprise system. Freedom is an ephemeral thing; all segments of our society must be free or eventually none will be free."[54]

The Taft-Hartley Act was vetoed by President Truman on June 20, 1947. The House of Representatives overrode his veto by 331 to 83, and the Senate by 68 to 25, far more than the necessary two-thirds margins. The reaction of the Carpenters was similar to that of the rest of the labor movement. The law was castigated as being "vicious, un-American and anachronistic," one that "set back industrial relations fifty years." It was characterized as opening the door "to all the old union smashing devices of the turn of the century: the open shop, the injunction, the company union, the blacklist."[55]

A twenty-point statement of policy on how the Brotherhood would operate under the law was sent to all locals. It was decided, among other things, that the national union was prepared to fight against any attacks on the union label, on the traditional refusal to work with non-union men or on nonunion material, and on jurisdictional boundaries. There was particular outrage against the required noncommunist affidavits:

> It is no secret that organizations like the Brotherhood of Carpenters have done more to combat Stalinism in this country than any other group, not excluding the FBI. Bill Hutcheson has probably thrown more sand into the wheels of U.S. Communism than any other one individual in the nation. To require a man like him to sign an affidavit stating that he is not a Communist is a little bit like requiring the President of the United States to swear he is not unemployed . . . Having no practical experience with the ways of Communists, Congressmen think in terms of legislative restraints and legislative wrist-slappings; the unions, on the other hand, having had to contend for years with the disruption, deceit and character-assassination that the Communists use as their stock in trade, think in hard, realistic terms. They know that the way to deal with Reds is not with the velvet glove but with the brass knuckle.[56]

As a direct consequence of the enactment of the law, the AFL established Labor's League for Political Education for the purpose of increasing labor participation in national and local election campaigns. A year earlier, this might have been opposed by the Carpenters as an abandonment of the traditional nonpartisan policy. But times had changed; the members were told that "your Brotherhood is going to help draw up the program and once an honest workable program is arrived at, we are going to support it to the fullest extent of our ability."[57] The Carpenters established their own Non-Partisan Committee for the Repeal and Defeat of Anti-Labor Legislation, to be financed by voluntary contributions, and all locals were urged to participate. Within a few months, 420 district councils and locals had set up corresponding local committees.[58] Neither the AFL nor the Carpenters specifically endorsed Harry Truman, but there is little doubt that the labor movement made a major contribution to his successful bid for election. National labor organizations spent almost $1.3 million in the campaign; the national office of the Brotherhood supplied $100,000 of this amount, and the local unions used the funds they collected for precinct work. Maurice Hutcheson urged members to register, and it was not necessary to tell them for whom to vote.[59] The results of the election were hailed as a victory for labor: 17 Senators and 106 House members who had voted for Taft-Hartley were defeated.

The conservative, anti-government philosophy of the Brotherhood's leadership was shaken by the events surrounding the promulgation of

Taft-Hartley as nothing before had done. The threats to unencumbered freedom to pursue jurisdictional goals, to boycott nonunion materials, to unrestricted picketing, hit at some of the principal weapons in the Carpenters' arsenal. Nevertheless, as the decade drew to a close, optimism pervaded the ranks of the Brotherhood of Carpenters. Membership was up. Between 1945 and 1950, despite a considerable price inflation, the real wages of union carpenters rose by 15 percent. The union had come through ten years of world upheaval with flying colors. It had lost a few battles, notably the one in Hollywood, but had won more. There were some clouds on the horizon, especially the potential impact of the new labor law. But the outlook for future progress was as favorable as it had ever been at the end of any previous decade, particularly the two preceding ones.

13

ADJUSTING TO A CHANGING
ENVIRONMENT: THE NINETEEN-FIFTIES

The decade between 1950 and 1960 was the first "normal" one for the United States since the 1920s, if one excepts the war that began when Communist North Korea launched its attack on South Korea in June 1950. The impact of the war on the American economy was not severe, and once it ended, things resumed a course that in retrospect seems remarkably stable. For example, although there was a burst of inflation—11 percent—between 1950 and 1953, the consumer price index advanced by only another 11 percent for the entire seven-year period 1953 to 1960.

Low unemployment provided trade unions with a favorable basis for expansion. Total union membership rose from between 14 and 16 million in 1950 to more than 18 million in 1960. The percentage of the labor force organized reached the all-time high of 25.2 in 1956, a figure that has never been exceeded. This was by no means labor's golden age, however. Were it not for the Taft-Hartley Act, even greater progress could have been made. Moreover, toward the end of the decade the internal government of trade unions was subject to the most searching inquiry in American history, and the revelations of practices followed by several labor organizations set the stage for legislated reform.

The less hospitable legal climate in which unions had to operate was not the only factor that led to a loss of momentum. Perhaps even more basic was the relative decline in blue-coller employment. The number of workers on manufacturing payrolls rose from 15.2 million in 1950 to 16.7 million in 1960. However, while manufacturing provided employment for 33.7 percent of all employees on nonagricultural payrolls in 1950, the figure was down to 30.8 percent by 1960. The rise of the service sector, which had always been more difficult to organize than manufacturing, contributed to the relative decline in union strength. The construction labor force held up better than manufacturing—an increase of 23.7 percent compared with 9.5 percent for manufacturing over the decade—but even that figure was exceeded by the 26.5 percent growth in the service sector labor force.

THE TWENTY-SIXTH GENERAL CONVENTION

The opening of the 1950 convention, at a time when the Korean War was in full swing, provided Hutcheson with the appropriate atmosphere to make what turned out to be his valedictory address to the assembled Carpenters. He blasted the few communists who might still be hidden within the Brotherhood ranks, and remarked that "we in the labor movement learned several decades ago that you could not depend upon or even take the word or pledge of a communist." He then went on in his peroration to provide a concise and accurate summary of the basic ideology of the Carpenters' Union:

> In all the years since the first general convention, the United Brotherhood has never been anything but a 100 percent American organization. The Brotherhood has never subscribed to the European concept of unionism which holds that capital and labor are mortal enemies engaged in a life or death struggle until one or the other perishes. It never begrudged an employer an honest profit. It always turned a deaf ear to the wildeyed theorists and followers of misty Utopias. Last but not least it has been one of the most outstanding foes of communism and all it stands for. Circumstances may alter, conditions may change, but the United Brotherhood will never be anything but an American institution devoted to the ideals laid down in the constitution and broadened and polished by succeeding generations. We have achieved understanding with our employers, understanding that was mature before some of the men who are now sitting in Congress trying to govern labor relations by laws and compulsion were even born.[1]

During the four years from December 31, 1945, to the same date in 1949, total membership showed a decline from 722,392 to 710,034. In fact, the membership picture was really better at the latter date. Some 26 percent of the members were in arrears in 1945 compared with only 7 percent in 1949, which meant a greatly improved financial situation. The proportion of membership provided by Canadian locals more than doubled, from 3 percent in 1945 to 8 percent in 1949, a change that was to provide fuel for Canadian autonomy demands. Another statistic is worthy of note: during the four years more than 8,000 apprenticeship certificates had been issued, far more than at any time in the past.

Special tabulations presented to the convention provide us with some interesting cross sections of the Brotherhood membership. Table 13.1 shows that Brotherhood locals were still relatively small, on the whole; almost half the members were in locals with less than 500 members, and 70 percent in locals with less than 1,000 members. The geographical distribution contained in Table 13.2 shows how much of a swing there had been to the western part of the country; the three states of California, Oregon, and Washington accounted for more than a quarter of the

Table 13.1. Membership in the United Brotherhood of Carpenters by size of local union, December 31, 1949

Members in locals	Percentage of total membership
10–49	2.9
50–99	6.2
100–199	11.7
200–299	9.4
300–399	8.9
400–499	7.2
500–599	10.7
600–999	13.3
1,000–1,499	13.9
1,500–1,999	6.2
2,000 and over	9.6
Total	100.0

Source: United Brotherhood of Carpenters and Joiners, Proceedings, Twenty-Sixth General Convention, 1950, p.114.

total membership, while the share of the previously dominant eastern locals was down. Part of this change can be attributed to the growth of membership in the lumber industry, but by no means all. A good many of the lumber locals were still in semibeneficial status; the three western states included 60 percent of all the semibeneficial members. Some had converted to full benefit status, however.

Pensions, which were to become a major issue as the years went on, came in for considerable convention discussion. The current pension of $15 a month was being paid to 14,600 people, which exceeded the income from dues and initiation fees allocated to this purpose by more than $30,000 a month. The convention defeated a motion to increase the pension to $60 a month but did agree to raise it to $30, to be covered by raising dues and initiation fees. However, the usual pattern was repeated; the membership turned down the dues increase in the referendum vote, and the pension remained at $15 for lack of funds. The Carpenters wanted higher pensions, but continued to refuse to pay for them. It should be made clear that from the start these Brotherhood pensions were never properly funded, and were paid out of current dues. They are not to be confused with the collectively bargained pension funds financed by employers that began to spread about a decade later.

A delegate from Local 36 of Oakland, California, introduced a resolution that would have recommended the inclusion of the following clauses

Table 13.2.   Membership in the United Brotherhood of Carpenters, by state, December 31, 1949

	Percentage of total membership	
State	Beneficial	Semibeneficial
California	16.3	34.6
New York	8.5	4.4
Illinois	7.3	3.4
Washington	6.2	12.7
Oregon	4.8	12.6
Pennsylvania	4.7	a
Ohio	4.6	a
Texas	4.4	a
Michigan	3.5	a
Others	39.7	32.3
Total	100.0	100.0

Source: United Brotherhood of Carpenters and Joiners, Proceedings, Twenty-Sixth General Convention, 1950, pp. 114–115.
a. Below 1 percent.

in all future agreements: "(1) That one out of every five men hired shall be fifty years of age or over when available; (2) That no member of the Brotherhood be denied employment due to race, color, or creed." It was defeated when Hutcheson opposed it on the ground that these were matters for local decision and should not be mandated by the national union.[2]

A heightened interest in political action was indicated by the adoption of a resolution suggesting that locals establish committees to raise political funds and registration committees to check registration rosters and make certain that members could vote.[3] Taft-Hartley was very much a factor in bringing about this action.

The only trace of internal dissension came in the form of challenges to the credentials of two California delegates for assisting communist organizations. One of them was seated without voice or vote; the GEB later found that he had attended meetings of an organization classified as subversive by the government, and "admitted his mind is not clear on the justice of our government's position in the present Korean situation." He was required to file a noncommunist affidavit, debarred from holding local office and voting at meetings, and put on probation for five years. The second man denied any association with the Communist Party, but he was nevertheless required to file an affidavit that he would not participate in any future communist activities.[4]

If any further evidence of its opposition to communism was needed, the Brotherhood asked the government to furnish it, on request, with the names of disloyal individuals among the membership. The GEB itself introduced a resolution pledging the Brotherhood to uncompromising opposition to communism and promised support of the Korean War effort. It called upon Congress "to protect our people against the sinister enemies, within and beyond our gates, who are plotting against our liberties, our Constitution and our lives."[5] Needless to say, the resolution was adopted unanimously. The Brotherhood was swept up in the prevailing anti-communist mood of the country, which was at its height at the time.

A resolution urging the AFL to lobby for socialized medicine, which had the support of the resolutions committee, was defeated when Hutcheson spoke against it. A few months later he amplified the arguments he had made at the convention in a speech to the House of Delegates of the American Medical Association, saying in part:

Socialization is like a wolf with a tapeworm; once it starts growing, it never can stop. Socialized medicine would only be the first bite out of our free enterprise system; it would not be many years before the carpenters would be feeling the teeth of socialization on the seats of their pants. I salute you today not only as doctors but as crusading citizens as well. We in the labor movement have our own cross of regimentation to bear [Taft-Hartley]. The fight you are making is part of the same war. It is a war against concentration of authority in a few hands in Washington.[6]

All the general officers and GEB members were reelected without opposition. Hutcheson's retirement was provided for by setting up the office of General President Emeritus, which he could assume with full membership on the GEB and at full salary.

JURISDICTION (6)

The Machinists Union, which had withdrawn from the AFL in 1944, decided to reaffiliate in 1950. When its application was received by the AFL executive council, Hutcheson complained that the union had been taking its jurisdictional cases to the National Labor Relations Board. He stated that if the Machinists were readmitted without a commitment to withdraw NLRB charges, the Carpenters would leave the AFL. The Machinists were told by letter that this was a condition of their reaffiliation..[7]

Nevertheless, after the Machinists had come back, they continued to file NLRB complaints against the Carpenters. Their excuse was that while it was against the policy of the national union to do so, the actions of locals could not be controlled. The AFL executive council advised the Machinists to adhere to the understanding that they would not go to the NLRB, provoking the reply that since the council seemed powerless to bring about compliance with jurisdictional awards, there was no alternative. President Hayes of the Machinists agreed, however, to continue negotiations with the Carpenters.[8]

Maurice Hutcheson, who had become general president in 1952, wrote Green that the Machinists filed seventeen NLRB complaints against the Carpenters, and that there could be no discussions until they were withdrawn. Green was directed by the council to inform the Machinists of its disappointment with their actions, and he received the following reply: "The International Association of Machinists does not intend to permit any government agencies to decide any of its jurisdictional difficulties. We are, however, willing to make every effort to settle these differences with other A.F. of L. unions. It is our intention to always protect the jobs and job opportunities of every member of this organization."[9]

It took another year of sparring before the two organizations could be brought to the negotiating table. In September 1953 it was agreed that each union would appoint a three-man committee to seek a solution to the controversy. To prevent negotiations from breaking down, outside consultants were brought into the bargaining sessions: Professor John T. Dunlop for the Carpenters and Father William Kelly for the Machinists. After twenty-six meetings stretching over another year, an agreement was finally reached on September 18, 1954. It set forth in considerable detail the division of jurisdiction between the two unions on both existing and new construction. It was also agreed that the Carpenters would not interfere in factories where the Machinists were certified or recognized as the bargaining agent, and vice versa. Stoppages of work were banned, and a procedure established for the resolution of future local disputes.[10]

There was general satisfaction with the resolution of what *The Carpenter* called, with justification, "the oldest and thorniest jurisdictional dispute in modern labor history."[11] But the rejoicing proved to be somewhat premature. In 1966 the Carpenters abrogated the agreement on the ground that new technology had rendered some of its provisions impossible to administer.[12] However, both sides appear to be observing those portions of the agreement that are still applicable, and the open hostilities of the past have not been resumed.

With respect to the building trades, a National Joint Board for the Settlement of Jurisdictional Disputes was established on May 1, 1948. The Building Trades Department, the Associated General Contractors, and a number of specialty contractor associations were the signatories to the agreement, which covered nineteen unions and seven specialty contractor groups.[13] After a year of operation the unions decided to return to the previous plan under which there was no employer participation, but when it became clear that the NLRB might not recognize decisions made by the unions alone, the Joint Board was reestablished.[14]

The Carpenters were not overly happy about participating in this scheme, but they had little choice. The basic reason was set forth in the report of their delegates to the 1951 Building Trades Department convention:

> No matter what the criticism of the [National Joint Board] has been or may be, however, it has justified its existence on one score alone. If it were not operating, the cases which have come before it would have been taken to the National Labor Relations Board and the unions involved in those cases would have been involved in long and expensive litigation. As it is, the N.L.R.B. has continued to refuse to process cases involving any building trades unions.[15]

The chairman of the Joint Board from its inception until 1957 was Professor John T. Dunlop of Harvard University. In an address to the 1954 convention of the Brotherhood, he pointed out that jurisdictional disputes were inevitable in construction because of certain structural features of the industry: subcontracting, technological change, the democratic process within unions, and the wide diversity of local conditions. He warned, however, that work stoppages over jurisdiction had been made illegal by the Taft-Hartley Act and were likely to remain so for the indefinite future. "Either this industry solves its jurisdictional problems within its own family or the government will. There is no other choice." He stressed that employers, because of their power to assign work, had to be involved equally with the unions in the adjustment process.[16]

When Dunlop resigned the chairmanship in 1957 after more than nine years of service, the Carpenters made it clear that they were by no means unhappy with the way the board had operated under his guidance:

> Filling the shoes of Mr. Dunlop is no easy matter. Throughout the early years of the Joint Board, when the problems were numerous and no backlog of experience existed, Mr. Dunlop performed yeoman service in keeping the Board functioning effectively. But for his patience, foresightedness, and determination, the Board might well have died aborning. Every move the Board made was plowing new furrows in virgin soil, and much of the credit for the solid growth of the Board in stature and recognition must go to the tireless efforts put forth by Mr. Dunlop.[17]

The Joint Board had three general functions in its task of making work assignments: to prevent work stoppages, to make decisions on individual jobs, and to promote and mediate national agreements. Contractors normally made initial work assignments, and objecting unions were obliged to file grievances with the board through their national presidents. The board processed a considerable volume of cases, and although it did not succeed in eliminating jurisdictional strikes, it was able to settle most of them promptly.

The Carpenters were involved in a substantial number of cases. In 1957, for example, it was party to 389 cases considered by the board, of which it was the initiating party in 198. Decisions favorable to it were rendered in 184 cases.[18] The Carpenter cases constituted 74 percent of all board decisions; the majority of the disputes were with the Sheet Metal Workers over the installation of metal cabinets, bookcases, and lockers, and with the Lathers over acoustical work.[19]

A number of national agreements were reached either through direct negotiation or through the intermediation of the Joint Board. There was a very detailed agreement with the Ironworkers dealing with the installation of monorails and package conveyers; with the International Brotherhood of Electrical Workers on luminous and acoustical ceilings; and with the Hod Carriers on the work of stripping panel forms. On the other hand, the Lathers Union was notified that the Carpenters no longer recognized an agreement between the two organizations that dated back to 1903 and had been renewed in 1950, initiating a new series of conferences.

As the foregoing data indicate, Carpenter jurisdictional disputes with other building crafts were by no means a thing of the past. They consumed a great deal of the time and energy of the national officers. The forum for their settlement had shifted, however, from the AFL executive council and the Building Trades Department to the Joint Board. Success was more a matter of law and precedent than of power, and the time when the Carpenters could expect to win them all had come to an end, as Maurice Hutcheson made clear to the 1954 convention:

> The adjustment of disputes over jurisdiction is one of the most difficult tasks facing your General Officers. In negotiating settlements of jurisdictional disputes it is necessary to reach some compromises. It is impossible to get everything we feel we should have. If we were able to settle on that basis there would be no need for negotiations. Our General Executive Board, in reaching their conclusions do so only after extensive surveys and agree only where agreement will benefit the majority of members.[20]

PRESIDENTIAL AND OTHER POLITICS

Although the Carpenters had earned their anti-communist credentials, they had no use for the strident know-nothing line that was associated with the name of Senator Joseph McCarthy. In an unsigned article entitled "Threat from Within," *The Carpenter* took sharp issue with the campaign being waged by the Senator from Wisconsin:

Of the cult that sees a potential enemy in everyone who disagrees with its views, Senator McCarthy is the high priest. By innuendo, finger-pointing and often without any real evidence, he has besmirched the good name of sincere citizens. Instead of using a rapier to pinpoint suspicious characters, he uses a blunderbuss loaded with buckshot. He aims it in a general direction and lets her fly. How many innocent people get hit seems immaterial. By their own standards, McCarthy and [Westbrook] Pegler are the only 100% Americans left in the nation.[21]

On another subject, an article prophetically condemned businessmen for their willingness to trade with the Soviet Union "to make an extra franc, pound, or dollar":

For the sake of a few immediate profits they are ready to accept the enslavement of millions of men and women living behind the Iron Curtain . . . Some of the industrialists of the West have long ago lost even the will to power which distinguishes creative from decadent capitalism. They seem to be determined to fulfill Marx's prophecy that capitalism digs its own grave. The Communist dictators could not ask for more.[22]

When the time came to decide on a policy for the 1952 presidential elections, the Carpenters, despite the continuing trauma of the Republican-sponsored Taft-Hartley Act, reverted to their traditional non-partisanship, although Maurice Hutcheson together with three other national union leaders personally supported Eisenhower. In a lengthy letter to Green, the GEB urged the AFL not to endorse any candidate, calling attention to the fact that support of one party could harm friends in the other.[23] However, the AFL, for the second time in its history, did abandon nonpartisanship; it endorsed Adlai Stevenson over Dwight D. Eisenhower.

In a comment on the Eisenhower victory, Maurice Hutcheson pointed out that many union workers had voted Republican (a poll taken in May 1952 indicated that 37 percent of manual workers intended to vote for Eisenhower against 54 percent for Stevenson). It was his view that the average skilled worker was earning as much or more than most white-collar employees, and was perfectly capable of making political decisions for himself. "The labor movement must never allow itself to be jockeyed into a position where it is hanging on the coat tails of any political party."[24] In 1952 the Carpenters were in a small minority within the AFL in opposing abandonment of nonpartisanship. They were later to pick up more support for the idea that despite changing circumstances, Samuel Gompers may have been right.

President Eisenhower was not unpopular with the Carpenters. On the occasion of a seventy-fifth anniversary dinner on October 23, 1956, he, together with three members of his cabinet, not only attended but remained for the entire evening. This political gesture was not necessary to keep the Carpenters neutral in the 1956 campaign, although once again the now united AFL-CIO endorsed Stevenson and spent a substantial amount of money on his behalf. A poll taken a few weeks before the election showed that 45 percent of manual workers and 39 percent of union members intended to vote for Eisenhower, despite the advice of the majority of their union leaders.

Non-partisanship did not imply that the Brotherhood was inactive politically. Its 1958 convention resolved unanimously that in order to protect labor's gains against anti-union forces, it was necessary to educate the membership to the need for electing friendly members of Congress and state legislatures. The GEB was requested to assign responsible individuals to every geographical district for the purpose of establishing political education committees at each echelon of the union structure.[25] Substantial donations were made to state labor federations in Idaho, Washington, and Kansas to help fight right-to-work laws.[26] However, when the AFL-CIO Committee on Political Education asked for a contribution, the GEB "reaffirmed its previous action not to participate in COPE activities, but to permit and encourage our local unions to participate if they wish to do so and we will assist them in every way possible."[27]

RELATIONS WITH THE AFL

William L. Hutcheson resigned the presidency of the Brotherhood effective January 1, 1952. He was succeeded by his son, Maurice A. Hutcheson. Prior to becoming first vice-president in 1938, the younger Hutcheson had been a journeyman carpenter for ten years, and in 1928 was appointed a general representative. He had been identified with the Brotherhood in one capacity or another since the age of 17 years, and had thirty-seven years of continuous membership on the date he assumed the post of general president. He was also elected to the position of fourteenth vice-president of the AFL.

On June 2, 1953, a committee set up by the then separate AFL and CIO to explore the possibility of unity between the two bodies reached a no-raiding agreement as a step toward unity. However, it was binding only on the affiliated national unions that signed it. William Hutcheson, who was still on the AFL executive council, although he had resigned the Carpenters' presidency, objected to the pact and left the council in protest. The Carpenters were not alone in refusing to sign; a year after the negotiation of the pact, only 65 out of 110 AFL unions had ratified the agreement.[28]

The Carpenters were not content with merely abstaining from the pact. On August 12, 1953, Maurice Hutcheson addressed a letter to George

Meany, who had succeeded Green as president of the AFL, which read in part:

> Having been advised that the Executive Council of the American Federation of Labor adopted a proposition of "no-raiding" between the A.F. of L. and the C.I.O. without giving any consideration to the disruptive conditions within the American Federation of Labor itself and as the record will show that for many years past the Building Trades Department, and its affiliates, have submitted to the Executive Council numerous complaints regarding disputes, and either through inability or disregard for the existing conditions the Executive Council of the American Federation of Labor has taken no action to remedy same; from the action just taken it indicates that they are more concerned with the affairs of the C.I.O. than they are with those of the Federation. We have no objection to "no-raiding" agreements between all organizations, in and out of the American Federation of Labor; however, if the American Federation of Labor is not able to control its affiliates, our organization being no exception, we fail to see where there is any benefit to the United Brotherhood of Carpenters and Joiners to continue paying per capita tax to the American Federation of Labor. Therefore, I am hereby notifying you of our withdrawal as of this date.[29]

In support of this action, the GEB issued a statement pointing out that some AFL unions were resorting to NLRB procedures against other AFL affiliates, including a great many against the building trades. Jurisdictional lines seemed to be breaking down: coal miners were seeking to represent shirtwaist workers; longshoremen, candy makers; steelworkers, bricklayers; machinists, electricians. The Building Trades Department had proposed the establishment of an umpire system within the AFL, but no action was taken. "Boiled down, it is a race to see who gets there first and obtains N.L.R.B. certification. Such a situation, in our opinion, amounts to outright surrender to a Government Board of the basic and inherent authority and power vested in the American Federation of Labor."[30] An editorial in *The Carpenter* concluded that "in the long run, any other course [than withdrawal] might have been unfair to the Federation . . . for a continuation of the Federation's present chaotic, dog-eat-dog attitude, insofar as jurisdiction is concerned, can lead to nothing except frustration, failure, and eventual disintegration."[31]

In the past, the AFL had responded to similar Carpenter threats by taking immediate conciliatory action, but the new president, himself a building tradesman by origin, counterattacked. To fill the post vacated by William Hutcheson, all AFL vice-presidents were advanced one step in rank. Despite the fact that the withdrawal of the Brotherhood would have cost the AFL $300,000 a year in dues, George Meany announced that in accord with standard AFL procedure, the Carpenters would be barred from all AFL bodies. He declared: "The Federation is not seriously hurt by the withdrawal. It is serious that a large organization has walked out and it's something we don't like to see happen, but when it does we have to accept it and go along as best we can."[32]

The walkout did not last long. After several meetings the Carpenters agreed to maintain their AFL membership, while the Federation issued a statement deploring jurisdictional conflicts and agreeing to set up a special committee with the mandate of working out a scheme to alleviate the conflicts. A commitment was made to expand the size of the executive council so that Maurice Hutcheson could be made a member at the next convention.[33] The AFL eventually adopted a no-raiding plan for its affiliated unions, and the Carpenters could claim some credit for bringing it about. In its report to the 1954 convention, the GEB claimed that the Brotherhood "has not at any time during its seventy-three years of existence been unaffiliated with the American Federation of Labor, and at no time during its existence has it ever been in arrears to that organization for per capita tax. It has been in continuous affiliation with the American Federation of Labor."[34] But this was a close call.

The Brotherhood did not display any great enthusiasm for the merger between the AFL and the CIO that took place in 1955. It favored unity, but with the proviso that it would not be bound to any program "which does not permit it to maintain its jurisdictional integrity and protect the advances it built up so painfully over the decades."[35]

A few years later the Carpenters had another run-in with the AFL leadership, although this time there was no threat of withdrawal. The issue was the expulsion of the International Brotherhood of Teamsters from the AFL-CIO in 1957 for alleged corruption. The Carpenters, together with the Laborers and the Operating Engineers, had entered into a pact with the Teamsters in 1955 to help organize the heavy and highway construction field, and were reluctant to break their ties with this powerful ally. With twenty other unions, the Carpenters voted against the expulsion at the 1957 AFL convention, but they were in a minority. The Brotherhood delegates took the position that expulsion was a poor mechanism for instituting reform, and challenged the legality of the expulsion process. "In effect, by appropriating power to pass judgment and levy punishment, the AFL-CIO assumed something approaching police powers, a right that was never vested in the organization by any membership vote. Regulating the morals of union members is a far cry from the purposes for which unions were created — namely to improve wages and working conditions."[36]

The question of continued cooperation with the Teamsters arose after the latter had become independent. The three building trades signatory to the so-called Four-Way Pact, including the Carpenters, decided to close down the Washington office that had been set up to administer the agreement, but the field staff continued in operation.[37]

Relationships with the AFL-CIO were not improved when the AFL-CIO executive council asked Hutcheson to explain charges of corruption that had been made against him, as well as his refusal to answer some questions put to him by the McClellan Committee.[38] A letter from Meany to Hutcheson stated:

The Executive Council of the AFL-CIO, meeting today, was gravely concerned at the public record of the McClellan Committee which records your refusal to answer questions relating to allegations of misuse of union funds. As you know, from your participation in determination of the Executive Council on such matters, such an attitude is incompatible with holding office in the American Federation of Labor and Congress of Industrial Organizations. The Executive Council, therefore, directs that at its next meeting you appear and give appropriate explanations of your refusal to answer these questions.[39]

In response to this demand, a resolution introduced at the 1958 Carpenters' convention empowered the GEB to withdraw the Brotherhood from the AFL-CIO to protect its interests in view of "certain courses of action and policies by the AFL-CIO and statements by leaders therof which, in the judgment of this convention, threaten and jeopardize the best interests and welfare of this United Brotherhood of Carpenters and Joiners of America and its members, and purport to discredit and impugn its leadership and to undermine and disrupt the effectiveness and solidarity of our Brotherhood and its organization."[40] The resolution was adopted with only two dissenting votes, but since the AFL-CIO took no further action against Hutcheson, this authority was never used.

After the AFL-CIO merger, the Carpenters, like most other building trades unions, affiliated with the newly established Industrial Union Department for those of its members who worked in factories. This affiliation did not last long; in 1957 the Brotherhood's per capita tax was withheld since the Carpenters "were receiving no benefit from the affiliation."[41] The cause of the rupture was a controversy between the construction and industrial unions over factory maintenance work. The building unions complained that several industrial unions had been putting pressure on their employers to stop contracting for construction work with employers of the building craftsmen, and instead to allot the work to inside men at the lower industrial wage level. This was an attractive proposition for industrial enterprises and for the inside men. The Industrial Union Department was actively promoting this scheme. In turn, the building trades unions and their contractors were urging the contracting out of the maintenance work.

The Building Trades Department condemned the action of the Industrial Union Department, and for a time there was some danger of full-scale warfare between the two wings of the labor movement. An agreement was eventually worked out between the building trades and most of the industrial unions, providing that the new construction should be within the jurisdiction of the building unions, while current maintenance work could be done by members of industrial unions. Joint teams were set up to adjust disputes, with the possibility of final reference to a special committee of the AFL-CIO.[42] To help industrial enterprises resist the temptation to use cheaper production worker labor, the building trades joined together to offer standard maintenance contracts to their contractors; there were occasional concessions on wage rates, and time and a half rather than double time was stipulated for overtime; payment for subsistence and travel time was dispensed with, and multiple shifts were permitted. All these concessions were designed to make the employment of building trades labor more competitive.[43]

THE TWENTY-SEVENTH GENERAL CONVENTION

The passing of the old order was marked by the death of William L. Hutcheson on October 20, 1953. The Brotherhood purchased a 65-acre tract of land in New Brunswick, New Jersey, to be maintained as a memorial forest in his honor. *The Carpenter,* in its obituary, emphasized his fight against the antitrust indictment in 1940. "The courage and determination of Brother Hutcheson thereby saved the labor movement from domination by bureaucrats."[44]

That the years 1950 to 1954 had been good ones for the Carpenters is indicated by membership figures. Total membership reached 823,574, a gain of 113,540 over the four-year period. Moreover, the proportion of members in arrears had fallen to 6 percent. Canadian membership did not rise as rapidly as that in the United States, declining from 8 percent to 6.7 percent of total union membership.

A resolution favoring a national health scheme was adopted without discussion, suggesting that rejection of such a resolution at the previous convention may have reflected the personal views of William L. Hutcheson rather than those of the Brotherhood membership. The convention approved a constitutional amendment giving local unions the right to require a member "who, by his actions or speech, arouses a suspicion of communistic leanings" to sign an affidavit that he was not a member of the Communist Party, refusal to sign constituting evidence of guilt. Along the same lines, resolutions opposing the admission of Communist China to the United Nations, urging the AFL-CIO to seek the expulsion of the Soviet Union from the International Labor Organization, and declaring that peaceful coexistence with the USSR was a snare and a delusion, were carried.[45]

A resolution proposed to delete from the constitution an old clause barring from membership anyone engaged in the sale of liquor. The resolutions committee made the following comment:

> The Committee considered that there are times when a member might want to go out on a Saturday night and tend bar somewhere, and we see no particular harm in that. On the other hand, you know as well as we all know that this particular clause is in our Constitution as a roadblock to the admission of certain undesirable elements into the Brotherhood. The Carpenters' Union is particularly free of gangsterism and racketeering, and we think that is one of the reasons that that particular element does not get into the Carpenter's Union.[46]

The resolution was defeated.

All officers were reelected without opposition. There was virtually no controversy at the convention, a very successful one for the new general president.

THE McCLELLAN COMMITTEE INVESTIGATION

A U.S. Senate committee set up to investigate improper activities in the labor-management field made some serious allegations against several national officers of the Brotherhood.[47] General President Maurice Hutcheson, Vice-President William Blaier, and Charles Johnson, a member of the general executive board, were among the targets of the committee's investigators.

The best way to approach this subject is by considering seriatim the charges levied against the officers of the United Brotherhood, and their replies. Congressional hearings, unfortunately, do not normally result in firm conclusions, unless they are followed by judicial proceedings. Since no criminal indictments were brought against any officer of the United Brotherhood as a result of the committee hearings, there was presumably no evidence of any unlawful activities. The only question is whether some of the practices could have been regarded as improper from an ethical standpoint, a question notoriously difficult to answer.

The Hutcheson Biography and Maxwell Raddock

In 1953 the general executive board had commissioned Maxwell Raddock to write and publish a biography of William L. Hutcheson for distribution at the 1954 Brotherhood convention. In 1956 Raddock was also commissioned to arrange and promote seventy-fifth anniversary regional celebrations for the Brotherhood. Raddock was the publisher of a newspaper called *The Trade Union Courier,* and evidence was introduced to the effect that in February 1958, the list of subscribers included several New York locals of the Carpenters' Union, which bought 5,000 copies of the *Courier,* and the Brotherhood itself, which took 5,500 copies.[48]

In 1950 William Green had notified all AFL central bodies that despite its repeated representations to the contrary, the *Courier* was in no way connected with the AFL, and the AFL executive council requested "that our affiliated unions withhold and discontinue giving aid or assistance to this publication."[49] Two years later George Meany addressed a similar letter to a list of organizations furnished by the Better Business Bureau, decrying the selling tactics of the *Courier* and adding: "For many years the American Federation of Labor has publicly and officially disavowed this type of activity by unscrupulous individuals. Let me emphasize, no businessman is doing the American Federation of Labor a favor by purchasing advertising space solicited in this manner."[50]

The 1952 AFL convention adopted a resolution condemning the *Courier,* and urging that the AFL, to protect the legitimate labor press from injury, ask all affiliated national unions and their locals to cancel any existing endorsements of the publication.[51] From this and other material it may be inferred that the *Courier* and its publisher were hardly in good repute with the trade union movement.

The original agreement between the Brotherhood and Raddock for the preparation of the Hutcheson biography called for 6,000 copies. As time went on the scope of the book widened, as did the prospective audience to which it was addressed. In the end, 87,100 books were produced at a total cost to the Brotherhood of $310,000, or about $3.50 a book.

Robert A. Christie, author of a history of the Carpenters' Union that had been published in 1956,[52] told the McClellan Committee that the Raddock volume had plagiarized between 5,000 and 6,000 of his words without acknowledgment, as well as substantial excerpts from other books. He called it "the worst history book I ever read . . . Because it is so pasted and glued together it is impossible for there to be any intellectual outline or substance to it."[53] Raddock defended the quality of the book, though he added: "If the researchers who were employed by me were careless or irresponsible in their work and supplied to me material taken virtually verbatim from other books without so indicating, that is highly regrettable. It should never have happened."[54]

Maurice Hutcheson, in a personal statement addressed to the Brotherhood members, said that the book had been authorized by the 1954 convention, and that the McClellan Committee had not provided any evidence that he or members of the general executive board had derived any personal profit from the undertaking. He observed that the Raddock book was longer than Christie's, and the cost per copy to the union, $3.50, was less than the $5.00 price for which the Christie book was selling.[55] William Blaier was also quizzed on the subject; when asked by

Robert Kennedy whether the union went to other organizations to find out what it would cost to produce the book, he replied: "Mr. Kennedy, no. We are not literary artists nor are we book reviewers; no sir. We had faith in Mr. Raddock, and I still have it, that he gave us a good product, and we have 80,000 books, and I think we have value received."[56]

In retrospect, it would appear that the general executive board made an error of judgment in entrusting the preparation of the Hutcheson biography to an individual whose operations had been repeatedly denounced by the AFL, and who had no professional qualifications as a biographer or historian. There is little doubt that it could have secured a better product by going elsewhere. There were no committee findings, however, that any Carpenter official benefited personally from the funds paid to Raddock for the book.

The Indiana Land Fraud Case

General Treasurer Frank Chapman had arranged for the purchase of several tracts of land in Indiana on behalf of himself, Maurice Hutcheson, and William Blaier, all of it with their own personal funds and without any union involvement. The State of Indiana subsequently purchased portions of this land for a highway right-of-way, resulting in a profit of about $75,000 to the participants. The profits were split five ways, with the assistant director of the state's right-of-way division receiving one-fifth, while a similar share went to the chairman of the State Highway Commission. At the time of the McClellan Committee hearings, Hutcheson, Blaier, and Chapman were under indictment in Indiana for bribing state officers.

The McClellan Committee attempted to demonstrate that Raddock had acted as an intermediary in trying to ward off the indictment. Raddock refused to answer any questions relating to the episode on Fifth Amendment grounds.[57] Hutcheson also refused to testify on this matter, but not on the basis of the Fifth Amendment: "On the advice of counsel, I refuse to answer the question on the ground that it relates solely to a personal matter, not pertinent to any activity which this committee is authorized to investigate, and also it relates or might be claimed to relate to or aid the prosecution in the case in which I am under indictment and thus be in denial of due process of law."[58] The charges of attempted "fixing" were repeated several years later by Robert F. Kennedy in his book on the McClellan Committee hearings,[59] but no proof of guilt was adduced.

In refusing to rely on the Fifth Amendment, a defense that the AFL-CIO had condemned as inconsistent with the responsibilities of trade union officials, Hutcheson left himself open to a charge of contempt of Congress. He was indicted for this offense, found guilty by a district court after waiving a jury trial, and sentenced to six months' imprisonment and fined $500. The case was appealed to the U.S. Supreme Court, which affirmed the conviction on May 14, 1962, by a vote of 4 to 2.[60] However, Hutcheson was placed on six months' probation after petition to the district court, and did not actually serve any portion of the sentence. An affidavit filed by George Meany attesting to his character contributed to the mitigation.

As to the principal case against the three officers, they were found guilty of bribery by a jury in 1960. The general executive board of the Brotherhood refused to believe in their guilt despite the verdict, being of the opinion that the matter was the consequence of strong anti-union attitudes in Indiana. Affirming its faith in the officers, the board concluded: "We unquestionably share the feeling of the defense attorneys that these brothers will be completely vindicated when the record of this case is considered in the calm judicial atmosphere of the Indiana Supreme Court where the hysteria fanned and fed by anti-union papers can exert no influence."[61]

Their confidence was not misplaced. In 1963 the Supreme Court of Indiana reversed the conviction of Hutcheson (the codefendants, Chapman and Blaier, had died prior to the reversal). In a unanimous opinion, the Court declared:

> The trouble with the prosecution's case is . . . that while the evidence as to whether Doggett [the Indiana highway official] was bribed is weak, it is weaker still and entirely insufficient on the point of whether the appellant agreed or conspired to bribe the said Doggett . . . The record is simply devoid of facts and circumstances from which inferences can properly be drawn that appellant entered into a conspiracy to bribe Doggett.[62]

Charles Johnson

The only damaging testimony adduced by the McClellan Committee against Brotherhood officers involved Charles Johnson, a member of the general executive board from 1945 to 1969, and head of the New York District Council of Carpenters. During the three-year period 1955 through 1957, Johnson received $224,600 in salaries and expenses from various Brotherhood units. Two brothers and a son were paid salaries and expenses of $223,500 as officials of New York City locals during the same period. Johnson also received commissions of $96,500 for petroleum products sold by an oil dealer to construction firms with which the Carpenters had collective agreements. "From the evidence submitted, the Committee can come to no other conclusion than that the business transactions between Walsh Construction, Merritt-Chapman-Scott, and Penn Products Co., were nothing more than payoffs to Johnson."[63]

Johnson was also charged with accepting a $30,000 payment for helping settle labor disputes at the Yonkers Raceway, with which the Carpenters also had an agreement.

Unfortunately, these allegations were never answered, since Johnson was excused from testifying before the committee for reasons of ill health. The charges were repeated by Robert Kennedy in his book, where Johnson was branded as a labor leader who had betrayed his trust.[64] However, Johnson was never charged with any violation of law. He remained a member of the general executive board for a decade after the committee hearings had concluded, and continued to play a leading role in New York building trades affairs.

The McClellan Committee was successful in unearthing some unsavory practices in the labor movement. Its findings led directly to the enactment of the Landrum-Grffiin Act, which has been instrumental in upgrading the conduct of internal affairs within American trade unions. However, a fair reading of the committee's hearings and findings leads to the conclusion that extensive and thorough investigation into the affairs of the Carpenters produced no evidence of wrongdoing against Maurice Hutcheson or other top officials of the Brotherhood, with the possible exception of the nepotism and conflict of interest charges levied against Charles Johnson. There was no evidence showing that Brotherhood funds had been diverted to the enrichment of its officers. There may have been mistakes in judgement, but the Carpenters' Union emerged relatively unscathed from this searching inquiry into its operations. None of its officers was forced to resign under the pressure of public opinion, as was the case with officers of other labor organizations. The general executive board concluded with some justification:

> No union in the nation has been investigated more thoroughly or completely than ours, including our general officers, and not one hint of the slightest kind or character of participation in the nefarious practices such as the use of sweetheart agreements, under the table contract negotiations and chartering of paper locals to forestall legitimate organization, etc., has been made. Therefore, we individually and collectively feel, and with considerable pride, that the integrity and leadership of all of our officers have been most conclusively established.[65]

INTERNAL ADMINISTRATION

Little has been said thus far about the manner in which the Carpenters' Union is administered. The existence of several studies written during the 1950s, plus the availability of some good data for the period, provide the opportunity to make some observations on this aspect of the Brotherhood's history.

Table 13.3.   Number of cases appealed, 1916-1958

Period	Appeals to the general president	Appeals to the general executive board	Appeals to the convention
1916–1920	840	100	14
1920–1924	621	54	14
1924–1928	481	38	7
1928–1936[a]	1466	122	8
1936–1940	467	42	3
1940–1946[b]	431	37	6
1946–1950	264	17	5
1950–1954	659	74	14
1954–1958	630	30	16

Source: Morris A. Horowitz, *The Structure and Government of the Carpenters' Union* (New York: Wiley, 1962), p. 92.
a. Covers an eight-year period.
b. Covers a six-year period.

The Carpenters had from the start an elaborate internal judicial system for dealing with infractions of rules and regulations. There are specific and detailed constitutional provisions governing charges brought against individual members at the local level, with guarantee of due process. In the 1950s, appeals from local judgments could be taken to the general president, with a further right of appeal to the GEB and the convention.

We do not know what proportion of local decisions are appealed. The total number of appeals for periods between conventions is shown in Table 13.3 for the years 1916 to 1958. A writer on union government noted that "the number of appeals is sufficiently large and stable to indicate that members and officers of subordinate bodies regard the appeals tribunals as fair and equitable and as one in which differences between locals and district councils and members can be equitably reviewed."[66] Relatively few cases were appealed to the GEB, but the same writer observed that this "is not important because the general president has gone over the cases, examined the trial minutes and the statements of both parties . . . His decision is as a rule based upon an adequate and fair review of the evidence."[67]

Only rarely has the decision of the general president been reversed. A 1962 study revealed that the last year in which the GEB had overruled the general president was 1914, while one had to go back to 1920 to find a convention reversal of a GEB decision. This could be interpreted as presidential domination of the GEB, but the following is a much more likely hypothesis: "The board members may realistically see the president's decision as one into which considerable time, thought, and effort, plus seasoned judgment have gone, and overruling such a decision would not mean substituting justice for injustice, but substituting the GEB judgment for the President's judgment."[68]

The nature of the cases appealed, which may be regarded as fairly typical of the kinds of cases that are heard at the local level, can be seen from Table 13.4. Of 137 appeals acted upon by the general president in 1957, some 113 involved fines ranging from $5 to $500. Of this number, 47 fines were upheld, 30 reversed, and the rest modified. The higher the fine, the greater was the chance of reversal or modification.[69] Seven cases were appeals from expulsion, and all of those were reversed. A good deal of heat is often generated at local levels over violation of working rules, and severe penalties are meted out. The national office is more concerned with preserving membership and can take a dispassionate view of the offense.

Table 13.4.   Appeals cases submitted to the general president, by issue, 1957

Issue	Number of cases
Violation of trade rules	46
Improper conduct	12
Failure to procure working permit	9
Failure to appear before local trial committee	9
Worked Saturday, Sunday, or holiday without permission	8
Worked during strike	8
Challenged election[a]	8
Piecework and subcontracting	6
Failure to obey business agent	5
Failure of union officer to perform duties	3
Crossing picket line	4
Improper assignment of work	3
Creating dissension	3
Dismissal of charges by local union	5
Working below scale	1
Working with nonunion men	1
Miscellaneous	6
Cases withdrawn	6
Total	143

Source: Morris A. Horowitz, *The Structure and Government of the Carpenters' Union* (New York: Wiley, 1962), p. 96.
   a. Five cases involved the same election.

A comparative study of union government yielded the following general conclusion about the Carpenters' judicial procedure: "The issue is not how one would rule if deciding a specific case, but whether the procedure is fair and adequate; whether evidence is carefully reviewed; and whether the rights of defendents are amply protected. On all these points, it is our opinion that the Carpenters' Union does an exemplary job."[70]

The general president and the GEB are often called upon to intervene in local situations that involve issues other than individual transgressions of rules. Some of these are political in nature, others may involve financial improprieties. In 1952, for example, a GEB committee investigation into the affairs of Spokane Local 98 indicated that a dissident group was causing turmoil by interrupting meetings and making threats of violence against local officers. The board decided that certain members, who were named, would be barred from attending meetings or holding local office, though their membership would not be terminated.[71] Local 1782 of Newark, New Jersey, was charged by the Essex County District Council with appointing delegates to meetings of subversive organizations, and the council asked that its charter be revoked. The board decided that this was too severe a measure; instead, it put the local on probation for five years, and barred the president from holding any office. Local 1778 of Columbia, South Carolina, displayed a charter of the Ku Klux Klan in its meeting hall. This was ordered removed, and the local told that "they cannot rent the hall to organizations ruled as subversive to the Constitution of the United States." The Susquehanna Valley District Council in New York was dissolved in 1955 and its four locals transferred to another district because it had not been operating satisfactorily.[72]

Almost every national union has the authority to put locals into trusteeships and to reorganize them for violation of the constitution or other regulations. The McClellan Committee hearings revealed that several unions had abused this authority. No such charges were made against the Brotherhood. The general office often consolidated locals in the interest of efficient management, and trusteed locals for a variety of reasons. At the end of 1969 some 67 locals were either under trusteeships or under less stringent supervision by the national office, out of about 3,000 locals holding Brotherhood charters.[73] By comparison, the Operating Engineers had one-fifth of their members in locals under trusteeship, some of them for many years; the Teamsters had 11 percent of their locals in trusteeship.[74] By these standards, the Carpenters do not appear to have made inordinate use of the power of trusteeship. There may have been cases in which either the imposition or the duration of trusteeships would have been barred by the Landrum-Griffin Act, but it seems fair to conclude that on the whole, this authority was employed by the national union primarily for the preservation of the organization.

Some might object to the arbitrary discipline to which left-wing locals and individuals were subjected; this is a policy for which the national leadership of the Carpenters has never made any apologies.

As far as dues are concerned, the Brotherhood has never been an expensive union. In an analysis of union dues structures made in the early 1940s, only three of sixty-seven unions studied had an average monthly dues rate less than the average hourly earnings of the members. The Brotherhood was one of the three.[75] Nor did the Carpenters pay high salaries, particularly in view of the sometimes arduous duties of its officers. Here is how the administrative situation was described in 1950:

> The Carpenters' Union operates a beneficiary and pension system for its members. Every one of its 821,000 members has a card on file at headquarters upon which his dues and assessments are registered quarterly. The union sends out 17,000 pension checks to its members monthly, and annually the union pays almost three million dollars in death benefits to the survivors of its members. To carry out its operations, the union uses IBM machines, and . . . there were 104 members employed in the office. The president, in addition, must decide all appeals, hear complaints, prevent dissension, and see that almost three thousand locals are operating effectively. For that he receives at present $20,000 per annum.[76]

At that time, twelve out of seventy-four unions studied were paying their chief executive officers over $20,000 a year, while eighteen others, in addition to the Carpenters, paid about $20,000. Unions comparable in size to the Carpenters—the Miners, the Steel Workers—were in the $35,000 to $50,000 range; only the Auto Workers were lower.[77] In 1955, pursuant to the authorization of the 1954 convention, the annual salary of the general president was raised to $24,700, and that of the general secretary to $19,500.

THE TWENTY-EIGHTH GENERAL CONVENTION

A year before the 1958 convention was held, the GEB had sent out for a membership vote some proposed major changes in the governing structure. It suggested that the referendum required to approve constitutional changes adopted by the convention be abolished as too costly and cumbersome. This included the election of all general officers. At the time there were seven geographical districts, each one represented by a GEB member. The board asked for authority to redistrict in line with membership distribution. Finally, it was proposed that the general office be moved from Indianapolis to Washington, D.C.[78]

All these changes were approved and went into effect in 1958. One of the board's first uses of its new authority was to increase the number of districts to ten, including two in Canada. It also purchased a site for a Washington headquarters, located on Constitution Avenue a few minutes' walk from the Capitol, at a cost of $1,991,000, or $39 a square foot.[79]

The years between the 1954 and 1958 conventions were relatively good ones for the construction industry, and the Carpenters managed to achieve a small increase in membership:

Membership category	December 31, 1953	December 31, 1957
Members in good standing	746,467	758,726
Members in arrears	52,614	62,025
Honorary and other members	24,493	19,043
Total	823,574	839,794

There had been a substantial increase in the organizing staff, and many new members had been recruited. However, the rate of turnover was high, as evidenced by the following figures:[80]

Year	New members	Suspensions and resignations
1954	94,310	104,680
1955	120,947	94,144
1956	121,289	91,080
1957	96,833	105,873

Hutcheson, who reported these figures, found it difficult to explain the high dropout rate. "In general, the whole attitude toward unionism is at the lowest ebb in several generations." In his view, the greatest challenge to the Carpenters' Union was to organize the hundreds of thousands of nonunion men working at the carpentry trade: "They pose a constant threat to our wage and working conditions. They literally make up a heavy anchor we must drag behind us as we move forward. The shadow of their non-union wages and non-union working conditions hovers over every bargaining session and contract negotiation we enter into. By the very nature of our competitive society, they set the pace and we can never get very far out of step."[81]

The committee on organization emphasized that the greatest problem lay in the home building industry. Local unions appeared to lack interest in organizing that branch of the industry. The committee expressed its belief that "immediate action is required to bring all of this work within the scope of our Brotherhood, even to the extent of chartering new Local Unions or to direct the establishment of competitive wage rates to accomplish this purpose."[82] This was to be a constant refrain for the next two decades.

General Secretary Richard E. Livingston had assumed office on March 1, 1957, following the death of Albert Fischer. (Frank Duffy had died on July 11, 1955, at the age of 92.) His report painted a dismal picture of the financial situation of the Home and the pension system. The average monthly cost per Home resident was $86.30, almost six times the pension amount. The pension account was in deficit; since 1937 there had been five referenda proposing increases to put the pension system on a sound financial basis, and all had been defeated.

The constitution had earlier been amended to permit women to join the Brotherhood, and by this time there were 8,864 of them in good standing. The ritual was amended to include the term "sister" as well as "brother."

A proposal was introduced that in effect would have required all locals to affiliate with state carpenter councils. The president of the California State Council argued for it on the ground that organization at that level was necessary to help rural areas not covered by district councils. The district councils were opposed, sensing a possible threat to their autonomy, and the resolution failed.[83]

Almost 20 percent of all registered apprentices in the United States, including the metal as well as the building trades, were training for carpentry work. The Carpenters had the largest cohort of apprentices in the country, and in view of the growing importance of this activity, the convention voted to establish an apprenticeship department at the national office and to sponsor regional and national apprenticeship contests. In recognition of the rapid increase in health plans, a health and welfare department to help locals formulate sound plans was also approved.

When it came time to elect officers, a number of state councils jointly introduced a resolution to the effect that all the incumbents be reelected unanimously. This produced the only real controversy of the convention. A delegate objected that this would make the nominating and election procedure meaningless, and urged that it be defeated. The standing vote was close, and a roll call was called for. Hutcheson suggested that the customary election procedure be followed and that no vote be taken, and his advice was followed. However, when the time came for nominations and elections, only one incumbent member of the GEB, Joseph F. Cambiano, was opposed, and he won handily. The convention supported Maurice Hutcheson not only by returning him to office, but also by unanimously endorsing a personal statement that he had made to the membership defending his stewardship.[84]

The 1950s were, on the whole, good years for the Brotherhood. Membership rose between 12 and 20 percent, depending on which membership categories are included. The real hourly wage rates of union carpenters had gone up by 30 percent; this had taken place despite restrictive legislation in the form of the Taft-Hartley Act and right-to-work laws enacted by a number of states. Vice-President John Stevenson, commenting on the adoption of an Indiana right-to-work statute, told the 1958 convention: "That kind of disturbed me, and because it disturbed me it disturbed my family, and after voting for nearly 40 years as a Republican, I kicked over the traces and voted straight Democratic at the last election."[85] The Brotherhood had not yet abandoned its traditional nonpartisanship in national politics, but it was moving in that direction.

There were some problems ahead. The first involved Canada: in 1947 the Canadian membership of the Brotherhood was 24,000; this had grown to 76,500 in 1959, making the Brotherhood the second largest union in Canada.[86] The North Ontario District Council had served notice at the 1958 convention that "elect your friends and defeat your enemies" was not appropriate for Canada, since the Canadian Labor Congress had decided to set up a new political party that all its affiliates would be obliged to support. Second, the nonunion home building sector of the construction industry was becoming an ever greater roadblock in the way of organization. Finally, the Carpenters' financial position was being threatened by the heavy burdens imposed by the pension system and the retirement home.

14

A DECADE OF PROGRESS:
THE NINETEEN-SIXTIES

In 1960 a special convention was called in order to revise the constitution to bring it in line with the requirements of the Landrum-Griffin Act. The Carpenters, and labor in general, were hostile to the new legislation. A Brotherhood publication termed it "vicious," a law "which puts dangerous new shackles on organized labor," the result of "five years of continuous smear-labor propaganda."[1] They believed that its real objective was to weaken unions. "As drunks, agitators, and Communists disrupt union meetings because the law gives them authority to do so . . . resentment is bound to develop."[2]

The special convention took place on the eve of the 1960 presidential election. Hutcheson called upon the delegates to step up their political action work, but at the same time he urged that the Brotherhood maintain a nonpartisan political policy. To emphasize its neutrality, the two presidential candidates, Vice-President Richard Nixon and Senator John F. Kennedy, both addressed the convention—a historic event, and indicative of the importance attributed to the Carpenter vote.

Nixon spoke first, and claimed that his father, though not a professional carpenter, had himself built the California house in which he (Richard Nixon) was born. He expressed pride in the fact that "the father of your present President, Bill Hutcheson, for many years headed the Labor Committee of the Republican Party." Recalling his service on the House Un-American Activities Committee, he praised the Carpenters' Union for its early stand on communism:

> One of the first unions to adopt a rule making it absolutely illegal and impossible for Communists either to belong to the union or to hold office in it were the Carpenters who, long before other institutions in this country and other leaders saw that their goals were not the goals of free trade unions. The Carpenters in 1928 took their stand, and for that you are commended, for the leadership that you gave not only to the union movement but that you gave to America as well.[3]

He claimed that he had been in favor of restoring the right of common situs picketing as far back as 1949, and pointed out that real wages had risen by 15 percent during the Eisenhower administration, when he was vice-president.

Senator Kennedy emphasized the achievements of past Democratic administrations in building homes, and also expressed his opposition to the ban on situs picketing. Nor did he neglect the communist issue: "The enemy is the Communist system and the enemy of the Communist system, the chief adversary of the Communist system, is our system."[4]

To emphasize the union's nonpartisanship, Hutcheson used exactly the same words in thanking the two speakers. The convention later considered a resolution asking the general president to appoint two members in every state to set up and direct committees on political education, and on the suggestion of California delegate William Sidell, it was referred to the GEB for action. The postconvention issue of the *The Carpenter* printed both the candidates' speeches, and concluded: "It is not our purpose to tell anyone for whom to vote. Rather, it is to emphasize the need for voting intelligently on November 8."[5] The GEB did contribute $36,000 to the AFL Committee on Political Education, but did not officially take sides.

After the political introduction, the convention got down to the business of rewriting the constitution. A consideration of some of the major changes affords us the opportunity of seeing what changes had to be made in even a relatively good constitution like that of the Carpenters in order to bring it into conformity with the code of good union government embodied in the Landrum-Griffin Act. However, some of the changes proposed or adopted had nothing to do with the new legislation, but were matters that had been debated in past conventions and were included in the general revision.

1. The constitutional clauses on penalties, charges, trials, and appeals were completely rewritten. The new procedure for presenting charges and conducting trials provided the individual involved with more safeguards. For example, a completely new provision made it possible for the members of the body that imposed a penalty to set it aside by majority vote.

2. Only those members who were expelled, suspended, or fined in excess of $50 could take an appeal to the general president. It was hoped that this would eliminate appeals on trivial offenses and reduce their volume.

3. Some delegates wanted to put a time limit within which appeals would have to be acted upon. As one remarked, "I know a case in point that I can bring to the attention of the delegates here where a man lodged an appeal against the sentence imposed upon him. When the appeal was finally answered the man had been dead for two weeks." Although this suggestion was not adopted, provision was made for an appeals committee of not less than five persons to whom the president could delegate the review of appeals, and thus speed the procedure.

4. Local unions had long been forbidden to use local funds for per-

sonal loans or gifts to members, or for political or religious purposes. It was proposed to add a ban on contributions to local credit unions sponsored by the local. A delegate objected that this would deprive locals of an excellent source of investment. The spokesman for the constitution committee pointed out that "the historical position of the International with respect to investments has been pretty clear. At all times this has been prohibited." The reason behind this was to prevent membership in the local from being jeopardized by failure to repay a loan to the credit union; "that is what we are attempting to do, to make sure that a credit union sponsored by a Local Union is not being used as an instrument to put pressure on the Local Union itself." The motion carried.

5. An interesting debate occurred on the proposal that "when a local union raises its dues, intiation fees or levies an assessment, a secret ballot vote shall be taken either by mail or by a special or called meeting." The issue of mail ballots for union voting has long been a controversial one, particularly in Great Britain. One delegate agreed that to permit a mail ballot would discourage attendance at union meetings: "We have to get these men out at the meetings. We should do everything we can to get them to the meetings." Another agreed:

> We have many people today that are in the union only because it is a means of securing a better livelihood. They use their dues to buy a job. Why give these people the opportunity to vote by mail when they are not even interested, attend not one meeting out of the year? It is not a hardship to ask a man to come down to his local once a year, to make this trip, regardless of what it is, to protect his bread and butter.[6]

The mail option was deleted. However, mail ballots have been employed on occasion in connection with the ratification of agreements, when national union officers believed that local factions were blocking satisfactory settlements.

6. The requirement that local unions or district councils must notify the national office and seek its approval when general wage or hour movements were contemplated was limited to situations in which financial aid from the general office was requested. This change, which enhanced the power of locals with respect to collective bargaining and strikes, was passed without discussion.

7. Prior to 1960, shop stewards in industrial plants controlled by the Brotherhood were appointed by the business agents. The new constitution provided for election by the employees concerned. One delegate objected to the change on the ground that it might lead to the election of stewards who were too close to the employer. "Many men who go into these shops become permanent fixtures, and as a result of becoming permanent fixtures they are liable to be swayed to the employer's side of the

fence, when voting on a steward or otherwise, because they are fearful of losing their jobs.["]7 The amendment was adopted despite this objection, bringing the Carpenters in line with the general practice of industrial unions.

8. A number of resolutions had as their purpose limiting the election of GEB members to delegates from their own districts, a perennial issue. After a heated debate, the resolutions were defeated; all GEB members continued to be elected at large.

9. A proposal to place business agents under the supervision of district councils met with strong opposition from the floor on the ground that it would impair local autonomy. It was adopted nonetheless.

10. It was proposed in several resolutions to prohibit the existence of segregated locals, as well as the alleged denial to blacks of admission to the Florida retirement home. The resolutions committee found that there was no truth to the latter allegation, and stated with respect to the former:

> We have many Local Unions whose membership is made up of more than one race or color and . . . we have a very small number of Local Unions whose membership is made up of only one race or one national origin. The existence of Local Unions made up of one race or one nationality is not a prima-facie indication of discrimination but it is a result of many reasons: Desire of the members of the particular Local Union, economic, industrial, proximity of residence to employment . . . Some of these Local Unions have been in existence for many, many years . . . They are proud of their organization, proud of their record, and desire to maintain their charter and identity . . . In summary, we feel that the United Brotherhood constitution has adequate and effective anti-racial discrimination provisions; the United Brotherhood has lived up to the spirit and intent of these provisions; the Brotherhood's position is clear and its record is enviable.[8]

11. No change was required in the previous constitutional provision stipulating that all officers were to serve a term of not less than two or more than four years. However, local officers were governed by the three-year maximum term mandated by Landrum-Griffin.

These were the principal constitutional changes proposed, discussed, and adopted or rejected by the special convention of 1960. One resolution that did not pertain to the main business of the convention was adopted; it deplored the expulsion of the Teamsters and other unions from the AFL-CIO, pointing to the valuable assistance given by the Teamsters and Longshoremen in honoring Carpenter picket lines.

Concern was expressed about the growing encroachment on building trades jurisdiction of industrial unions engaging in plant construction work, particularly in missile bases and the aerospace industry. The convention ended with a remark by Hutcheson that was well taken:

In every way, shape, and manner, [the Convention] was completely free, open and democratic. Every delegate was given full opportunity to have his say. Every resolution received a fair hearing. Every decision was made by the vote of the delegates, after unrestricted opportunity to debate. Nothing was suppressed or buried or barred. No one can ever challenge the fact that this Convention has been a model of democracy in action.[9]

THE CANADIAN PROBLEM

Reference has already been made to the growing demands by Canadian members for a greater degree of autonomy.[10] Some of these demands were made at the 1960 convention in connection with the rewriting of the constitution. Before considering them and subsequent developments during the decade, a few pages on the history of the largest of the Canadian affiliates, the Ontario Provincial Council, are in order to provide some background on carpenter organization in Canada.[11]

The first Carpenter local to be established in Canada was Local 18, of Hamilton, Ontario, chartered in 1882. This was followed by the organization of Local 27 of Toronto at a meeting addressed by Peter J. McGuire. Some 400 members of this local participated in a labor parade on July 22, 1882, the largest ever held in Canada up to that time. It engaged in a strike the following spring, but lost. "Work was backward and the city was flooded with immigrants dumped to crowd every branch of the trade." Eleven additional charters were issued during the next three years, though all but one lapsed soon thereafter. Strenuous employer opposition and the competition of the Knights of Labor, who were very active in Ontario, made progress difficult.

One of the most prominent figures in this early history of the Canadian Carpenters was Harry Lloyd. He came into Local 27 with a group of millwrights who had actually been organized before the Carpenters. He became the tenth president of the United Brotherhood in 1896, and was a delegate to AFL conventions. Among the other early leaders were Tom Moore of Local 713, Niagara Falls, Charles Moat of Local 27, and Arthur Martel of Montreal. Moore proved to be one of the outstanding figures of the Canadian labor movement. He was appointed a Brotherhood general organizer in 1911 and organized all the local unions in the Niagara Peninsula, which became the Frontier District Council. In 1918 he was elected president of the Trades and Labor Congress (TLC) of Canada, which was the Canadian equivalent of the AFL, and served in that position until his death in 1946.

As of December 31, 1912, there were twenty-nine local unions of carpenters with a membership of 1,837 operating in the Province of Ontario. In that year the first of five meetings of an informal Ontario

Provincial Conference was held; it was replaced in 1913 by a regular provincial council. During the First World War, Moore greatly expanded membership in Ontario by vigorous drives to organize woodworkers in shops and factories producing war products.

When Moore became head of the TLC, he was succeeded as general organizer by Jim Marsh, who served for some years as president of the Ontario Provincial Council until he became Deputy Minister of Labor for Ontario, a post he held until he died in 1945. Another president of the Provincial Council, Fred Hawse, eventually moved to the position of Director of Apprenticeship for the Province of Ontario, where he served for twenty years.

On the eve of the Great Depression, which did not affect Canada as severely as it did the United States, there were forty-two locals and three district councils in the Province of Ontario, with a total of 3,571 members. This represented a decline from the high level of 7,451 members in 1920, mainly due to the loss of furniture workers and mill men.

In the early 1930s, organizing efforts were begun among the bushworkers of northern Ontario. There had been earlier unsuccessful efforts by the Industrial Workers of the World and the One Big Union, a Canadian counterpart of the IWW. Some independent locals had continued in existence, but they did not flourish:

> The independent unions were smashed three or four times mainly because they were a small, poorly financed, loosely organized union, but the bushworkers proved to one and all during these trying times that they wanted a union and were prepared to fight to the death if necessary (many of them were killed, including two organizers who were shoved under the ice and found in the spring of the year).

The independents received a Carpenter charter in 1936. The working conditions faced by the men were among the worst in the country:

> Tarpaper shacks with no windows, log camps built by each bushworker or sawed dugouts which were musty, vermin ridden hovels were the lot of the bushworkers. No thought was given by the Companies at attempting to keep the worker clean or respectable . . . Horses were treated much better because a horse cost the Company money—a man did not. Food consisted of a diet of rotten salt pork and beans, generally boiled. Wages were very low and men could be and were beaten for their pay . . . Many or most of the bushworkers were landed immigrants and so neither spoke English nor understood the English language.[12]

Building a union among these men was a slow business, not least because of the intense opposition of the large lumber companies. Finally in 1946, after a long strike, the union won collective agreements with most of the employers. Within the next fifteen years, their hovels were replaced by well-built bunkhouses with toilet and shower facilities; they had access to good cafeteria food and medical facilities; their hours of work were reduced to forty-four per week. The bushworker "is no longer the dirty lout that he was in early years. One can no longer recognize a bushworker by the smell of his clothes." There was probably no group of members in the history of the Carpenters' Union that benefited more from unionization than the bushworkers of Ontario.

The Canadians were justly proud of their achievements in organizing construction, factory, and lumber workers, and as Canadian nationalism deepened, their views on a number of issues began to diverge from those of their brother carpenters in the United States. The Canadian Labor Congress (CLC) had been formed in 1956 by a merger between the TLC and a new federation of industrial unions, the Canadian Congress of Labor, that had been established in 1940 as a parallel to the CIO in the United States. The CLC was moving toward the creation of an independent labor party, and the Canadian locals wanted the right not only to discuss politics at union meetings, but to be able to back a labor party if they saw fit. Their main argument was that in Canada, unlike the United States, members of parliament were bound to vote along party lines, so that the formula of electing individuals friendly to labor had no meaning if a party as a whole took anti-labor positions. Delegate Edward Thompson of Local 878 urged the 1960 convention to give the Canadians more leeway:

> Mr. Chairman, what are we going to do here today? Are we going to crucify Canada? Is that what we are doing? Now let us give them some consideration up there. They know their business up there. Do not turn everything down that Canada brings in. I think this is a good proposition. I am telling you, Jack, and I am telling you, Maurice, too, I think it is good. Bring it in. Let Canada have a little say in this Brotherhood, more than they have got, because you have turned down everything they have brought up.[13]

The Carpenters' leadership, backed by a majority of the delegates, would not allow any special constitutional dispensations for Canada. They feared that encouragement of separatist tendencies might lead to an eventual split between the U.S. and Canadian branches of the Brotherhood. They were able to defeat all the Canadian resolutions submitted to the convention, but the controversy was by no means over; it had just begun. The next phase came in the form of a dispute with the International Woodworkers' Association (IWA) over the right to organize loggers in Newfoundland.

After the 1930s, the only major jurisdictional dispute between the Carpenters and the Woodworkers involved the efforts of both organizations to organize the loggers and lumbermen in Newfoundland. The Brotherhood began an organizing campaign there in 1955; the IWA, the following year. In 1957 an independent organization, the Newfoundland Lumbermen's Association, offered to join the United Brotherhood. It was decided not to accept this group as such, but rather to seek to organize the members in newly established locals. The IWA claimed that it had established collective bargaining relationships with two pulp companies. In fact, the IWA had lost a strike there, and the two locals that it had formed were disbanded after having been decertified. The IWA set up a new local and began to organize once more.

In 1961 the Brotherhood sent an investigating committee to Newfoundland to see whether there was a basis for organization. Among others, it called upon President Claude Jodoin of the Canadian Labor Congress, who assured the committee that the Brotherhood had the right to enter Newfoundland. It conducted a mail ballot among as many loggers as it could find, and ascertained to its satisfaction that there was a majority sentiment for joining the Brotherhood.[14]

The IWA then filed a charge with the Canadian Labor Congress to the effect that the Carpenters were interfering with its established bargaining relationships, in violation of a CLC no-raiding clause similar to that of the AFL-CIO. The Brotherhood maintained that the IWA no longer held any bargaining relationships, and that the province was open to organization. Unable to bring about an agreement between the two unions, the CLC proposed that it issue a charter to an independent union, and that both the Brotherhood and the IWA refrain from any further organizational efforts for a year. At that time, the Congress would arrange for a choice to be made between the two.

The Carpenters rejected this solution on the ground that they would lose a promising organizational momentum. Moreover, the Newfoundland Federation of Labor, an arm of the CLC, had taken a strong pro-IWA stand, putting in question its neutrality. After the CLC executive council had ordered the Carpenters to halt organization, they retaliated by notifying the CLC that they would stop paying per capita tax until the order was withdrawn. Sixty Brotherhood delegates to the April 1962 convention of the CLC walked out in protest.[15]

Things remained at a standstill until 1964, when an agreement was reached containing the following provisions:

1. The competitive activities of the Brotherhood and the IWA were to remain on a status quo basis, and complaints filed by both parties were to be withdrawn.
2. The CLC organizing code would be observed by both sides.
3. The CLC constitution would be adhered to.
4. An IWA-UB consultative committee would be formed to deal with future disputes.[16]

Thus, the Brotherhood won the right to remain in Newfoundland. One of the two locals there, No. 2564, reported that it had 4,500 members in 1964.[17] President Jodoin of the CLC delivered a laudatory address at the 1966 Carpenters' convention, and the Brotherhood continued to play an active role in CLC affairs.

The 1962 convention was not a favorable time for renewed Canadian autonomy demands to be raised, in view of the friction between the Carpenters and the Canadian Labor Congress. Canadian resolutions failed of adoption, leading Secretarty McCurdy of the Ontario Provincial Council to complain:

> I would express my very great disappointment that, once again, nothing has been done on the political resolutions coming from Ontario . . . This undoubtedly is due to the fact that our American brothers are not aware of the political structure in the Province of Ontario, as it differs from the way you operate in the United States. [In Canada] a candidate from any of the parties may not, cannot, and he dare not depart from the policy in that party. So, supporting the candidate is not satisfactory as far as the Dominion of Canada is concerned.[18]

This statement provides a good illustration of why American trade unions were able to follow successfully the kind of political policy that would have been difficult, if not impossible, in western Europe or Canada. The system of strict party discipline that prevails in almost every democratic nation apart from the United States, and the tendency of voters there to cast their ballots for the official nominee of a party rather than for an individual politician, makes it impracticable to pick and choose among the "friendly" candidates from the contending political parties, to "support your friends, defeat your enemies." The internal controversy that the Carpenters and other American unions feared might be generated by permitting political discussion and party support at the local level has largely been avoided by the tendency for trade union workers in countries other than the United States to vote fairly uniformly for labor parties. Secretary McCurdy of the Ontario Provincial Council was not a radical. There was no issue of ideology involved; it was simply the different perceptions on the two sides of the border of how effective political action could be combined with effective trade union action in the economic sphere.

The next step in the Canadian situation was the formation in 1965 of the Canadian Council of Carpenters to coordinate the legislative activities of the district councils and locals. Every provincial council joined it. A resolution was introduced at the 1970 Brotherhood convention asking that the new council be recognized as an autonomous section of the Carpenters, in view of the fact that Canada "is the only major country in the world that does not have an independent trade movement," and of "the growing concern amongst the Canadian people at the erosion of

Canadian independence due to economic, political and trade union control by the U.S." The Canadian Labor Congress had adopted a policy of encouraging self-government of the Canadian sections of international unions, a move that was by no means confined to the Carpenters. Moreover, a Canadian parliamentary commission had spelled out the following six-point guideline for Canadian branches of American unions:

1. Canadian members and locals of international unions should be recognized as the Canadian section of the international.
2. The Canadian section of the international union should have the machinery and the authority to deal with all matters of concern to the Canadian members and locals. As a general principle, international headquarters should not put a Canadian local into trusteeship without the advice of the Canadian headquarters.
3. In particular, Canadian members and locals of international unions should have complete authority with regard to their collective bargaining program, to the settlement of disputes, and to the conclusion of collective agreements, none of which should require approval in the United States.
4. Canadian officers of international unions should be selected by Canadians either by delegates at the conventions or by the Canadian membership.
5. As far as possible, machinery should be provided so that control over expenditures and staff in Canada rests with the Canadian section and its officers. Financial statements in reasonable detail giving members a clear view of the financial operations of the union should be made available to all members.
6. Generally, all steps should be taken to give Canadian sections of international unions full authority to deal with all matters, whether they involve the internal administration of the Canadian locals or general social and economic policy or collective bargaining, without any control from outside.[19]

Although these guidelines did not have any formal legislative mandate behind them, they did indicate the growing strength of Canadian separatism. To follow the guidelines in full would have meant the creation of a virtually independent body within the Carpenters' structure, with little authority over Canadian members left to the national headquarters in the United States. The convention referred the entire matter to the GEB for further consideration, accompanied by a warning that if the Carpenters, and other unions as well, did not act, " [Canadian] Federal legislation will be brought down that will bring these requirements about."[20] Hutcheson, from the chair, promised only that the GEB would give the Canadian locals the same consideration as it gave to U.S. locals, "and you cannot expect any special favors." Further developments in what continues to be a problem will be discussed in the next chapter.

INTERNAL MANAGEMENT UNDER LANDRUM-GRIFFIN

The management of internal union affairs became more complicated after the enactment of the Landrum-Griffin Act. It is problematic how much the final results of actions taken by the general office were changed, but the procedures employed were more careful, if one is to judge from the proceedings of the GEB.

A few examples will serve to indicate the new approach. Local 101 of Baltimore, which had once been under trusteeship for many years, again ran into financial problems. In the past, in all likelihood a trustee would have been appointed to manage its affairs. In 1960, however, a subcommittee of the GEB was delegated to conduct a careful investigation, and eventually came up with a package of recommendations including the following: (1) better cost control and regulation of the time spent by business agents; (2) possible disaffiliation from state and local AFL-CIO bodies in the interest of economy; (3) establishment of a more reasonable level of strike pay; (4) election of officers all at one time to avoid too frequent elections; (5) examination of investments to see whether higher yields could be secured; and (6) an increase of monthly dues to 1½ hours of pay. It was believed that the local could put its affairs in order by following these recommendations.[21]

The Spokane district council also found itself in financial difficulty. A GEB committee found that the council had not raised its per capita tax since 1954, despite the fact that wages had risen from $2.60 to $3.63 an hour in the interim. The GEB made the following recommendation: "The only way recurrent financial crises can be avoided is by identifying per capita taxes with journeyman wages . . . It is our recommendation that a formula be worked out for automatically increasing per capita taxes on a percentage basis of journeyman wages as these increase over the years." Two locals not affiliated with the council were directed to join it.

Local 425 in El Paso, Texas, was being subjected to a great deal of internal turmoil. On only one occasion in the previous ten years had a business agent succeeded himself. The local had substantial assets, amounting to $144,000, and controlled 65 percent of the commercial and industrial work in its area. It was recommended that a general representative attend local meetings to maintain order; that there be no more than two meetings a month; and that initiation fees be reduced in order to organize the home building market, which was nonunion. The importation of transient Mexican labor was having a demoralizing effect on the local labor market, so the board recommended that the AFL-CIO look into the situation.[22]

Archie Anderson, a member of Local 7 in Minneapolis, had been charged with affiliation with communist groups, and with falsifying his application for membership by denying his communist sympathies. He was expelled from membership by the Twin Cities district council and appeared personally before the GEB, which upheld the expulsion after he refused to answer some key questions. However, he was reinstated to membership by a federal court on the ground that he had not been provided the right to be confronted by witnesses and to cross-examine them. A new trial was held at which twenty-one witnesses were heard. The trial committee found that he had attended Communist Party meetings in 1948 and 1949, and was in fact a member of the party. He was expelled once more, and the GEB concurred. The result was the same as it would have been a few years earlier, but the due process requirements were much more stringent.[23]

Roscoe Proctor of Local 35, San Rafael, California, was charged with falsely stating in his membership application that he was neither a member of nor sympathizer with the Communist Party. He was tried by a committee of the local, which followed the legal requirements carefully. He refused to testify on the advice of counsel, and was found guilty. In his appeal to the GEB he claimed that his constitutional rights has been denied and that the questions contained in the membership application interfered with his political beliefs. The GEB sustained his expulsion, and prevailed.[24]

OTHER GENERAL EXECUTIVE BOARD ACTIONS

With the growing complexity of the carpentry trade, the GEB found itself increasingly involved in settling jurisdictional disputes among locals. For example, Local 2212 of Newark, New Jersey, had until 1939 belonged to the Upholsterers' Union. It claimed jurisdiction over resilient floor coverings, venetian blinds, slipcovers, and draperies in fifteen New Jersey counties, a claim disputed by other locals. In 1959 Local 2212 extended its jurisdictional claims to include shops producing sink tops and Formica coverings. A GEB investigating committee found that shops set up to fashion and install sink tops had gone into making cabinets as well. The committee recommended that this practice be discontinued because it constituted unfair competition for cabinet shops with which the Brotherhood had long enjoyed friendly relations. For the rest, the Essex County district council was given jurisdiction over shops that manufactured fixtures, counters, sink tops and table tops; Local 2212 was awarded the right to install plywood under resilient floor coverings in one- and two-family houses; while in buildings where mechanics affiliated with other building trade unions were employed, the district council was to control the work, except where the only organized tradesmen were members of the Stationary Engineers. The claim of

Local 2212 for the manufacture and installation of venetian blinds, curtains, and draperies was sustained, except for installation in commercial and industrial buildings.[25]

This was by no means the only internal jurisdictional dispute the GEB was called upon to settle. When the Glen Canyon Dam in northern Arizona, an isolated area, was being built, the GEB had to apportion jurisdiction among the surrounding district councils. The general office has also helped recruit men for large isolated projects, and at times has participated in negotiating agreements for such projects.

The board was continually being called upon to appropriate funds for special purposes, and a few are worth recording. A $100,000 scholarship fund for children of Brotherhood members was established in honor of Frank Duffy.[26] A more surprising gift was one of $50,000 to an AFL-CIO memorial fund in honor of Eleanor Roosevelt. "Organized labor has particular cause to revere her memory. She staunchly supported the goals and aspirations of the union movement. While there may be room to question some of the positions she took, her motives were beyond reproach."[27] The Bookbinders Union, which had been waging a long strike against the Kingsport Press, was given $6,000; $50,000 was given in support of 140,000 AFL-CIO members who were on strike against General Electric; and another $50,000 to help the Steelworkers Union in a struggle against the copper mining companies.[28]

The board emphasized its support for the long-standing Carpenter policy of maintaining good relations with former members who had become construction superintendents. It made a special ruling that they would continue to be eligible for membership even if they went through Carpenter picket lines to do their normal work in the event of a strike. If they wished to maintain their membership they would be exempt from strike assessments, but they could not hold union office.[29]

THE TWENTY-NINTH GENERAL CONVENTION

In his report to the 1962 convention, General President Hutcheson told the delegates that the four years since the previous regular convention had not been good ones in terms of employment. "Automation, whatever its promise for the future, has made strong progress in eliminating jobs in all fields, not excluding construction. New techniques and new machinery, month by month, have been chipping away at the number of man-hours required to do a given job, or to erect a given building. Consequently, while construction dollar volume has remained at very high peaks, the employment situation for our members has not kept pace."[30]

The following membership figures bear out these observations:

Membership category	December 31, 1957	December 31, 1961
Members in good standing	758,726	682,570
Members in arrears	62,025	56,544
Honorary and other members	19,043	3,539
Total	839,794	742,653

A good deal of the construction expansion was taking place in home building, and the Carpenters had failed to make any substantial inroads into that portion of the industry. Hutcheson was quite clear about this problem:

> The days of organizing large blocs of men working for a large contractor or industrial plant are past. With a few exceptions, here and there, the large employers have been organized. However, there remains a vast pool of small operations that need organization and should be organized. The housing field is particularly in need of attention in many areas. Contractors in the housing field are generally small and hard to reach; yet the housing field must be organized if our Brotherhood is to grow and prosper. The housing contractor who starts out on a small scale often branches out into building multiple unit structures, and, eventually, high rise apartments. From this, it is merely a step to general contracting. If he operates non-union as a house builder, it becomes increasingly difficult to organize him as a general contractor.[31]

This is an excellent statement of a theme that was to be echoed on many occasions during the next two decades. The problem applies to all the building trades unions, not only the Carpenters. In one sense, however, the construction unions are more fortunate than their industrial union confreres: the latter have organized their industries to practical limits, and have no place to go in the face of declining blue-collar employment. The building trades still have wide organizational frontiers, but the going is rough. For example, the use of NLRB elections as organizing devices is not available to the construction unions as a practical matter.

Another topic discussed by Hutcheson was that of appeals from local decisions, which continued at a high level. Of 731 appeals reviewed from 1958 to 1962, 450 had been processed and decisions rendered. A number of appeals were disallowed because the amount of the fine involved did not exceed $50. Despite the problems created by large volume, Hutcheson felt that the $50 limit imposed by the 1960 convention was working serious injustice upon many members by barring them from appeals in which matters of principle were involved, and urged that this limit be removed. The convention agreed, and the change was made.[32]

The Building Committee reported that the new Washington headquarters had been completed at a cost of almost $5.5 million. Cabinet woods of walnut, oak, cherry, teak, redwood, and butternut were used throughout. On the basis of personal inspection, I can attest to the fact that this is one of the most beautifully finished buildings in Washington, a fitting tribute to the art of the carpenter.

One of the principal bones of contention at the convention was a proposal to increase the per capita tax. This was strongly opposed by representatives of furniture workers; a delegate from one local declared that while his union had a rate of $2.40 an hour, one of the highest in the United States, the average for the 375,000 unorganized furniture workers was only $1.65. Further organization would be impeded by higher dues. The counterargument was made that while in 1907, with a per capita tax of 25 cents, it took the average carpenter 38 minutes to earn that amount, only 23 minutes were required to earn the tax in 1962. The addition to the per capita went through.[33]

The GEB reported that standard contracts had been signed with sixty-five national contractors who had maintenance agreements with industrial firms. As a result, a good deal of the work that had been going to industrial unions was won back by the building trades. As a further step toward the same end, a Construction Industry Joint Conference, chaired by John Dunlop, with Maurice Hutcheson as labor chairman, was established on April 7, 1959. The president of the Associated General Contractors, which was affiliated with the Conference, appeared before the convention and told the delegates: "We want to thank you as an organization for getting us so many qualified contractors . . . You have given us more contractors than any other source . . . You are the natural source of our membership because if you are a good carpenter you are a good contractor. You know the business. You are not like some of those boys who finish college and have a desire to be a contractor."[34] This helps explain the attitude of the Brotherhood toward former members who moved into contractor ranks.

For the first time in some years, there was a successful challenge to the administration slate in the election of officers. William Sidell, secretary of the Los Angeles district council, was selected by a caucus of Eighth District delegates to run for the GEB against Joseph Cambiano, the incumbent. The delegates felt that Cambiano, at the age of 78, was no longer up to the job of representing the district adequately on the board. Sidell spoke to Hutcheson before the election took place, and was told that while it was his privilege to run, the officers felt duty-bound to support Cambiano if he wanted to retain his board seat. Sidell won by a margin of 981 to 802 votes, but Hutcheson did not hold this successful challenge against him. Impressed by Sidell's ability, Hutcheson appointed him second general vice-president in 1964, and Sidell eventually succeeded to the presidency. There was also a challenge to the incumbent board member for the Third District, but that failed by a substantial margin.[35]

JURISDICTION (7)

The Carpenters were one of the most active clients of the National Jurisdictional Disputes Board. For the four-year period 1958 through 1961, they were involved in about 70 percent of all decisions handed down by the board. Of the total in which they were a party, the Carpenters won 48 percent and lost 28 percent, while the rest were split. Most of the controversies were with the Lathers over acoustical ceilings, metal studs, and partitions; with the Sheet Metal Workers over cabinets, shelving, and bookcases; and with the Painters and Glaziers over mirrors, shelving, and door frames. Hutcheson told the delegates at the 1962 convention that "too many cases involving our local unions are going to the Joint Board on Jurisdictional Disputes. We want to keep peace within the family of labor, but not entirely at our own expense."[36]

From 1962 through 1965, the Carpenters won 39 percent of their cases and lost 43 percent, with the rest split. The Brotherhood withdrew from Joint Board participation in 1964 because of the large number of decisions going against it, and stayed out for almost a year. It returned to the board in March 1965, when a revised plan was adopted providing for a new Appeals Board, headed by an impartial umpire, to which appeals from the National Joint Board could be taken. The new plan also spelled out more precisely the criteria that the Joint Board was to use in making decisions, including agreements of record, past trade practices in general, and prevailing practices in the locality.[37]

The national office stressed the importance of getting work assignments prior to the start of work. The fact that a majority of the cases involving the Carpenters were initiated by other unions indicated that local unions had been following this policy with some success.[38]

During the four years 1966 through 1969, the Carpenters won 43 percent of its cases and lost 46 percent before the Jurisdictional Board. The GEB had the following comment to make on the experience of this period, with particular reference to initial job assignment:

> Our membership had the original assignment in a vast majority of the cases going before the Board and we are not unaware that this is due primarily to the efforts and the alertness of our business representatives in the field. We do, however, emphasize that this increased effort must continue and every effort be made to continue to receive original assignments for our members prior to the actual start of work on projects rather than waiting until the work commences and being confronted with another craft having secured an assignment

on work which should properly have been awarded by the responsible contractor to members of the United Brotherhood .[39]

A dispute with the Lathers over the installation of acoustical ceilings, which lasted for many decades, was resolved in 1966. The Lathers claimed jurisdiction by virtue of a 1903 agreement and Building Trades Department decisions in 1908 and 1920. In preparing for hearings before a panel empowered by the agreement to render a national decision, the Carpenters contacted 1,000 contractors and received replies from 600, most of them to the effect that carpenters were normally assigned to do the work. The Lathers refused to participate in panel hearings and filed for an injunction against the panel in federal court, but could not get it. The Carpenters argued their case before the Jurisdictional Board after submitting eighteen volumes of briefs, and were supported by employer representatives from all over the country. The panel noted in its decision, which went in favor of the Carpenters, that "this record is undoubtedly the largest compilation of briefs, statements, and evidence ever presented to a jurisdictional tribunal in the building and construction industry."[40]

National jurisdictional agreements were concluded with a number of unions, including the Plumbers, the Boilermakers, and the Laborers. The negotiation of these agreements, plus the processing of cases before the Jurisdictional Board, became one of the major preoccupations of the national union. It was even more evident than in the past that the president of the Carpenters' Union, as well as that of any other union, had to protect the jurisdiction of his union in order to survive politically.

But the necessity of being alert on the jurisdictional front is a two-edged sword. For every case a union wins, another may lose, unless there is some way to split the difference. If decisions begin to run against a union, resentment against the Jurisdictional Board builds up, and relief is sought in changes either in board procedure, or in its chairman, or in the union officers. Aggressive direct action in the form of work stoppages tends to tarnish the image of the building unions in the eyes of contractors and customers, and is counterproductive. Contractors attempt to get out from under union contracts. Jurisdictional disputes may well have been a more important factor than high labor costs in impeding union organization. This dichotomy between what is politically necessary within the union and what is feasible outside complicates life for construction union officers.

Management groups were constantly complaining to the general office about the actions of locals in jurisdictional matters. The general president recognized the problem, and stated that even though there might be as little as fifteen hours of work involved in a particular dispute, the national union would still have to take action.

POLITICAL ACTIVITY

The nonpartisan political stance of the Carpenters' Union was finally abandoned in the 1964 presidential election. Lyndon Johnson was running against Barry Goldwater, who had an unblemished anti-union record in the Senate. Maurice Hutcheson made the following statement, which appeared in *The Carpenter:* "Let me make my position completely clear. I am rejecting the Republican Platform of 1964 as an insult to the intelligence of the voters. I am repudiating the Republican Presidential and Vice-Presidential candidates as avowed and deliberate enemies of our trade union movement. I am talking as a Republican, who finds it impossible to vote Republican this year."[41] A notice sent to all locals urged that the membership vote for Lyndon Johnson, who, in addition to having a generally pro-labor record, had done the Carpenters a special favor by signing a bill providing tax exemption for the orange business associated with the Lakeland Home.[42] After the election the general treasurer, Peter Turzick, claimed that of the sixty-five congressmen and Senators who received financial help from the political fund of the Carpenters, all but ten had been elected.[43]

To coordinate future political work, a new body, the Carpenters' Legislative Improvement Committee, was set up in 1966. The GEB recommended that every board member contribute 1½ percent of his salary, and every general representative 1 percent, to build an adequate fund for political activity.[44] The goal of the CLIC was to collect a million dollars from voluntary membership contributions.

The 1968 presidential election was a repetition of 1964. A statement signed by every member of the GEB was sent to all Brotherhood members endorsing the Humphrey-Muskie Democratic slate. The statement was prefaced by the remark that the general office had never tried to tell any member how to vote, but that it was "a responsibility of leadership to present information to the members that can enable them to determine in their own minds what is best for themselves, their families, their union, and, of course, their nation." The voting records of Humphrey and Nixon were compared, and the conclusion reached that "this is no time to jeopardize our gains by turning over control to Mr. Nixon or Mr. Wallace, neither of whom has shown any real concern for the cause of organized labor."[45] By the GEB count, Nixon had cast 59 out of 60 votes against labor on issues considered important for labor, while George Wallace, when governor of Alabama, had followed anti-labor policies. From 1966 through 1970, the Legislative Improvement Committee spent over $200,000 in contributions to individual and party election funds.

In taking this political stance, the Brotherhood allied itself with the majority of the American labor movement. However, there was no deviation from its strong anti-leftist position. A resolution introduced at the 1966 convention supporting the Johnson administration's Vietnam policy was passed unanimously.[46] When the radical Students for a Democratic Society announced that it was planning to infiltrate the labor movement, locals were urged to exercise great care in granting temporary working permits to students, for:

> these punks, and there is no other word to adequately describe them, are not interested in bettering the living and working conditions of American wage earners. Their aim is to destroy the whole American economic and social structure. Because they have been able to intimidate white-livered, weak-kneed college authorities, they think they can do the same in the world of work. Union leaders are not as gutless as college administrators.[47]

The Carpenter echoed these sentiments, noting that "some union leaders may be shy on some of the deeper academic subjects, but they all have solid acquaintance with the facts of life—with holding a job, butting heads with an arrogant employer, fighting off insidious assults of the Communists from whose book SDS is taking a page."[48] Needless to say, neither the SDS nor any other contemporary radical group gained as much as a toehold in the Carpenters' Union.

THE THIRTIETH GENERAL CONVENTION

The year 1966 was a good one for organized labor. Maurice Hutcheson, in his opening address, noted that in construction, "the agreements negotiated this year have far exceeded anything in past years, not excepting the war years when the scarcity of labor was unprecedented." Between 1962 and 1966 carpenters' real wages rose by 9.2 percent, while unemployment, starting at 5.5 percent in 1962, fell to 3.8 percent in 1966.

As the following figures demonstrate, this very favorable economic climate was accompanied by an increase in membership, but not as substantial as might have been expected:

Membership category	December 31, 1961	December 31, 1965
Members in good standing	682,570	718,048
Members in arrears	56,544	59,247
Honorary and other members	3,539	2,473
Total	742,653	779,768

The failure of the Brotherhood to capitalize on the opportunities afforded by high employment and rising wages was the main theme both of Hutcheson's opening talk and his formal report. While approximately

100,000 new members were coming into the Brotherhood each year, between 90,000 and 95,000 were leaving. This could not be ascribed to the high propensity of young people to change jobs, since the average age of those who left the union was about 35, with almost five years of membership. Hutcheson ascribed lagging membership growth to the reluctance of many locals to admit new members. "Sometimes these men are kept on permit indefinitely. Sometimes they are turned down by the examining committee, even though they may have years of experience . . . Often older members feel concerned that if newer members are taken in they will provide competition for jobs when conditions slip backward a little." After calling this a shortsighted policy, Hutcheson issued the following warning:

> I am inclined to believe that the existance of so much nonunion work is responsible for a good deal of the membership loss each year . . . Nonunion jobs operate six and often seven days a week. While the rate may be lower, the long hours make the weekly paycheck average out fairly high . . . At the present time there is a relative scarcity of qualified mechanics. Nonunion contractors must set wages relatively high to man their jobs. But let there be even a minor setback and nonunion wages would soon take a tumble. The nonunion employer has absolute control over the wages he pays; therefore, he sets wages at the lowest possible point that will permit him to man the job . . . This will put pressure on union wages which will be difficult to resist successfully.[49]

George Meany, President of the AFL-CIO, spoke to the convention. He praised Maurice Hutcheson for his work as chairman of the AFL-CIO committee on social security, "which gave all of America such able leadership in the fight for Medicare, a fight we won after long and bitter struggle, and I wanted to take this time to publicly thank him for his service to the whole of the trade union movement."[50] This was in sharp contrast to the uncompromising opposition of William Hutcheson to government-sponsored medical care programs two decades earlier.

The number of appeals to the general president for the years 1962–1966 increased somewhat over the previous four-year period, perhaps as a result of eliminating the $50 minimum limit. A total of 866 appeals were filed; 706 were acted upon; and 36 were appealed to the GEB, most of them involving disciplinary proceedings.

The convention spent a great deal of time on the apprenticeship program. The average age of working carpenters in 1966 was fairly high, 45.6 years. Even though the carpenters' apprentices numbered 30,000, by far the largest group of any single trade in the country, Hutcheson noted that "our apprenticeship programs are barely turning out enough journeymen to replace those members who die in a given year." Vice-President Finlay Allan pointed out that "from a strictly selfish point of view, any union has a big stake in seeing its members are as competent as possible.

It is easier to bargain for members who can do a good job and produce a profit for the boss. If negotiations get really tough, an employer is going to think twice before trying to replace workers whose skill and knowledge will be seriously missed."[51]

In 1967 the Brotherhood signed an agreement with the Department of Labor for an expansion of its apprenticeship program. For a subsidy of $2.9 million over a two-year period, the Brotherhood agreed to train 2,000 men, many of them from minority groups, in carpentry skills, and to provide a pre-apprenticeship program of 26 weeks. The following year, the Brotherhood became a prime contractor at fourteen Job Corps Centers, where the programs were devoted mainly to building recreational facilities in national parks.[52] It is fair to say that no union in the United States was more interested in building up an apprenticeship program, viewed by the Carpenters as a necessary element in the maintenance of union strength and the prevention of debasement in skill standards.

Turning back to the 1966 convention, several proposals to amend the constitution were defeated. One would have reinstated the election of officers by referendum. Another called for compulsory retirement of officers at age 65, and evoked the following comment from the resolutions committee: "The principle of compulsory retirement at any given age is undemocratic. This principle is reflected in negotiated pension plans under our respective bargaining agreements. If it is not applicable to our rank and file membership, it should not be imposed on our fulltime officials."[53]

Despite considerable opposition from the floor, a pension plan for officers and other Brotherhood employees was approved. Apart from a discussion over the merits of rival candidates for a vacated Canadian seat on the GEB, and a challenge to one incumbent, the election of officers took place without incident. General President Hutcheson ended the convention with a plea for more political activity, particularly to achieve the Carpenters' principal legislative goals, the outlawing of state right-to-work laws and the legalization of situs picketing.

AFFIRMATIVE ACTION

On June 21, 1963, the presidents of eighteen building trades unions agreed on a policy of nondiscrimination in employment. The first union to act officially to implement that policy was the United Brotherhood of Carpenters. The GEB issued a statement which not only pledged that any qualified worker would be accepted as a member, but went beyond the Building Trades commitment by directing the elimination of racially segregated locals:

Any such unions as may exist have been kept segregated not at the behest of our Brotherhood, but because of the insistance of their own members, who are members of minority groups. We have decided that to be consistent our Brotherhood cannot maintain segregation at the request of the same minority groups who are justifiably anxious to wipe out all segregation. However, with the best will in the world to be helpful in correcting any discrimination, we cannot under any circumstances permit the imposition of quotas, either by a governmental agency or by any outside private organization. We consider quotas totally undemocratic and out of line with sound trade union practice.[54]

In reporting this statement, the *New York Times* noted that "the carpenters' union is considered one of the more enlightened building trades unions on questions of race. It has not been a focal point of the attack on discrimination in the building trades by Negro groups. Where discrimination has existed, it has been a matter of local practice without sanction of the national union."[55]

In response to an equal opportunity questionnaire from the Labor Department, Hutcheson informed the Secretary of Labor that the Brotherhood had no data on racial patterns in local unions, since such records had never been kept. He pointed out that employers shared with unions the responsibility for any discrimination in admission to apprenticeship programs, and concluded: "We are interested in wiping out all discrimination based on race, creed, or color in job opportunities, and you may rest assured we will continue our efforts to achieve this goal."[56]

The Brotherhood had no problem complying with equal employment access regulations; that had been its policy from its very foundation in 1881. The determination to eliminate segregated locals, however, did mark a change from its refusal to take such a step eight years earlier. The general office proceeded to enforce the policy, which did not always prove easy because of the resistance of some black locals to mergers. William Sidell, when he became general president, was obliged to visit several cities in order to persuade members that it was in their interest to consolidate. In Baltimore, for example, a black local of long standing insisted that it wanted to maintain its identity and its own officers, and even threatened legal action if the national office persisted. Eventually it was prevailed upon to admit some whites, but it did not give up its charter.[57]

On the whole, the 1960s were good years for the Carpenters' Union. The economy remained surprisingly stable in the face of the military demands created by the war in Vietnam. One can only look back in wonder at the fact that for the entire decade, consumer prices rose by only 31 percent, an average annual increase of only slightly more than 2.5 percent. The average unemployment rate over the entire decade was 4.8 percent, and in 1969 it had fallen to what many would regard as the irreducible minimum level of 3.5 percent. The money wage of union carpenters rose by 72 percent for the ten-year period, which meant a real wage increase of more than 30 percent. Politically, apart from the final two years of the decade, the federal government was controlled by administrations friendly to the labor movement, although labor had not been able to persuade Congress to enact the amendments to the Taft-Hartley Act governing situs picketing and right-to-work laws that were at the top of its legislative agenda. As Hutcheson told the 1970 convention, "The past four years have been among the most prosperous years the construction industry has ever experienced. Wage rates have climbed far faster than ever before in history."[58]

This proved to be the calm before the storm. The years ahead were to be much more turbulent for labor and management, and for the economy in general.

15

BATTLING INFLATION AND RECESSION: THE NINETEEN-SEVENTIES

The 1970s started out calmly enough. Although unemployment rose to 4.9 percent of the labor force in 1970 from the remarkably low figure of 3.5 percent a year earlier, no one was talking about recession. There was concern about the 6 percent increase in the cost of living during 1970, but this was not yet perceived as a harbinger of worse ahead. The volume of construction activity remained high. No one could have foreseen how quickly the economic structure of the country would be affected first by inflation, then by recession, and finally by a completely new phenomenon for the United States, a combination of inflation and economic stagnation—stagflation.

The construction industry faced some special problems. As Quinn Mills has pointed out, "The inflation of the 1965–70 period had involved, in construction especially, so much distortion of traditional wage relationships that governmental intervention was accepted as merited by the requirements of equitable administration of stabilization."[1] Union wage rates in the building trades rose much more rapidly than those in manufacturing, and there were many strikes in construction, leading to special treatment for this industry.

THE THIRTY-FIRST GENERAL CONVENTION

The years 1966 to 1970 were characterized by a substantial increase in Brotherhood membership. The total 1970 membership of 821,000 did not come up to the 1958 peak of 840,000, but the number of members in good standing, 759,600, was an all-time high. The increase was achieved despite the continuing pattern of an annual membership loss of 90,000 due to retirements and withdrawals.

The financial problems involved in carrying the pension scheme and the Home were becoming increasingly critical. The pension amount had been doubled, from $15 to $30 a month, by the previous convention, and despite a similar percentage increase in the per capita tax going to support the programs, the central fund was dropping steadily. The most recent applicants for admission to the Home were largely in their late seventies or early eighties, and more than half required hospitalization.

For the Home to qualify as an extended-care facility under Medicare would have required many structural changes in the physical facilities, so the Brotherhood had to bear the full cost of care. From 1967 to 1970, the cost of the Home per occupant had risen from $2,333 to $3,203 per annum.

The convention committee responsible for dealing with the Home and pensions nevertheless recommended that the Home be retained, and despite considerable opposition from the floor, this recommendation carried. However this victory for those for whom the Home issue was a sentimental one proved short-lived. Two years later the GEB ordered a referendum on the matter because of the continuing financial drain; it was costing more per *month* to maintain an individual at the Home than the *annual* pension currently being paid. By a vote of 59,480 to 24,049, the membership agreed that the Home should be discontinued.[2]

The pension proved to be a much more stubborn issue. Few members had been personally involved with the Home, but every member had a potential claim to a pension. There was a strong belief that the pension was an entitlement paid for out of dues. The finance committee of the convention recommended a referendum on the phasing out of pensions by exempting from the scheme anyone joining the Brotherhood after July 1, 1971, but this proposition was defeated resoundingly. The convention did agree to raise the monthly per capita contribution to the pension fund by 25 cents, but this was a drop in the bucket.

By 1973 the fund's quarterly income was $3.2 million, compared with expenditures of $4.5 milion. The GEB determined that an additional $1.25 monthly tax would be necessary to keep the fund from insolvency, and submitted this proposal to a referendum. The proposal was defeated by a very close margin; the board thereupon decided that there would have to be a quarterly adjustment of the pension amount based on financial ability to pay; in March 1973 it was reduced from $30 to $22 a month.[3] The members were thus put on notice that they would have to pay for the benefits they wanted.

The 1970 convention received a report from the General President's Committee on Contract Maintenance, a group composed of the presidents of thirteen national unions that had been in existence for fourteen years. Its purpose was to facilitate the performance of construction work in industrial plants by building tradesmen. The agreements signed through the intermediation of this group contained a commitment against work stoppages, and concessions on overtime rates, shift work, and subsistence and travel pay. A number of problems had arisen: some locals were attempting to enforce all the terms of local agreements, even those in violation of the agreements; contractors sometimes found it difficult to adhere to jurisdictional lines for industrial work, and were given too little leeway; and the growing disparity between construction and in-plant wages was leading some firms to conclude that it was more economical to have the work done by their own employees. As of 1970 there were ninety-two such contracts in effect, but no new ones had been reached for some time. The conclusion was, "We have thus far only made a small ripple on the surface."[4]

The Brotherhood's Apprenticeship and Training Department, which had been initiated in 1966, now included seven full-time field coordinators and forty-six full-time instructors. Both Hutcheson, in his opening address, and Sidell, who was in charge of the program, complained of harassment by government bureaucrats who were attempting to substitute selection of apprentices by lottery for the selection system that was based on merit. Both men expressed strong opposition to random selection, as well as to the racial employment goals that the government was seeking to install. The four-year period of apprenticeship, which had been criticized as being of excessive length, was considered by Carpenter officials to be essential to the development of well-rounded skills.[5]

One of the more interesting debates at the convention involved a constitutional amendment submitted by the GEB providing that if a local engaged in a jurisdictional work stoppage contrary to the order of the general president, or was involved in a work stoppage that had lasted more than thirty days, the GEB would have authority to investigate and reach a binding agreement on behalf of the local. The following explanation was set forth in support of the proposal:

> Because of the seriousness of the problem in our industry of unnecessary work stoppages and strikes, it is generally felt that some solution must be found and the necessary authority granted to the elected leadership to implement the solution once it is found. It is generally recognized that the major problem facing us today is not the alleged high wage increases but is the continual and frustrating stoppages of work for one reason or another . . . It is becoming increasingly more difficult to evade the pitfalls of legal procedures as a result of illegal and other types of work stoppages that naturally result in damage suits, and sometimes costly legal maneuvering . . . We must find some way to protect our friendly contractors . . . such as the National Constructors' Association, against our mutual enemies, the open shop, ABC and non-union, non-competitive employers.[6]

This bid for greater centralization of authority in the national office met with strong opposition. A delegate from a Canadian local felt that help from the general office was already too slow in coming, "because the job is finished, and everything is dropped or forgotten until the next time, so the only recourse left to us is to pull the men, to stay away from the job, or to harass the sub or the general [contractor], whichever the case may be." Others maintained that the GEB could not be familiar with local situations and that its intervention would only imperil future negotiations, emasculate local committees, and induce contractors to hold out for thirty days. In the end, the proposal was defeated. This episode illustrates the difficulties faced by a national union, in an industry characterized by decentralized collective bargaining, of exercising the degree of control over local unions essential to the continued economic health of the organization. Delegation of greater authority to the national office appears to members to be undemocratic; the failure of the national office to intervene in order to correct local maladministration is taken by outsiders as a sign of weakness. Striking the right balance is not an easy matter.

Another controversial amendment brought in by the GEB related to the membership rights of construction superintendents who had come up through the Carpenters' ranks. The board wanted to confirm a ruling it had made earlier giving these men the right to continue their membership, to be exempt from strike assessments, and to go through picket lines to perform their normal functions. An additional issue was whether they should be permitted to hold union office; apparently many of them did so. Hutcheson explained that the amendment had been brought in because many superintendents were being fined for going through picket lines, and unless they were protected, they would be lost to the union, with unfortunate consequences for Carpenter success in securing initial work assignments. The delegates voted against the right of the superintendents to hold office, but they were evenly split on the balance of the proposal, which was turned back to the GEB for further study. There was apparently a good deal of resentment against the superintendents; as one delegate declared, "They negotiate with their contractor, they have their own little deal under the door and . . . they haven't got the trade union movement in their heart."[7] The board later affirmed the substance of the original resolution on membership rights.

All the incumbent officers were reelected unanimously. However, a resolution to establish the office of General President Emeritus, which would carry full pay for life, met with unexpected opposition. A number of speakers, though not criticizing Hutcheson, the intended beneficiary of the proposal, contrasted this generosity with the reduction of pensions for retiring members. The voice vote was so close that the chairman had to call for a standing vote, which was in the affirmative. From the transcript of the proceedings, it is clear that the delegates were not at all reticent about expressing their views on this delicate question. (Hutcheson had left the chair, but he was present during the discussion.)

NATIONAL AGREEMENTS

The trend toward the consummation of national agreements accelerated during the 1970s, reflecting the rapid change in construction technology.

Prefabricated housing provided an opportunity to win back some of the on-site jobs that had been lost, through the negotiation of contracts with the manufacturers of prefabricated materials. In a major policy statement, the GEB proclaimed that all prefabricated houses, regardless of the materials used, were within the Carpenters' jurisdiction, and that local agreements were to conform to uniform national standards.[8] In order to prevent jurisdictional disputes in this portion of the industry, a tri-trade agreement was reached with the Plumbers and the Electricians providing for a master contract for work within the factory and the apportionment of site work on the basis of established jurisdictional lines.

A national agreement was reached with a subsidiary of Monsanto, the large chemical firm, giving the Carpenters exclusive bargaining rights for employees engaged in installing artificial turf, a new material used in the burgeoning sports industry.[9] The Wall and Ceiling Contractors' Association agreed that carpenters should receive the majority of work assignments—though not all, for the Lathers were still doing some of the work.[10] After a conference with employers in the resilient floor industry, during the course of which it emerged that 65 percent of products used were being installed by retail shops, none of which employed Carpenter members, it was agreed that collective bargaining in this industry should be placed in a separate category, and a moratorium on wage increases was decreed until an appropriate bargaining mechanism could be developed.[11] The Brotherhood joined in the national Building Trades Nuclear Power Construction Stabilization agreement, which superseded all existing agreements for new projects in this industry. "Although the Board was not in complete agreement with its contents, it was of the opinion that it would be in the best interests of the Building Trades to combat the open shop and the Environmental Protection Agency."[12]

In order to protect specialized contractors operating in more than one labor market, the GEB imposed some national rules to facilitate manpower mobility. These contractors were permitted to employ for five days two of their specially trained employees without local working permits. For work lasting more than five days but not requiring more than four men, the contractors were required to secure 50 percent of their men locally. Moreover, any member working outside his home jurisdiction for five days or less could no longer be required to pay for a working permit. The board justified these exemptions from normal working rules in the following terms:

No longer can we maintain a posture adopted years ago for a total local autonomous structure, as all evidence points to a diversified, specialized, mobile industry which requires our members to move about with a reasonable amount of mobility. Too long have we been criticized due to strict employment requirements, indiscriminate economic assessments directed at vast numbers of our members in the form of work permits or assessments paid, in many instances, several times during the period of one month because the nature of the industry demanded our members move in and out of numerous jurisdictions.[13]

This by no means settled the problem. In southern California, for example, some local unions continued to be reluctant about clearing outsiders for work:

The Board reaffirmed its concern over the need to find a suitable solution to this problem as it is having a detrimental effect, not only in the Southern California area, but in other areas of the country, where contractors signatory to agreements are cancelling authorization for the renegotiation of such agreements due to the failure of our affiliates at the local level to amicably find the solution to the problems of manpower mobility.[14]

All of this would have been quite familiar to Peter J. McGuire, who had to wage a never-ending fight against local exclusionist policies.

WAGE CONTROLS

To a country that had been accustomed to a stable price level, the 6 percent increase in the consumer price index during 1971 seemed like runaway inflation. There was a general belief that trade union construction wages were a major cause of the inflation; from July 1, 1969, to July 1, 1971, they rose by almost 25 percent, compared with an increase of 12 percent for manufacturing employees during a comparable period.

The Nixon administration responded by suspending the operation of the Davis-Bacon Act on February 3, 1971. However, this action was soon revoked and superseded by an executive order of March 29, 1971, the contents of which had been agreed upon by unions and contractors, with the concurrence of Secretary of Labor James Hodgson. Under the terms of this order, a tripartite Construction Industry Stabilization Committee was set up, with four representatives each for labor, management, and the public. In addition, craft dispute boards were established within branches of the industry to advise on sectoral wage standards and to scrutinize collective agreements to determine whether they were in conformance with the standards established. Specific wage guidelines were laid out, and any agreements in excess of the stipulated amounts, with certain exceptions, were made unlawful.

Hutcheson's reaction to this drastic government intervention in the labor market was predictable. "Between a President who seemingly has little respect for blue collar workers, minority groups which are determined to break down the whole concept of apprenticeship training, and a press that pictures construction workers as feather-bedding millionaires, the need for solidarity is unprecedented. To me, it all adds up to one thing. We in the building trades need to mobilize our political strength more effectively than we ever have before. It is in the political arena that most of our future battles lie."[15] Sidell, who was a labor member of the Stabilization Committee, declared: "To say that as an organization we dislike the burden of wage controls and our required expenditure of money and manpower is a drastic understatement."[16] In August 1971 wage controls were extended to all of industry, including the industrial segment of the Brotherhood's membership.

The Stabilization Committee was successful in bringing down the rate of wage increases in construction. Union wages in that industry rose by 11.7 percent from 1971 to 1973, as compared with 14 percent for manufacturing. In fact, the committee received the full cooperation of the construction union presidents, who were themselves concerned lest wage increases get out of hand. Many of them regretted the committee's termination with the lifting of controls in 1974, but found it impolitic to say so openly. However, the Carpenter Craft Board, at least, proved to be a useful innovation, for the GEB called for its continuation after the controls were lifted as a means of facilitating the negotiation of agreements and the avoidance of strikes.[17] This body has remained in existence on a voluntary basis.

THE THIRTY-SECOND GENERAL CONVENTION

The 1974 convention was opened by General President William Sidell, who had succeeded Maurice Hutcheson on the latter's retirement in 1972. Sidell had been elected to the GEB in 1962 and appointed a vice-president two years later. The son of a cabinetmaker, Sidell was born in Chicago and then moved to Los Angeles with his parents. When he completed his formal education, he became an apprentice in Local 721. He held successively the posts of warden, recording secretary, organizer, assistant business representative, business manager, and president of the 4,500-member local. In 1957 he was elected secretary-treasurer of the Los Angeles District Council of Carpenters, the largest council in the country. It would be difficult to find anyone whose career was more closely linked to the carpenter's trade and the union.

In his opening address and his report to the convention, Sidell outlined recent developments that had occurred in the organization and the outstanding problems that faced it. They are best considered seriatim, with some additional annotations where appropriate.

1. Jurisdictional disputes. The greatest problem was the fragmented nature of the industry: "There simply is too much duplication and this

results in the greatest cancer of all, the jurisdictional dispute. The jurisdictional dispute foments work stoppages, worker insecurity, confusion and concern, decreasing if not actually destroying productivity."

2. Centralization of authority in the national union. "I see us playing a greater role at the General Office level, directing, guiding, and giving greater service . . . This in itself is indeed a departure from the traditional stance of our Organization. Whether we wish to recognize it or not, there has been over recent years a voluntary abdiction of certain traditional autonomous activities."

3. Recruitment of new members. This was the union's top priority. To stimulate organizing activity, Sidell had initiated two formal programs: the Coordinated Housing Organizing Program (CHOP) and the Voluntary Organizing Committees (VOC). The first was aimed at the housing market, which was a union stronghold prior to World War II but had since become mainly nonunion. During the war years, when there was little housing construction, many union carpenters had moved into defense and industrial construction work and were reluctant to return to housing, where wages were lower and jobs uncertain. The task before the union was to organize the half-million carpenters who were employed by nonunion contractors, as well as to halt the incursion of these contractors into the union sector. The VOC program was aimed at the industrial sector. It was estimated that there were 1.5 million industrial nonunion workers within the Carpenters' jurisdiction. For both programs, locals were urged to set up three-man organizing committees, and the national office was to supply manuals, spot radio announcements, audio-visual material, and other informational matter. Regional conferences of full-time Carpenter officials were instituted at which the new organizing programs and techniques were explained. A system of regular reporting was instituted so that the national office could keep track of local efforts.[18]

4. Industrial councils. A new structural unit, the industrial council, was instituted as a means of assisting local unions in their normal contract administration activities, as well as to coordinate bargaining on a regional level. The Southern Industrial Council, which dated back to 1967, had eighty-nine affiliated unions at the close of 1973. Similar councils had been set up for the mid-Atlantic states and for Indiana. These steps reflected a growing awareness on the part of the Carpenter leaders that the potential for considerable growth lay outside the construction industry.

5. Pensions. Few developments were of more immediate significance for the well-being of the members than the spread of area-wide multiemployer pension plans. In view of high employment turnover, single-employer pensions were of little benefit to a construction carpenter. By 1974 some 550,000 members were covered by such plans, half of them

vested. The Brotherhood itself had set up a Carpenter Labor-Management Pension Fund which was designed to make it possible for locals in nonurban areas to secure coverage.[19]

All of these initiatives implied a greater future role for the national office if the Brotherhood were not to stagnate. Not only the growing industrial sector but the organization of nonresidential construction required coordination on more than a local labor market level. The local unions and the district councils were often not happy about this trend, and they resisted it, but effective management of the carpentry trade could not continue on the basis of the traditional union structure. The general office was not reaching out for power for the sake of power; it was simply recognizing the shift that had already occurred, as Sidell observed.

The report on contract maintenance work provided another example of the need for national coordination. The number of contracts in effect had reached a peak of 105 in 1968, but dropped to 53 by 1971. In an effort to arrest the downward trend, the General Presidents' Committee had been authorized by the twelve participating unions to freeze wage rates at any location where the gap between in-plant and outside rates had become dangerously high as a result of the construction wage explosion of 1969–1971. The Carpenters were also losing maintenance work because of jurisdictional disputes and other causes of delay that were unpalatable to industrial firms. "Industrial owners were just not bothering to give our contractors an opportunity to bid the work because they would not tolerate inefficiency, work stoppages and high construction rates and excessive premiums." But the breakthrough came in 1972, when an agreement was reached with Westinghouse on the maintenance of generators which spread to other power-generating operations. A national agreement was also reached with the National Erectors' Association, which did contract maintenance work in the steel industry.[20]

The 1970 convention had amended the constitution to give the GEB final authority to decide all appeals other than those involving suspension or expulsion from membership, geographical jurisdiction, consolidation, and the formation of district councils. The general president was given full authority to decide all cases involving elections. As a result, the number of appeals to the convention declined from 68 in 1970 to 21 in 1974. In fact, most of the appeals were being handled by the newly established Appeals Committee; of the 774 filed during the four-year interconvention period, 713 were decided by the committee and only 61 went to the GEB.[21]

Some important constitutional changes were enacted by the 1974 convention. An amendment to the Obligation undertaken by each member on joining the union committed him to observance of local bylaws in addition to the previous requirement to abide by trade rules. There had been some legal challenges to local disciplinary cases on the ground that the constitution contained no such requirement, although in fact it had always been assumed that members would have to honor the bylaws of their locals.

Another innovation was the grant of authority to the GEB to bring charges against individual members. Up to that time all charges had to be initiated at the local level, which meant that particularly in political situations, the national union could not act as long as the offenders controlled the local vote. Moving in the same direction was a constitutional amendment proposed by the GEB that codified and elaborated previous practice with respect to the authority of the national union to establish, dissolve, and determine the geographical jurisdictions of local unions and district councils. Although the other proposals passed with little opposition, this one ran into trouble. One opponent asserted:

> It covers too much area. It would literally give dictatorial powers to the general reps. in a particular area that they have jurisdiction over. The general rep. would come in; he would dissolve a Local Union or merge a Local Union, a District Council, or even a State Council, without a single vote being cast, and I believe that the membership in these areas should have a voice in this, and I think that local and area problems are best handled by the Local Business Agents, the District Council personnel, the State Carpenters Councils' personnel. They understand the problems. They understand the geographical boundaries better.[22]

Sidell made his position clear:

> If we have Locals combine in a given area where there are ten members in one Local and eleven members in another Local and twelve members in another Local, sure, I know it is their desire to stay there and have that Local stay in existence, but it also has to be looked at from the standpoint that somewhere in this great world of ours and in this great organization we have to put something together here whereby we can move. We can't let this thing flounder out there without any guidance or direction.[23]

The proposal carried by the narrow margin of 1,142 to 1,048. Local separatism was still a force to be reckoned with.

The constitution was also amended, without debate, to permit local unions to spend their funds for political purposes provided the expenditure was based "not on party politics or the candidate's political affiliation but is based on the candidate's position as a supporter and advocate of objects, principles and legislative goals of our trade." The GEB had actually made a ruling to this effect at the beginning of 1974, and this constitutional change was designed to forestall any legal challenges to the new policy.

The 1974 convention went through the constitution systematically and made more changes than any convention since the Special Convention of

1960, which had been called for this specific purpose. It was evident that more than ever, a trade union constitution must be a "living document" in order to anticipate or react to the growing volume of legal challenges. In the past, much had been left to custom and practice; now there was a need, in an era when the number of lawyers and lawsuits was proliferating rapidly, to pin things down.

WIDER BARGAINING UNITS

There have been recent pressures for wider-area bargaining units in construction for a variety of reasons. At the national level, there has been concern about "leapfrogging" by strong local unions, leading to frequent strikes. Local union leaders have felt themselves obliged to improve upon, or at least match, the improvements gained by contiguous locals even where the underlying economic situation was different. Competition among locals has also tended to raise costs unduly and open the way for nonunion contractors.

Some outstanding examples of the tendency toward wider units were provided by the Carpenters. In 1974 the Central Illinois District Council of Carpenters, covering thirteen central Illinois counties, was formed, with six locals joining. In 1975 the council negotiated its first contract, replacing six previous bargaining units. Research workers who looked into the situation reported that "although union respondents were reluctant to discuss intraunion problems with us, it is clear that the surrender of local autonomy was opposed at least by the larger locals. Pressure for the enlarged bargaining unit came from the international."[24]

An even greater unit expansion took place in Wisconsin, also in 1975, with the establishment of the Greater Wisconsin Bargaining Unit. The unit includes two independent locals and three district councils, all of which had constituted separate bargaining units. The agreement of 1975 was signed with the Wisconsin chapter of the Associated General Contractors. "The local and council officers knew that the international favored larger bargaining units, and the existence of an almost statewide Michigan carpenters' bargaining unit was evidence that such units could be established and made to work."[25] The increased authority to determine geographical boundaries vested in the general office by the 1974 convention has helped to consolidate bargaining areas.

POLITICAL ACTION

The carpenters had no difficulty remaining neutral in the 1972 presidential election. Almost the entire labor movement refused to endorse the

Democratic candidate, George McGovern, who was considered to be too far to the left. The GEB simply urged all members to vote after studying the background of the candidates. Judging by the results, a great many of them voted for Richard Nixon.

The 1976 election was another matter. The building trades had been getting along well with Gerald Ford, who succeeded Nixon after the Watergate affair, particularly because Ford had committed himself to approving legislation allowing situs picketing, a goal long sought by the construction unions. The bill that Labor Secretary John T. Dunlop had managed to shepherd through Congress contained additional provisions that would have given national unions more control over local bargaining. As Sidell put it, "The second part of the vetoed bill would have raised collective bargaining from an essentially local level to larger areas and even a national level in an effort to stabilize the construction industry." The purpose of this portion of the bill was to prevent locals from pricing themselves out of the market by providing a greater input of advice from the national union. When President Ford, under pressure from the right-to-work lobby, vetoed the bill, leading to Dunlop's resignation, Sidell wrote: "Secretary Dunlop was the one bridge remaining between the Ford Administration and organized labor, and this tie has been severed in the name of political expediency."[26] The GEB unanimously endorsed Carter for the presidency, and Sidell advised the members: "The Carter-Mondale slate must be elected if America is to move again and if we are to have jobs for our members."[27] In view of the close margin by which Carter was elected, it is quite possible that Ford might have won had he signed the bill. The building trades would certainly not have opposed him, and might indeed have endorsed his candidacy. A relatively small switch of votes in a major industrial state might have provided him with the necessary margin of victory.

In 1980, although the labor movement was dissatisfied with many of the policies of the Carter administration, the Republican candidate seemed less attractive. In consequence, virtually the entire labor movement, including the Carpenters, endorsed Carter for reelection. General Secretary John S. Rogers wrote in his *Quarterly Circular*:

> We cannot permit the reactionary, conservative forces which have consumed the Republican Party to take charge of both the White House and the Congress . . . For the first time in modern American history, workers have a clear choice. Our vote and support can only unequivocally be directed toward the reelection of the Carter-Mondale ticket. Both Reagan and Anderson have demonstrated, both in act and deed, their total and consistent opposition to the goals and aspirations of workers.[28]

After the Republican victory, the Brotherhood leadership oberved ruefully that the union policymakers and many of the members differed in their views of the campaign issues. "Let us hope this break does not remain, and that the membership recognizes that the leaders of their unions do, in fact, have a good understanding of what is happening."[29]

One notable aspect of the 1980 campaign, showing how far the Carpenters had departed from their traditional political neutrality, was the major role played by the Carpenters' Legislative Improvement Committee (CLIC). Common Cause estimated that the Carpenters spent $555,000 in support of congressional candidates, fourth highest among trade unions (the United Automobile Workers, the Machinists, and the AFL-CIO itself provided more), and tenth highest among all private organizations.[30] After the election all locals were urged to continue to participate in the union's political fund-raising program, under which half of all receipts are retained by the locals and the rest forwarded to the general office for national political use.[31]

CLIC operations are by no means confined to political campaigns. During 1981, for example, it was heavily involved in efforts to prevent scuttling of the Davis-Bacon Act, and succeeded in helping defeat amendments to military construction bills that would have had such an effect. Another piece of legislation that it has been active in defending is the Occupational Safety and Health Act. Although environmentalists may be unhappy, Brotherhood members in the lumber industry were undoubtedly pleased by CLIC efforts to expand the National Forestland acreage that can be used for multiple purposes, as well as in winning some degree of income security for Brotherhood members who lose their jobs through the expansion of the Redwood National Park in northern California.[32]

JURISDICTION (8)

While exhorting locals not to engage in jurisdictional strikes, the national office was constantly involved in arbitrating local disputes through the national jurisdictional machinery. The data in Table 15.1 show how substantial an amount of litigation was involved. On April 1, 1971, the Brotherhood was suspended from the National Joint Board for failure to accept and implement some of the board's decisions, but it rejoined on June 1, 1973, when the procedural machinery was revised and the National Board was replaced by an Impartial Disputes Board.[33] The large number of adverse decisions sustained by the Brotherhood from 1970 to 1973 reflects the fact that the Brotherhood did not participate in these cases, which are thus unrepresentative of its success rate.

As the data indicate, the policy of the Brotherhood in bringing cases before the various disputes tribunals has generally been a defensive one, reflecting its success in securing initial work assignments. Even after its suspension was lifted, the Brotherhood continued to be very selective in

Table 15.1   Cases decided by the jurisdictional disputes board involving the
United Brotherhood of Carpenters, 1970–1980

| | Total requests | | Decisions | | | |
Year	By Carpenters	By others	Total	Won by Carpenters	Lost by Carpenters	Split
1970	89	160	249	80	161	8
1971	16	245	261	30	231	0
1972	0	223	223	7	213	3
1973	0	172	172	35	133	4
1974	0	84	84	37	39	8
1975	21	105	126	57	61	8
1976	27	105	132	68	57	7
1977	34	64	98	56	38	4
1978	11	33	44	24	18	2
1979	20	57	77	36	38	3
1980	27	86	113	43	68	2

Source: United Brotherhood of Carpenters, *Proceedings*, Thirty-Second General Convention, 1974, p. 167; Thirty-Third General Convention, 1978, p. 194; Thirty-Fourth General Convention, 1981, Reports of the General Officers, p. 23.

bringing cases before the board, and perhaps because of this policy, it won more cases than it lost during most years since 1974.

The cases that found their way to the board represented only the tip of the iceberg. Most of the disputes were settled by meetings at the job site; the following figures present the number of jurisdictional assignments of general representatives over the years 1970–1980:

1970	2,108
1971	1,856
1972	1,648
1973	1,632
1974	1,650
1975	1,403
1976	1,300
1977	1,294
1978	1,169
1979	1,382
1980	1,296

The downward trend in the need for assignments was attributed to "the conscientious application of existing agreements and cooperation between business representatives in the field and the Jurisdictional Department at the General Office."[34]

A number of national jurisdictional agreements were reached during the decade: with the Electrical Workers on luminous and acoustical suspended ceilings and automated interoffice transporters; with the asbestos workers on refrigerated storage tank installation; and with the Plumbers on Brown-Bouverie turbine installation. A mutual assistance pact was signed with the International Woodworkers' Association in 1973, providing that either union was free to organize a nonunion plant, and in the event that both organizations were engaged in the same campaign, meetings were to be held to determine which should continue. There was to be exchange of information and joint conferences in collective bargaining, as well as mutual respect of picket lines.[35] (This agreement did not apply to Canada.) There were discussions of a possible merger between the Brotherhood and the Woodworkers, but these terminated without result. The mutual assistance pact remains in effect.

The IWA is one of two existing unions with jurisdiction that conflicts directly with that of the Carpenters. The other is the Furniture Workers' Union, with which the Brotherhood also engaged in fruitless merger discussions during the decade. It has been difficult for small unions to survive in the atmosphere of the early 1980s, and the trend among unions to merge reflects the need for the financial power that can come only from larger aggregates. However, the IWA and the Furniture Workers are still committed to an independent existence.

A jurisdictional dispute that was rivaled in duration only by that with the Machinists was finally settled in 1979 when the Wood, Wire and Metal Lathers International Union joined the Carpenters' Union. Under the terms of the merger agreement, local unions of the Lathers with sufficient financial resources were to continue their separate identities, but the smaller ones were to be consolidated. A Lathing subdivision was created within the Carpenters, which the former general president of the Lathers was to supervise in his capacity as assistant to the general president of the Carpenters. The New York local of the Lathers, No. 46, was exempted from the merger and joined instead with the Iron Workers. Some years earlier, when the Iron Workers were out of the AFL, Local 46 had taken over reinforcing rod work, and had held it ever since. A majority of its members were rod men, and the Carpenters had no desire to dispute these jobs with the Iron Workers. Apart from that one local, however, the remaining 11,000 members of the Lathers became Carpenters, with all the obligations entailed, and the not unimportant right of working at any job within the Carpenters' jurisdiction, which greatly widened their potential job opportunities.

In his address to the 1978 convention, Vice-President Campbell made the following observation about jurisdictional strikes: "The jurisdictional disputes and the unauthorized walkoffs are two of our worst problems. We have been lucky in some cases where this has happened, the owners have not gone to the courthouse and brought suit against us. In case some have forgotten, there was a case in recent years that cost the Brotherhood $1,173,734.12."[36] He was referring to a suit brought by the Noranda Aluminum Company of New Madrid, Missouri, arising out of a jurisdictional strike, which the carpenters had lost after carrying an appeal all the way up to the U.S. Supreme Court.[37] This was the largest judgment ever rendered against the Brotherhood in a single case. Controlling jurisdictional strife had become a matter of survival.

THE INDUSTRIAL DEPARTMENT

The growth of the nonconstruction portion of the Brotherhood's membership necessitated the creation of a new body in 1979 to cater to their special needs. Its main function is to provide training and support for local union officers and staff in the industrial sector. Thus, an extensive training program was developed for shop stewards, whose work is quite different from that of business agents on construction jobs. Government funds were obtained to finance an industrial safety program in the lumber and wood products industries. The kind of information that industrial unions furnish to their regional and local bodies—financial analysis of employer firms, contract analysis, and general business data—is among the services provided by the Industrial Department of the Carpenters.

Some first steps have been made in the direction of coordinated bargaining, where a number of locals bargain with separate units of the same corporate enterprise. The Industrial Department also makes certain that the legislative interests of its industrial members are not overlooked, as, for example, in the problems of exporting logs and tariffs on imported plywood.[38]

The Brotherhood joined the AFL-CIO Industrial Union Department in 1980, paying per capita tax on 200,000 members. By this time, fully a quarter of its members were in industries other than construction, and included such diverse occupations as peat moss gatherers, musical instrument makers, and builders of fiberglass. About 150,000 were employed in the lumber and related industries.

PENSIONS AND OTHER BENEFITS

The payment of dues-financed pensions in the face of the long-standing unwillingness of the membership to support them financially had become an impossible drain on the resources of the Brotherhood. Resolutions introduced at the 1974 convention to patch up the financial structure of the system went down to defeat. Because of lack of funds, the monthly pension was reduced to $14 in 1978.

Over the four-year period 1974–1978, the Brotherhood paid out $75 million for pensions and other benefits. The number of pensioners was increasing, while the dues-paying membership was declining. It was estimated that in order to finance a $30 pension, the per capita tax going into the fund would have had to be raised from $1.45 to $5.03 per month.

The constitution committee introduced a proposal at the 1978 convention to eliminate pensions, but softened the blow by recommending an expanded program of funeral donations. The term "semibeneficial" as applied to industrial locals was to be eliminated, and in its place, a separate schedule of benefits provided.

It was not an easy matter to sell this program to the membership. Delegate Wigner of Local 792 complained that while the members' pensions were being discontinued, the tax to support pensions for officers and employees was rising. "So in effect what you are doing if we vote these two in is sending us back telling our 30 year members, 'Sorry, fellows, the pension program for you is out. However, we now have a pension program for this office girl which we'll pay for by sending Washington each month 10 percent of her salary.'" Delegate Spano of Local 1102 declared that "I really, honest to God, can't conceive, I can't conceive going back to my local, or any other Rep in this hall going back to his local and tell these old-timers that, 'In three months, you guys are cut off.'"

General President Sidell made an impassioned plea in favor of the committee proposal. Pointing out that maintaining pensions would entail a huge increase in the per capita tax, he continued:

> Let me tell you, if you go back to your membership and tell them that you have put another $3.58 on their dues structure, I'm going to tell you that you'll lose a lot more members than you ever thought of losing [by abolishing the pension]. Plus the fact we are struggling for our life. We are struggling for our life to try and ward off the non-union elements in this country. And I'm telling you that if our old-time members were able to rationalize the effect of this to the extent that they thought this organization was going to go down the tubes, I'd tell you right now, they would be willing to sacrifice their pension.[39]

The convention finally voted to accept the committee report. The Carpenters found it difficult to give up a pension scheme that had been in effect for half a century. Since its inception, more than $221 million had been paid out. The death and disability program, which had been initiated in 1881 and which was to continue, had cost the union $129 million. By comparison, strike benefits paid by the general office had come to only $10 million from 1886 to 1978.[40]

The costs of benefit unionism had become too high. Moreover, the retired carpenter with no means of support apart from his union pension had become nonexistent. Apart from Social Security, there were now employer-financed pension schemes covering 80 percent of the members. The Carpenters' Labor-Management Pension Fund, which was initiated in 1971, had accumulated assets of over $20 million by 1980, and paid out over $1.5 million from 1978 to 1980. This scheme is entirely separate from the many pension plans that have been set up on a regional basis. The Brotherhood has finally become what Peter J. McGuire had wanted it to be a century earlier—a protective union.

CANADIAN INDEPENDENCE

The autonomy demands of the Canadian locals mounted in intensity during the 1970s, and were not resolved by the end of the decade. The prime mover was the British Columbia Provincial Council, with less support from the rest of Canada. In 1972 the GEB was asked to recommend changes in the constitution that would, among other things, permit the Canadians to elect their own board members, guarantee that all Canadian dues be spent in Canada, and provide separate conventions for the Canadian locals. These proposals were all rejected by the board; the furthest it would go was to establish a Canadian research director to develop statistics and monitor legislation in that country.[41]

This did not satisfy the separatists. The Canadian Labor Congress adopted a code of minimum self-government standards for all Canadian locals of American unions that went far beyond what the Carpenters were prepared to allow. A resolution that would have expanded Canadian autonomy went down to defeat at the 1974 convention.

In 1978 the Canadians asked that they be permitted to elect a delegate to conventions of the Building Trades Department, rather than having him appointed by the general president. A Canadian section of the Building Trades Department had been established, and this request was part of a concerted move by all the building unions in Canada. The proposal went down to defeat.[42]

The separatist movement would not stay down, however, and later in the same convention a number of resolutions were introduced asking for several additional concessions: (1) all Canadian officers to be elected by Canadians alone; (2) policies dealing with Canadian affairs to be determined by the Canadian officers and members; (3) a Canada Conference that had been set up ten years earlier, but that had become inoperative due to internal dissension and had recently been reactivated, to be strengthened. The conference was composed of two representatives from each province who were officers of their locals or district councils. What the Canadians wanted was an organization with its own full-time officials — in effect, a miniature national union. As one Canadian delegate put it: "It's of the utmost importance . . . that we have the election of Canadian Officers to speak on behalf of Canadians at the National level of our country. We need a national structure to be able to do that . . . We not only should do that which has been done so well by the Brotherhood in the United States, but more important, we must do it."[43]

The effectiveness of the autonomy demands was reduced by the fact that while they were supported by District 10 (the West), they were opposed by District 9 (the East), a split that in the past had rendered the Canada Conference ineffectual as an instrument for promoting Canadian separatism. The convention instructed the GEB to seek ways to strengthen the Canada Conference within the existing structure of the Brotherhood.

This did not resolve the problem. The Carpenters' internal dispute became part of a larger one involving all the Canadian building trades. In 1981 the U.S. building unions, with about 400,000 members out of the 2.3 million affiliated with the Canadian Labor Congress (CLC), were suspended from that body for failure to pay their dues to the Congress. The dues were withheld in retaliation for the refusal of the Congress to take action against its Quebec branch, the Quebec Federation of Labor, which had granted affiliation to 12,000 members of a unit of the International Brotherhood of Electrical Workers that had seceded from the IBEW. The building crafts were also opposed to Congress support of the New Democratic Party, a socialist group, and objected to the system of voting at Congress conventions that accorded them a voice less than proportionate to their membership.[44]

General President Konyha, who succeeded Sidell in 1980, noted that the building trades had tried for years to obtain fair representation at CLC conventions, but representation continued to be based on the number of local unions affiliated with a national union rather than on its total membership. For example, the public service unions had 45.6 percent of the total vote but only 27 percent of membership; the building trades had 15 percent of the members but only 7.2 percent of the vote.[45]

After suspension of the building unions, the CLC established construction branches in a number of localities in Canada, which could conceivably become the nucleus of independent Canadian national unions, whereupon the building trades announced that they planned to establish a new central body, tentatively named the Canadian Federation of Labor. The Carpenters, with 87,000 members, constituted the largest single group within the Canadian building trades.[46]

A renewed effort was made to bring life into the Canada Conference of Carpenters, which had led a fragile existence since its establishment in the 1960s. A meeting was convened in March 1981 at which a new set of bylaws was adopted. It was also agreed that the Conference would be financed through per capita taxes.[47]

But this was not enough, as the debates at the 1981 Brotherhood convention showed. A half dozen resolutions called for resumption of per capita payments to the CLC, the election of Canadian GEB members by the Canadians themselves, and more Canadian autonomy in general. The complaint was made that the Canada Conference was without any authority to decide policy and that it needed salaried officers to function effectively. In the end the resolutions were defeated, but the threat of secession still hung over the Brotherhood.[48]

THE THIRTY-THIRD GENERAL CONVENTION

The opening speech of the general president to the 1978 convention was in marked contrast to the one he had delivered four years earlier. Indeed, it was the most pessimistic note that had been struck at any Carpenters' convention since the Great Depression. President Sidell said in part:

> These past four years have not been kind to us. We have had record unemployment, a decline in membership, defeats at the hands of the judiciary, our most prized legislation gone down in either veto or filibuster, our membership reeling from the hard blows of inflation, the unfair competition of cheap foreign imports and the staggering effect of illegal aliens in our ranks . . .
>
> We lost a total of more than 68,000 members over the four-year period since the last Convention. And I might say that we lost 28,000 additional members since January 1st of this year . . . Our situation today stands in sharp contrast to that of four years ago when we reported having reached the second highest level of membership in the history of our Brotherhood . . . This period of high unemployment and almost total economic stagnation coupled with sharply increased employer resistance and more intense anti-union activity by right-wing groups of every description, including Merit Shop Contractors, the National Association of Manufacturers, Right-to-Work Committees, and many other anti-union forces, all combined to generate one of the most difficult challenges faced by the trade union movement in this century.[49]

A few statistics serve to justify the dark picture Sidell painted. For the four-year period prior to 1976, the Carpenters had won 56 percent of NLRB representation elections in their industrial sector; after 1976, only 48 percent. Half a million construction jobs were lost in 1975 and another 90,000 in 1976, as a result of the worst recession to hit the United States since the 1930s. The unemployment rate among the Carpenters' industrial members varied between 13 and 25 percent during the recession.

Sidell stated that some locals were not responding to the national union's CHOP and VOC organizing drives, partly because they wanted to maintain a monopoly of the available jobs for their current members.

"In 1974, we were primarily concerned with organizing residential construction in the United States and Canada. Today we are concerned with open shop conditions in every segment of our construction jurisdiction . . . We have experienced one of the most devastating blows that I have ever seen and can recall ever reading in the history of the Building Trades Unions in this nation."[50]

The membership loss was indeed serious. While the loss in total membership was in fact 68,587, the number of members in good standing fell by 94,831. The membership gains of the previous fifteen years had been swept away. Part of the problem was the continuing high turnover; as many as 100,000 members dropped out of the Brotherhood each year. Some of these people left the industry; there were about 8,000 deaths each year; and retirements and decertifications also took their toll. One practice, followed in a few areas, that was stopped by the national office in order to facilitate the return of former members was the requirement of new examinations. The theory behind this requirement was that skills might have become rusty with disuse; the real cause was often local desire to restrict membership.

The committee on organization outlined a number of factors that were hindering progress. One of them was government training programs. Many of the trainees were outside the apprenticeship program and often went into nonunion work. It was recommended that a careful study be made of CETA, Model Cities, and other government programs to determine their impact on the Brotherhood. Aliens working illegally were also taking jobs from union carpenters. Some locals were adding to the problem by exacting double dues from traveling members. The main factor, of course, was simply the recession.

The number of contractors with whom the union had entered into maintenance agreements rose, as well as those with whom the general office signed construction contracts. However, the activities of the Associated Building Contractors, a strong anti-union organization, had made some inroads into the ranks of former union contractors. "Their operations have become more sophisticated, and the amount of money they have for pursuing their open shop activities has increased rapidly."[51]

A resolution was introduced by the Cleveland district council that would have made the concurrence of affected local unions a requirement for effectuation of national agreements. A voice vote was inconclusive; the proponents of local autonomy were still strong. The resolution was finally defeated after Sidell had warned the delegates that to block national agreements would be tantamount to inviting nonunion contractors in. "All of those people in those areas that feel they have a solid position and that they will never become non-union in their area, gentlemen, I want to tell you that this is not so. I want to tell you that in areas that were 100 percent union at one time are now dwindled to 50 percent union and 40 percent union, and I'm talking about heavy construction, I'm talking about commercial work, and I'm not even talking about residential work."[52]

Several constitutional amendments were adopted. The delegates agreed to raise the per capita by $1.50 a month over a three-year period. The only objection came from a British Columbian delegate, who charged that the Brotherhood was spending most of its money in the United States. The acceptance of this increase with so little opposition reflected the current rate of inflation; the membership had become used to seeing prices move up rapidly.

The GEB offered a constitutional change making working permits valid for thirty days, in order to eliminate full-month charges for portions of a month. The district councils were also authorized to waive working permit requirements entirely. The aim of this amendment was to prevent the dropout of traveling members who decided to work non-union when confronted with onerous conditions for securing working permits. A proposal opposed by the general office that would have provided for filling GEB vacancies by election rather than appointment, thus reducing the power of the general president, was defeated.[53]

Considerable progress was noted in the spread of so-called pro-rata pension agreements, which established reciprocity between negotiated pension plans so that members who moved from one area to another did not lose their pension rights. The general office undertook no liability for these agreements, but simply acted as a clearinghouse to enable local plans to maintain contact with one another. As of March 1, 1978, pension funds in twenty-eight states had become participants in the program. This represented a major step forward in helping to provide economic security for retiring members, and far outweighed in importance the Brotherhood pension plan that had been terminated.

All the incumbent officers were reelected without opposition. The convention ended with a plea by General President Sidell that Carpenters face up to the threat of the open shop through intensification of the CHOP and VOC organizing campaigns, and a promise that the general office would use the greater financial resources made available to it to assist local efforts.

ROUNDING OUT THE CENTURY

The economic climate in which the Brotherhood had to operate did not improve after the 1978 convention. The GEB noted in June 1980 that "the current economic downturn has already resulted in an abnormal decline in membership, which in turn has had a detrimental effect on General Office income.[54] Unemployment in the construction industry reached 18.3 percent in August 1980. The Brotherhood lost 8,000 members during the first seven months of 1980 as a result of the economic decline.[55]

A change in the top leadership of the Carpenters' Union occurred during these years. In 1978 R. E. Livingston resigned the post of general secretary and was succeeded by John S. Rogers. Rogers had served as an official of Local 1837, Babylon, New York; as a national union representative; and as secretary-treasurer of the Suffolk County, New York, district council. Coming to Washington as assistant to the general president in 1969, he was appointed to the GEB in 1974. He also held the posts of president of the New York State Council of Carpenters and vice-president of the New York State AFL-CIO.

In 1980 General President Sidell resigned for reasons of health and was succeeded automatically by the first vice-president, William Konyha. Konyha had begun working with his father, a home builder, at the age of 14, and became a member of Local 1180 in Cleveland. After serving in the Seabees during World War II, he was elected president of Local 1180; he became a general representative of the Brotherhood in 1952, president of the Ohio State Council of Carpenters, and in 1970 was elected to the GEB.

Mr. Konyha retired in 1982 and was succeeded in turn by Patrick J. Campbell, who, like his predecessor, had served with the armed forces during World War II. After serving an apprenticeship, he became a journeyman carpenter and was elected to the presidency of Local 964 in Rockland County, New York. A few years later he became a general organizer; he was elected to the GEB in 1969, rising to the positions of second and first vice-president before becoming general president. When working in New York, he was president of the New York State Council of Carpenters and vice-president of the state AFL-CIO.

Like all their predecessors, these general officers of the Brotherhood had served apprenticeships, worked at the trade, and moved up the administrative ladder from local to district to national levels. Their lives were devoted to the trade and to the union. This has tended to be true of most American trade unions, but for none more than the Brotherhood of Carpenters.

A great deal of emphasis continued to be placed on the apprenticeship program. The Brotherhood had 25.9 percent of all apprentices listed by the Department of Labor for fifteen trades at the end of 1978, the largest number of any union.[56] The Carpenters' Apprenticeship and Training Department developed the Performance Evaluated Training System, a visual education method designed to help an apprentice learn at his own pace through color slides and other devices. Expansion of the apprenticeship program and improvement in training were regarded as one of the most effective means of combating the incursion of nonunion work, since the Brotherhood could thus offer a superior quality of labor and help sustain a higher wage level.

Above all, resources were devoted to organizational work. The organizing staff was expanded by 1981, double the number three years earlier. The CHOP program was widened to include commercial and industrial as well as residential building, while the VOC program concentrated on manufacturing employees. A new technique of targeting specific areas for intensive attention was developed; in the fall of 1980, for example, thirty staff representatives were assigned to Florida to work with local and district council organizers. Florida is a right-to-work state where anti-union employer groups are strong, and it is hoped that in this way the concerted open-shop movement can be overcome. Brotherhood organizing activities have been supplemented by parallel efforts on the part of the AFL-CIO Building Trades Department, which established an organizing department in 1978.

As a means both of holding current members and attracting new ones the Brotherhood engaged in a national advertising campaign for the first time in its history. During 1980 and 1981, three national television commercials were shown, two of them encouraging listeners to call a WATS line number for further information about the union. This program is part of a general campaign by American trade unions to improve their public image. It is expensive, but the union officials involved judge it to be effective. This is the Carpenters' evaluation:

> There is little room for doubt about the success of our venture into the national media to advertise our Brotherhood and its services to members. We attract many organizing leads from both the construction and industrial sectors; we helped to improve the image of unions generally and our union in particular; we built a host of new friends and many members here told us that a renewed sense of pride and interest in our organization was kindled when they watched and listened to our ads on national radio and TV.[57]

THE THIRTY-FOURTH GENERAL CONVENTION

The Brotherhood returned to Chicago to celebrate its centennial. Twenty-five hundred delegates representing almost 800,000 members assembled in the city where a hundred years earlier, a handful of delegates speaking for 3,000 carpenters had founded the Brotherhood. The occasion was a festive one. A play, titled *Knock on Wood,* was written especially for the convention and played to large audiences. A 234-page history of the organization was presented to each delegate at registration.[58]

One of the high points of the convention was an address by the honorary co-chairman of the Centennial Observance Committee, President Ronald Reagan. Despite cool relationships between the White House and the labor movement, the President was well received and was given a standing ovation both at the beginning and the end of his speech. Reagan told the delegates that he had represented the Screen Actors Guild at the bargaining table for twenty years, and referred to the jurisdictional dispute that took place in the movie industry during the 1940s:

> I remember an inter-union squabble in which I faced your big Bill Hutcheson. It's kind of ironic to look back on that because Bill was a Republican; in fact, he served on the Republican National Committee. And at that time I was a Democrat.
>
> If there is one challenge that I have for those in organized labor today, it is that they should, in the footsteps of Bill Hutcheson, recognize that organized labor should not become the handmaiden of any one political party.[59]

The membership picture had not changed much since the 1978 convention. Total membership stood at 762,000 on December 31, 1980, representing a decline of 19,000 over the three previous years. More than 100,000 people had been initiated during that period, but the dropout rate was even greater, a reflection of the doldrums in which the construction industry found itself. The committee on organization pointed out that from 90 to 95 percent of the home building industry remained non-union, and that promoters and developers, who employed no workers directly and who were replacing contractors, were a major deterrent to organization.[60]

The geographical distribution of the membership as of December 31, 1981, is shown in Table 15.2. The shifts that took place over three decades can be seen by comparing these data with those in Table 13.2. California's share of the beneficial membership rose, but there was a sharp decline in its semibeneficial membership, reflecting a depressed lumber industry; otherwise, the changes were relatively small. Arizona and Florida together accounted for only 4 percent of total beneficial membership, indicating the failure of the Brotherhood to make much progress in the Sunbelt.

The Canadian membership proportion rose to 10.8 percent of the total. Most of the Canadian members were concentrated in British Columbia and Ontario, which were at odds on the quest for autonomy. Apart from the Canadian question, there were only three issues that caused any controversy at the convention, two involving money. The first had to do with a constitutional amendment restricting the amount any local could charge a member from another jurisdiction for a working permit to the excess by which the dues of the issuing local exceeded those of the transferee's home local. The constitution committee justified this proposal as follows:

Table 15.2. Geographical distribution of the U.S. membership of the United Brotherhood of Carpenters, December 31, 1980

	Percentage of total membership	
State	Beneficial	Semibeneficial
California	19.8	27.0
Illinois	8.2	a
New York	7.0	10.3
Ohio	4.7	a
Washington	4.3	7.6
Pennsylvania	4.1	2.6
Texas	3.7	2.8
Michigan	3.7	3.6
Oregon	3.5	6.8
Other	41.0	39.3
Total	100.0	100.0

Source: United Brotherhood of Carpenters, *Proceedings*, Thirty-Fourth General Convention, 1981, Reports of the General Officers, pp. 78–79.

a. Less than 1 percent.

The provision for members working out of the jurisdiction of their home locals or district councils has been a source of difficulty for many years. It is likely that more controversy and ill-will between local unions, district councils and individual members has been generated over these provisions then any other provision of our laws. The problem grows worse each year as our membership becomes more mobile and tends to specialize in particular skills involving our trade, which, in turn, involves traveling over long distances.[61]

The principal opposition to the change came from local unions on the periphery of large metropolitan areas. For example, a representative of the Ventura County District Council of Carpenters stated that from 300 to 400 Los Angeles Carpenters were normally working in Ventura at any one time, and would be making no contribution toward servicing the area. The secretary of the Los Angeles Council replied that "what a lot of people want [the permits] for is to keep a little kingdom. They don't want anybody in their own local union. They want to keep that so it gives them a phony billing permit, or high dues fee to keep them out. You cannot organize on this kind of a system."[62] That was the main argument for the change — it would facilitate organization and reduce the temptation of union men to work nonunion jobs away from home. The amendment eventually carried by a margin of 56 percent.

The second money issue had to do with a proposal to raise the contribution to the pension fund for Brotherhood officers and staff from 10 to 13 percent. Several delegates opposed the increase on the ground that it meant a further drain on the union's financial resources, but the convention approved the proposal.[63]

The third issue was another GEB proposal to facilitate organization by eliminating mandatory initiation ceremonies and permitting individuals to be inducted into membership immediately after signing application forms. The objections were that this deprived locals of the opportunity to screen prospective members, and that taking a public oath to abide by the union's obligation was likely to mean more to a new member than merely signing a piece of paper. President Konyha pointed out that in places where building sites were distant from local union halls, it was difficult to get new men in. "They would have rather signed the application there, raised their hands in front of the Business agent that was organizing them."[64] Again, the GEB prevailed.

All incumbent officers were reelected without opposition in what was one of the most harmonious conventions in the history of the Brotherhood. General President Konyha, in his closing address, stressed the importance of electing to Congress and to state legislatures friends of the labor movement, and defeating its enemies, echoing the sentiments that had been advanced by his predecessor, Peter J. McGuire, almost a century earlier.

16

A LOOK BACKWARD AND A LOOK AHEAD

In the course of a century, the United Brotherhood of Carpenters and Joiners of America grew from a small organization founded by a devoted band of carpenters into one of the giants of the trade union world. It has had a major impact on the construction industry and has affected the lives of its members in many ways, not least in the conditions under which they worked.

It is difficult to determine with statistical precision what portion of the advance in wages and the improvement in working conditions is attributable to union activities. Growth of the nation's capital stock and productivity would have undoubtedly resulted in higher wages and better conditions even in the absence of unions. However, virtually every employer, whether he bargains with unions or resists them, is convinced that unions do make a substantial contribution toward skewing income distribution in the direction of the employee. The same is certainly true of the millions of workers who have been willing to finance their unions by paying dues.

The Bureau of Labor Statistics began to publish a series of union wage rates in construction, by trade, in 1907. Prior to that date, the only systematic series are for the building trades as a whole for union and nonunion men alike. These data suggest that from 1881 to 1907, money wages in the building trades rose by 57 percent.[1]

Scattered wage data that appeared in *The Carpenter* enable us to gain a rough impression of how the union carpenter fared over the same period. These data are presented in Table 16.1. The cities shown were selected on the basis of data availability for the years in question, and are not necessarily representative of the entire country. Nevertheless, they do suggest that in the quarter of a century after the establishment of the Brotherhood, its members fared somewhat better than did the average construction worker.[2]

The data situation is better for the last three quarters of a century. Table 16.2 contains, by five-year intervals, indices of hourly wage rates for union carpenters since 1907, in both money and real terms. These data indicate that carpenters' real hourly rates increased fourfold from 1907 to 1978. If the pre-1907 data are to be given credence as a rough approximation to what actually occurred, it would appear that a union carpenter at the close of the first century of the union's existence was

Table 16.1   Hourly wage rates of members of the Carpenters' Union, selected cities, 1881 and 1907[a]

City	Hourly rate (dollars)		Index of wage increases (1881 = 100)
	1881	1907	
Boston	0.25	0.4375	175
Chicago	0.30	0.55	183
Cincinnati	0.25	0.45	180
Cleveland	0.225	0.45	200
Denver	0.30	0.50	167
Detroit	0.20	0.35	175
New York	0.325	0.625	192
Philadelphia	0.25	0.50	200
Pittsburgh	0.25	0.50	200

Source: *The Carpenter*, various issues, 1881, 1906–1908.

a. Daily wage rates of carpenters for these cities covering the period June 1889 to September 1891 are available in the Aldrich Report (U.S. Senate, 52nd Cong., 1st sess. *Retail Prices and Wages*, July 19, 1892, pt. 3, pp. 1928–37). They cannot be compared directly with the hourly rates shown in this table because cities were moving from a ten-hour to a nine-hour day during this period.

earning almost five times as much for every hour he worked as did his brother carpenter when the union first saw the light of day.

Wage improvement was by no means constant over the century. The decade of the 1920s saw a substantial rise, despite the weakness of the union in the face of a successful employer offensive against it. A relatively tight labor market, due to a high level of construction activity as well as the cessation of large-scale immigration, undoubtedly contributed to this result. But the union was able to prevent money wages from being cut in the face of declining consumer prices.

Money wages fell during the Great Depression, for the first and only time in the period covered by the tables. Since 1935 money wage rates have risen steadily, although real wages declined during World War II, when wage controls were in effect, as well as during the most recent quinquennium, with its combination of inflation and unemployment.

It is necessary to adjust for hours worked in order to translate hourly wage rates into daily or weekly earnings. In 1881 the standard working day for carpenters was ten hours, as against eight hours at the present time. This would reduce the daily earnings increase over the century to a factor of four, rather than five. As for weekly earnings, the customary working week for union carpenters during the first few years of the union's existence was five ten-hour days, with a nine-hour day on Saturday, for a total of fifty-nine hours a week.[3] When compared with the present forty-hour week, the improvement factor would be reduced to three over the century if weekly earnings are the basis of comparison. Of course, the additional nineteen hours of leisure time constitute a major improvement in a worker's living standard.

It is difficult to determine the most interesting comparison of all, the change in annual earnings over the century. Construction has always been a seasonal industry, and few building workers are employed the year round in most parts of the country. The present situation has been characterized as follows:

Construction earnings are quite seasonal, with a third quarter peak and a first quarter trough. The seasonal pattern is much the same in South and North, though the intensity of seasonal fluctuations is much greater in the North. Seasonality permeates the construction labor force, and extends to all income classes. Income levels do not identify a large group of stable, year-round workers immune from seasonal fluctuations. Earnings outside the construction industry dampen the fluctuations quite imperfectly, and leave a great deal of instability in total earnings.[4]

Table 16.2 Indexes of money and real hourly wage rates for union carpenters, 1907–1978 (1907 = 100)

Year	Index of hourly wage rates for union carpenters	Consumer price index	Index of real hourly wage rates for union carpenters
1907	100	100	100
1910	116	100	116
1915	127	109	117
1920	243	214	114
1925	276	188	147
1930	324	179	181
1935	275	147	187
1940	343	150	229
1945	386	193	200
1950	594	258	230
1955	757	286	265
1960	949	317	299
1965	1148	338	340
1970	1632	415	393
1975	2367	576	411
1978	2787	698	399

Source: Bureau of Labor Statistics, Handbook of Labor Statistics, various issues.

This is undoubtedly a good description of what conditions were like in 1881 as well; the only question is whether seasonality was greater then than now. One thing we do know is that in the past, daily wages were reduced for those who could find winter work because of the impossibility of working ten hours a day during that season.[5] But we do not have any comparative data on the number of days that carpenters worked per annum. Some progress has been made in winter construction technology, but the fact that during the decade 1969–1978 insured unemployment among construction workers was running from three to four times higher in February than in August of each year indicates that the winter seasonal shutdown, particularly for outside men, is still with us.[6] In all probability, carpenters work a greater number of hours per year in 1981 than they did in 1881, but the difference cannot be stated with any precision.

Thus far, we have been concerned only with earnings exclusive of fringe benefits. But earnings are by no means the sole index of worker welfare. The leisure time gained by a reduction in working hours may be a more important constituent of welfare than earnings. Indeed, when carpenters were working a ten-hour day, they were continually urged by the union leaders to demand shorter hours even at the expense of a proportional cut in earnings. "Man was not born for a mere brutish existence of work and eat and sleep. There is something higher or else civilization is a farce."[7] An alternative strategy advocated was to work only eight hours in winter, accepting a decline in income, but then to attempt to make up the income loss by demanding higher wages when the spring season began. The one-third reduction in the workweek that has taken place since 1881 may have made a greater contribution to the living standards of the carpenters than increased earnings.

A hundred years ago, the carpenter had no protection against the vicissitudes of illness and old age other than what he was able to save out of his earnings. This explains the attraction of "benevolent" unionism, which provided a minimal degree of protection. The recent spread of employer-financed pension schemes,[8] on top of Social Security, serves to assure the carpenter of a decent retirement income. Medical insurance is also being provided to many carpenters. There are no separate data on the value of fringe benefits to carpenters. However, unionized building tradesmen receive fringe benefits that cost the employer from 10 to 20 percent of base wage rates.[9]

Another aspect of Brotherhood activity of great benefit to its members was protection of the job area through strong defense of jurisdiction. Competition among rival unions for jobs carries potential danger for the viability of any labor movement, permitting employers to play one union off against another and weakening all. Almost from the beginning, the United Brotherhood maintained that all carpenters belonged in one organization, and since it has always been the largest union in its field, insisted that it had precedence over its rivals. It absorbed a number of unions catering to workers in the carpentry trade, including the Amalgamated Wood Workers, the Amalgamated Society of Carpenters, and more recently, the Lathers.

As technology changed, new jobs were created which the Carpenters claimed belonged to their trade by virtue of the skills involved and the tools employed. The Brotherhood was persistent in its claims; some of the classic jurisdictional disputes in American labor history were its controversies with the Sheet Metal Workers and the Machinists. Had it not been for the aggressive pursuit of jurisdiction, the job area allotted to union carpenters would have been much smaller than it is today.

At the same time, particularly within the last two decades, the national office has been working toward elimination of the local disputes over work assignment that give rise to costly jurisdictional strikes. These stoppages, which can be very damaging to the union, are often caused by insufficient local knowledge of awards rendered by national arbitration boards. The national office is not shy when it comes to assertion of jurisdiction; local unions need not be concerned that their claims will be neglected if they allow their disputes to go through normal channels instead of resorting to direct action.

The development of national agreements with large contractors who operate throughout the country, and participation in the building trades' maintenance contracts, are other means to the same end: expansion of the job pool available to union carpenters. Local unions have not always understood the necessity of making concessions to contractors when not forced to do so by immediate local circumstances, but in the long run, these uniform agreements have meant more jobs for the entire membership.

The Brotherhood can claim one of the best anti-discrimination records of any of the older American unions. Black and other minority workers were admitted to membership from the very start of the union, and the Brotherhood may have been the first union to elect a black vice-president, in 1884. It did permit separate black and white locals to be organized in the South when the alternative would have been no organization at all, but the national officers continually urged upon their local unions equal opportunity for all members. There was often local resistance, particularly in the South, and it certainly would not be accurate to say that the pattern of segregated locals that prevailed in that part of the country resulted in equal opportunity. However, the general office eventually ordered desegregation of all locals before the federal government mandated such action by executive order.

The Brotherhood made an important contribution to the efficiency of the construction labor market by facilitating labor mobility. One of the main purposes of those who founded the national union was to make it possible for carpenters to move from city to city in search of employment, to be welcomed when they arrived at a new place and directed to vacant jobs. The unvarying policy of the national union that members who had valid traveling cards had the right to work anywhere in the union's jurisdiction often met with opposition at the local level, particularly when jobs were scarce. But the national union insisted that walls could not be erected around local labor markets, and took appropriate action when monopolistic tendencies became flagrant.

Another major contribution toward market efficiency was the promotion of apprenticeship programs as the preferred means of entry to the trade. The union has always insisted that all construction carpenters receive a broad training, and there is only a single journeyman classification in the construction locals. A formal apprenticeship is the best means of ensuring the requisite skills. A recent study sponsored by the Department of Labor reached the conclusion that graduates of apprenticeship programs "worked more steadily, learned the trade faster, were more likely to be supervisors, and acquired supervisory status faster" than those trained by informal means. The union motivation in furthering apprenticeships was well summarized:

> An overriding objective is to protect wages and meet employers' manpower needs in such a way as to give them an incentive to continue dealing with the union. In achieving this objective, the union views apprenticeship as a means of turning out a cadre of well trained craftsmen who will have strong attachment to their unions and crafts. Unions realize that they can maintain their competitive position only if their members are more productive than the alternatives available to the employer. Moreover, business agents have considerable difficulty placing poorly trained journeymen and keeping them employed. They therefore tend to prefer apprenticeship to other types of training.[10]

Although the core of the Brotherhood membership has always consisted of building tradesmen, the union extended its jurisdiction to mills and factories fabricating wood products at a very early stage of its existence. The recent establishment of a seperate industrial department as a formal part of the union's national structure acknowledges the fact that the industrial membership has become an increasingly important part of the organization.

THE PROBLEMS AHEAD

As the United Brotherhood of Carpenters and Joiners approaches its second century, it faces a number of problems, some long-standing, some new. There is usually a tendency to look upon current difficulties as being more serious than any encountered in the past, if only because we know how past problems were solved. But there is some justification for the view that the Carpenters, in common with the rest of the American labor movement, will experience during the next decade challenges equal in intensity to those of crisis periods in the past.

The Bureau of Labor Statistics estimates that the employment of carpenters in 1990 will be between 1,183,000 and 1,228,000, depending on varying assumptions about economic activity, compared with 979,000 in 1978. The average increase of empoyment for all occupations between 1978 and 1990 is estimated to be between 15 to 25 percent, so by this measure, the range of 20 to 30 percent growth for carpenters suggests a relatively favorable outlook for the future.[11] If all the new entrants to the trade could be recruited into membership, the Brotherhood would be in good shape. But many of these new people will go to work for nonunion contractors; that is the essence of the problem.

At various times in the past, the labor movement has been confronted with organized anti-union campaigns: the open-shop movement of the early part of the twentieth century, the American plan of the 1920s, company unionism during the 1930s. Another attempt appears to be shaping up in the drive for a "union-free environment," and the Carpenters cannot expect to remain immune to its impact.

Precise figures on the proportion of nonunion construction work are not available on a national basis. There is a good deal of variation by locality and by the nature of the product market. In Chicago, Los Angeles, and San Francisco the proportion is small, while in Atlanta, Baltimore, Dallas, and Washington, D.C., more than two-thirds of all construction is being done by nonunion contractors. But even in union cities, much of the residential building is nonunion. Los Angeles, where not many years ago virtually all homes were being built by union carpenters, now has the largest single chapter of the anti-union Associated Builders and Contractors. Even where commercial and industrial building is still largely union, the residential sector is often unorganized; Denver, Kansas City, and Portland, Oregon, provide good examples.

Part of the current problem lies in the growth of militant anti-union employer groups. The most important of these is the Associated Builders and Contractors (ABC), which in 1978 had 13,265 member firms employing over 440,000 workers. The Associated General Contractors consisted mainly of union contractors until a few years ago, but recently its open-shop membership has grown rapidly—more than 40 percent by July 1979. Many of its members are "double breasted"—that is, they operate union or nonunion, depending on the locality and the job.

The spread of nonunion construction has been facilitated by the increasing differential between union and nonunion labor costs. It is difficult to make direct comparisons between the two, since higher union wage rates are often offset by the higher productivity of union labor, with little difference in the resultant unit labor costs. One major cost difference lies in fringe benefits: almost all union workers receive such benefits, at a cost of 10 to 20 percent of base wage rates. Many nonunion workers receive no fringe benefits at all, and where they do, the payroll cost is about half that in the union sector.[12] And this is by no means the only relevant cost factor. Insistence by the building unions that all workers except apprentices must be paid the standard journeyman rate precludes the subdivision of jobs into skill categories. Jurisdictional disputes, strikes involving fairly wide areas, and the lack of flexibility in managing jobs due to some union working rules and jurisdictional boundaries also tend to widen the cost gap.

While the greatest task facing the Brotherhood is maintenance and expansion of organization among building carpenters through unionizing the residential industry and resisting encroachments in commercial building, it is not the only one. Something will have to be done to satisfy the home-rule ambitions of the Canadian members in order to avert a split. The interests of the Canadian carpenters, particularly in the sphere of politics, do diverge from those in the United States, and this will have to be taken into account. It is not exclusively a Brotherhood problem, for all the international unions with Canadian members are being subjected to similar pressures. But the Carpenters' Union is one of the largest in Canada, and has more to lose if a break were to occur.

The construction industry, particularly in the nonresidential sector, has been moving steadily toward a national orientation. This is reflected in the growing importance of the national and project agreements negotiated by the general office. This structural trend is likely to continue, and it will entail further involvement of the national union in collective bargaining arrangements. The Carpenters' locals have insisted on maintaining their independence in bargaining, and have often resisted the concessions that were essential to consummate national agreements. The Brotherhood is more decentralized than most labor organizations, and this is a favorable element in ensuring democratic union government. However, the economic development of the industry is going to require an enhanced role for the general office, and statesmanship will be required at both the national and local levels to create a new balance of power that combines the necessary efficiency in collective bargaining with the preservation of local autonomy.

The movement of national union headquarters to Washington after World War II reflects the growing importance of the federal government in industrial relations. The Carpenters, more than any other American union, opposed this trend, arguing that while a favorable federal administration could assist labor, the potential damage that could be inflicted by a hostile one was necessarily enhanced. An obvious example, known to every building tradesman, is the Davis-Bacon Act. Throughout most of its existance it has helped maintain union working conditions, but it could become a powerful instrument for the promotion of the open shop. The inference is that the Carpenters, together with other unions, must step up their political activities. This does not necessarily entail unvarying support of one political party; the traditional AFL policy had a good deal to commend it. It does mean the expenditure of more funds and the greater personal involvement of more members to ensure that those who make our laws, and those who administer them, will help advance the interests of organized labor.

There are other clouds on the horizon: the possibility that the American economy will continue to be characterized by a combination of inflation and unemployment; a possible deterioration in the image of organized labor held by the American people; changes in technology that will require broadening of the carpenter's skill, initially through a strengthening of the apprenticeship program, and some redrawing of jurisdictional boundaries. None of these problems is insurmountable, but they all will require hard work.

The United Brotherhood of Carpenters is today one of the most powerful trade unions in the world, a far cry from the 36 delegates, representing 4,800 workers, who assembled in Chicago in 1881 to create a union. It is useful to look back at what has happened since then; history does not enable us to predict the future, but it does teach us some important lessons. The most significant is that the Brotherhood has been confronted with some daunting obstacles in the past, greater than anything it faces at present. There were some temporary setbacks, but the union was always able to overcome them and emerge stronger than ever.

It would be difficult to improve upon Peter J. McGuire's statement, written in 1881, explaining the need for organization:

> Without organization or union, workingmen become the victims of each other—one will accept work for less than reasonable wages, for fear another may step in and take it . . .
>
> We organize to elevate our trade and craft from the degradation of low wages and long hours . . .
>
> We organize to abolish piece work and the system of botch work it produces, and thus advance the standard of skill and proficiency among our members . . .
>
> We organize to assist each other to secure employment for mutual aid in case of sickness or disability, to protect each other's rights and redress our wrongs, and to obtain adequate legislation that will secure our wages from the unscrupulous sharks that infest our trade . . .
>
> We organize to solve the labor question! We propose a peaceful and orderly solution of that question.[13]

Much of what McGuire wrote a century ago still applies today. Despite all that has happened over the century, the key to progress now, as then, lies in organization. If there is one thing to be learned from the Brotherhood experience it is that a well-administered, well-financed, and solidly structured organization yields large returns to the working people who sustain it.

Appendixes
Notes
Index

Appendix A

Peter J. McGuire's Valedictory: Testimony before the Industrial Commission, April 20, 1899 [1]

McGuire's testimony before the Industrial Commission was in essence a summation of his views of the American labor movement. The commission had been established by Congress to look into the industrial relations problems of the country, among other things, and held extensive hearings. [2] McGuire was questioned on a number of subjects, and his views merit summary.

Discrimination.   When asked whether Carpenter membership was open to all, McGuire replied: "We do not ignore anyone on account of color, race, creed or religion. We have whites and blacks together in some cases, in the South we cannot mix whites and blacks ordinarily in one union. We have colored unions in the South, seventeen colored unions."

The boycott.   "The Boycott is an answer to the blacklist. The boycott has been talked of so much, it is an element in human nature that existed before the labor organization, of non-intercourse of those you don't care to have anything to do with."

Strikes.   "Until we asserted our power by strikes enforced on us, employers would very rarely meet us . . . We find that strikes have a highly educational effect. Much as we regret them and deplore the necessity of entering into them, they have taught the employers to be more reasonable and not precipitate a closing down of the operations . . . Strikes, further, in our trade have been instrumental in breaking down the system of piece work and lump work."

Piecework.   "It is nothing more than sweating in the tailoring trade, or those trades where there are a number of middlemen in between the man who does the work and the party who sells it or who uses it. We found in many cities four and five parties in interest between the workingmen and the party who had the work done, each one taking their slice of the amount of the contract in the first instance. We have broken that system down in every city but seven."

Extent of organization.   McGuire claimed that the Brotherhood controlled more than 70 percent of the carpenters in the leading cities. The fact that the union represented only 9 percent of all those in the country whom the census listed as carpenters he ascribed to the fact that most of them were scattered in small towns and villages, as well as to his belief that many who described themselves as carpenters had never worked at the trade. It might be added that the census included as carpenters those doing carpentry work in industrial plants and mills, few of whom were then organized by the Brotherhood.

Arbitration.   "If the workingmen universally were to declare through their organizations that they would never enter into a strike . . . the employers would never meet them in conference or conciliation . . . Voluntary arbitration we advise and accept, if possible . . . Once the government trespasses upon the domain of private industry to adjust disputes, it is tantamount to the demand that they ought to own the industry, and you reach collectivism."

Wage determination. "I believe that wages are regulated by competition between the employed and unemployed, between the scarcity and the plentiness of labor seeking work . . . The minimum is always tending downward and downward where there is active competition, unlimited and unrestricted, but where there is organization, the tendency is upwards."

Apprenticeship and quality of work. "The rules of our organization are that a boy or person engaging himself to learn the trade of carpentry shall be required to serve a regular apprenticeship of four consecutive years, and shall not be considered a journeyman unless he has complied with this rule and is twenty-one years of age at the completion of his apprenticeship. The same number of years are required from those who enter the trade over the age of twenty-one. They are classed as improvers, not apprentices . . . But it is more the exception than the rule to find the apprenticeship system in vogue now . . . Many of the contractors will not be bothered teaching a boy a trade for there are so many botch workmen that can be picked up who are preferable to skilled mechanics because they can deliver a larger quantity of work, regardless of the quality of the work, and with the trade cut up as it has been in some instances, one good skilled man can work in three or four botches by laying out the work for them and giving them assistance. Then again there are so many bosses in the trade without capital, shops which simply have an office, have no place to train an apprentice or hold an apprentice."

Hours of work. McGuire asserted that forty-two cities in which carpenters had gained an eight-hour day in 1886 never departed from this standard. He opposed Saturday half-holidays, however, on the ground that this might impede the spread of the eight-hour day, pointing out that this was precisely what happened in England.

Profit sharing. "In theory, gentlemen, profit-sharing may be very agreeable; but to be practical, the employee should have the right to look over the books and examine the status of the business regularly. Employers running their business on an individual basis are not likely to submit to that . . . At best, profit-sharing is a makeshift to meet the labor question, and it is a recognition that after all, the worker is entitled to something more than the market rate of wages."

Immigration. "Our people are on record in favor of the restriction of immigration, very strongly so." He particularly opposed the admission of temporary migrants from Scotland and England who worked half a year and returned to their homes. "All the immigration that is coming over . . . congests our larger cities in the tenement-house districts, goes into Pennsylvania and Ohio and a few other states and reduces the price of labor there in the mining industry and in railroad labor." McGuire favored an educational test for prospective immigrants and the elimination of contract labor, the padrone system. But he added: "There are a number of our people who favor total suspension of immigration. I am not one of that class. I think there ought to be an outlet to this country for any man who desires to seek freedom here and leave his native home to get it."

Labor leadership. "It will not do to place a man at the head of a labor organization who is an inefficient or incompetent workman, for his fellows around him will pull him down in the estimation of others, and the bosses with whom he deals will say, 'Well, if you cannot send us a better man than that, we are not going to have any dealings with your organization.'"

Manual training. McGuire came out strongly for manual and technical training in the public schools, if only to give students "an idea of the nobility of labor, where many now come from the public schools with a detestation of manual labor and with a dislike for anyone who is a manual laborer . . . We have a surplus and overstock of well-educated boys and girls with good commercial education, good general education, but lacking in respect for manual labor; and they are the most helpless class in the world, as conditions become more desperate . . . and the very pride that is in them . . . keeps them from organizing."

Socialism. "I am not prepared, and I think the conservative men in the trades union movement are not prepared to say that we are ripe for collectivism, when our civil-service laws are not effective as they might be . . . How much worse would it be if the Government service was to cover the railroads and the telegraphs and the multitude of industries." But if the civil service were more efficient and could provide protection from party patronage, "we would be more disposed to favor the extension of public employment and public ownership."

This summary does not begin to do full justice to the richness of the testimony of this remarkable leader of the Carpenters. McGuire was not only a tireless organizer, but clearly also a man of considerable intellectual attainment, who had developed over the years a consistent social and political philosophy. But he was a poor administrator, and it was this failing that eventually brought about his downfall.

Appendix B

Report on the Situation at Bogalusa, Louisiana, by the President of the Louisiana State Federation of Labor (December 1919)[1]

The Great Southern Lumber Company who owns the Lumber Mills and the Pulp and Paper Mills at Bogalusa, Louisiana, are perhaps the largest lumber producers in the United States. They claim that the Saw Mill located at Bogalusa is the largest mill in the world; they are also connected with several large enterprises; they are interested in the large mill located at Virginia, Minnesota which they claim to be the next largest mill in the world.

About three years ago, they put a very large pulp and paper mill at the Bogalusa plant, and about that time the workmen at Bogalusa began to try to organize. They asked for organizers and several attempts were made to help the people there. About this time, a young man named Rodgers, an organizer for the Carpenters and Joiners, went to Bogalusa and while there was arrested as a suspicious character. He was released after getting the news to some of his friends in New Orleans. However, they claimed that he was a dangerous character and filed charges against him in Federal Court and while he was in jail at Bogalusa, the Bogalusa officers had put dynamite caps and fuse in his grip. This grip was produced in Federal Court as evidence, but their case was so flimsy and so crude that the Federal authorities dismissed it without trial. Later, James Leonard, at that time vice-president of the State Federation of Labor and an organizer of the A.F. of L., went to Bogalusa and was told by the authorities there that they would not permit any organizer to come there and organize the men. Mr. Leonard left Bogalusa and returned to New Orleans.

This, however, did not stop the desire of the workers at Bogalusa who were in touch with the State Federation, and later on W. M. Donnells was sent there as an organizer for the Carpenters and organized the carpenters of the place. Then in rapid succession, the organization of all lines followed until we had seventeen local unions at the place with a splendid central union. Seeing that the men had organized in spite of their efforts to thwart it, the company became furious and tried to intimidate the members of the locals. Finding that this would not work, they then started a systematic system of discharging all white union men and putting non-union negroes to work in their places and at the same time making a great deal of noise and trying to work up a spirit of antagonism to the organization of negroes, even telling the farmers and planters that we were trying to organize the negro farm laborers. This forced the hand of labor and a campaign or organization was then begun to organize the negroes in the employ of the Great Southern Lumber Company.

This brought on quite a little feeling. The company called a mass meeting of the citizens where several public men, among them a Congressman, made speeches opposing the organization of negroes. Donnells spoke at that meeting and defended the right of Labor to organize. Seeing that the men were determined, the company then entered into an agreement to the effect that they would stop discharging the union men if they would cease organizing negroes. This arrangement was made with the understanding that no union man should be discriminated against or prejudiced in any way because of his membership in a union. This arrangement had not been made thirty days when the company immediately started discharging both white and colored union men and issued an ultimatum from Mr. W. S. Sullivan, the vice-president and general manager of the plant, that he would not recognize any union man and that he would not meet nor confer with anyone representing union labor and instructed his office to so inform Donnells and others. This agreement was made in April of 1919 and from that time on things happened fast at Bogalusa. Mr. Sullivan who is vice-president of the Great Southern Lumber Company, is also mayor of the town of Bogalusa.

Mayor Sullivan then placed about thirteen of his henchmen that had not joined labor, on the police force of the town. They were furnished guns and automobiles; they were augmented by a number of deputies appointed by the sheriff of the parish and then began a reign of terror in the town. They tried to get rid of all the leaders by terrorizing them and by offering them bribes to leave the place. Finding this would not work, they sent their employment man to Chicago and other cities to secure three thousand negroes with the intent of placing non-union negroes in the industries there and forcing the union men to leave. They failed to get any man in Chicago. I was informed by reliable parties in Chicago that they did not offer sufficient wages and that the men were informed that no labor trouble existed. However, the men knew that they were wanted as strike-breakers and would not go. On failing to get men, they immediately began arresting men, both black and white on all kinds of trumped up charges and taking them to the county seat about twelve miles away. The automobiles furnished the police and deputy sheriffs were used for the purpose of taking the men to the county seat; but the men, when discharged for lack of evidence, had to get back to Bogalusa any way they could. In addition to this, several men were beaten by these same gunmen, others were ordered to leave while some of them were offered bribes to leave.

Previous to this, a committee had been appointed, two by the company and two by the men to investigate wages and working conditions in the lumber industry throughout the state and eastern Texas and Western Alabama and Mississippi. This committee reported that the Great Southern Lumber Company was paying less wages than any mill west of the Mississippi River. One of the men representing the company was a sawyer who had at that time never joined the union, however, when he was selected by the company to represent he accepted and when the report was made, he was accused by the company of not making a fair report. He then joined the Sawyers' Union and was soon made president of the union. They then tried to induce him to leave. He owned his own home in the town and also a small farm just outside of the city. He was told by the henchmen of the company that he had better sell his property and leave the place. He refused to do this, and while attending a meeting he was called from the hall when seven of the gunmen attacked him and placed him in an automobile and run him five miles out of town where they took him out of the car and there proceeded to beat him into an almost unconscious condition, then dictated a letter which they compelled him to write to his wife telling her to sell all their property and leave at once as he was not coming back. This man, whose name is Ed. O'Bryan, was then taken to a station on the Northeastern Railroad and placed on the car bound for New Orleans and was told by the gunmen that they were Department of Justice agents and that he was under arrest by the Federal Authorities as an I.W.W. agitator. They had in the meantime painted a sign on the man's back which read, "I am an I.W.W." and when placed on the train they found Brother Donnells on the same train; they also told him that both he an O'Bryan were under arrest by the Federal Authorities as I.W.W. agitators. They held guns, both of them, and would not allow them to speak to each other. At the first station out of New Orleans, two of the gunmen got off of the car while one still stayd on. On reaching the yards in the city, this man also got off and left O'Bryan's wounds treated they went to the office of the Superintendent of the Department of Justice and filed complaints from which nothing has yet been heard.

The president of the Colored Timber Workers' Union was another one they offered a small sum of money to leave and sell his property. He owned a home and some livestock in the place all told valued at about $3,500. They offered him $2,000 to sell this property and leave; he refused to do so. They went that night to his house and shot it to pieces and searched for him. However, he had told the white labor people of the offer to leave and they had gotten him away where they could not find him. They then blamed the white labor people for getting him away and then gave out a statement to the press that Lum Williams and another labor sympathizer had paraded the negro Dacus up and down the street while they were heavily armed and had defied the authorities to arrest him. I am informed by a number of people who are not members of labor that this is a false statement as nothing of the kind was done, and that the gunmen who claimed to have a warrant for the arrest of Dacus had nothing but a trumped up charge. That was their excuse for going to Lum Williams' place on the following day where they murdered Williams who was president of the Central Trades Council together with three others. The claim of the gunmen that the union men had arms in the building was untrue as there was not a gun in the building. They drove up in their automobiles and without warning began to shoot. Williams was the first to appear at the door where he was shot dead without a word being spoken by either side. Two other men who were in his office at the same time were shot down and the bodies of the three men fell one on top of the other in the doorway. The other men attempted to leave the building by the back door where two of them were shot down while coming out with their hands above their heads. The only shot fired by any man connected with the labor people in any way was fired by a young brother of Lum Williams who shot Captain LeBlanc in the shoulder with a 22 caliber rifle after he had shot his brother to death. This Captain LeBlanc was a returned soldier and was placed in command of the gunmen in Bogalusa. One of the men wounded at the back door of the building where the killing occurred, was taken to the sanitarium where he died three days later, but no one was allowed to see him while he was alive. Young Williams was arrested immediately and charged with intent to kill while the thirteen gunmen who did the murder, were not arrested till three weeks later when the grand jury took action and bound them over to await the final action of the regular session of the grand jury in May. They were immediately released on a bond of $40,000 each and have returned to Bogalusa where they are still armed and defying the law of the state. They have been continually arresting negroes for vagrancy and placing them in the city jail.

It seens that a raid is made each night through the section of the town where the negroes live and all that can be found are rounded up and placed in jail charged with vagrancy. In the morning, the employment manager of the Great Southern Lumber Company goes to the jail and takes them before the city court where they are fined as vagrants and turned over to the lumber company under the guard of the gunmen where they are made to work out his fine. There is now an old negro in the hospital at New Orleans who they went to see one night and ordered him to be at the mill at work next day. The old man was not able to work and was also sick at the time. They went back the next night and beat the old man almost to death and broke both of his arms between the wrists and elbow. This old man was taken from the hospital and went to the county seat and appeared before the grand jury and the papers made a big thing of it and said we were trying to stir up race trouble. The State Federation has taken the matter up with labor throughout the state and we intend to fight the thing to a finish. However, we are badly handicapped for funds to fight the combined forces of the entire lumber industry as they have organized an organization to fight us and now we have a man named Boyd who was editor of the Lumber Men's Journal traveling through the southern lumber states forming local organizations with the sole purpose of defending the Great Southern Lumber Company and fighting any attempt on the part of labor to organize the lumber industry in the south. I have it from reliable sources that they have succeeded in lining up the hardwood lumber people also in this anti-union organization. They are holding meetings in all the towns in the southern lumber states. We have employed the Hon. Amos L. Ponder as an attorney to defend young Williams for the shooting and to prosecute the thirteen gunmen. We are having some investigating done and hope to be able to bring them to justice along with those who are responsible for the many outrages against humanity and justice. However, they are still terrorizing the people that live in Bogalusa and just last week Brother Donnells in company with Brother Donnelly was on their way to Bogalusa to hold a meeting. Brother Donnelly is now president of the Central Body at that place. On arriving at the depot in New Orleans one of the gunmen met them there and told Donnells that if he went to Bogalusa he would be murdered and made several threats. They had him arrested on two charges, one for threatening to kill and one for carrying concealed weapons. He was released on bond in each case and no doubt no effort will ever be made to have him appear for trial in New Orleans.

Appendix C

Membership of the United Brotherhood of Carpenters and Joiners of America, 1881–1981

Year	Members in good standing	Total membership[a]
1881	2,042	—
1882	3,780	—
1883	3,293	—
1884	4,364	—
1885	5,789	—
1886	21,423	—
1887	25,466	—
1888	28,416	—
1889	31,494	—
1890	53,769	—
1891	56,987	—
1892	51.313	—
1893	54,121	—
1894	33,917	—
1895	25,152	—
1896	29,691	—
1897	28,269	—
1898	31,508	—
1899–1900	68,463	—
1901–1902	122,568	—
1903–1904	161,205	—
1904–1905	161,217	—
1905–1906	170,192	—
1906–1908	178,503	—
1910	200,712	232,467
1912	195,499	244,388
1914	212,160	261,049
1916	212,833	261,722
1920	371,906	400,104
1924	322,044	350,391
1928	295,250	346,136
1932	134,059	242,005
1936	291,732	301,875
1940	—	319,848

Year	Members in good standing	Total membership[a]
1945	437,074	722,392
1946	—	586,274
1947	—	737,514
1948	—	671,191
1949	—	710,034
1950	677,180	760,908
1951	—	800,770
1952	—	822,828
1953	746,467	823,574
1954	—	804,343
1955	—	825,799
1956	—	852,038
1957	758,726	839,794
1958	—	806,572
1959	—	806,640
1960	—	788,702
1961	682,570	742,653
1962	—	741,066
1963	—	742,130
1964	—	766,025
1965	718,048	779,768
1966	—	798,891
1967	—	788,944
1968	—	793,421
1969	759,606	820,924
1970	—	814,365
1971	—	814,250
1972	—	825,775
1973	788,962	849,829
1974	—	843,789
1975	—	823,802
1976	—	794,345
1977	694,131	781,242
1978	—	768,762
1979	—	785,932
1980	—	762,163

Source: United Brotherhood of Carpenters, Research Division; *Proceedings,* General Conventions.

Notes: 1881–1908: average for July to June; 1910–1928: as of June 30; 1936–1980: as of December 31.

a. Includes members in arrears.

Appendix D

General Officers of the United Brotherhood of Carpenters and Joiners of America, 1881–1981

General Presidents

1881–1882	Gabriel Edmonston
1882–1883	John D. Allen
1883–1884	J. P. McGinley
1884–1886	Joseph P. Billingsley
1886–1888	William J. Shields
1888–1890	D. P. Rowland
1890–1982	W. H. Kliver
1892–1894	Henry H. Trenor
1894–1896	Charles B. Owens
1896–1898	Harry Lloyd
1898–1899	John Williams
1899–1913	William D. Huber
1913–1915	James Kirby
1915–1951	William L. Hutcheson
1952–1972	Maurice A. Hutcheson
1972–1979	William Sidell
1980–1982	William Konyha
1982–	Patrick J. Campbell

General Secretaries

1881–1901	Peter J. McGuire
1901–1948	Frank Duffy
1948–1956	Albert E. Fischer
1957–1978	Richard E. Livingston
1978–	John S. Rogers

NOTES

1. THE PREDECESSORS OF THE UNITED BROTHERHOOD

1. George E. McNeill, *The Labor Movement* (New York: M. W. Hazen, 1892), p. 71.

2. John R. Commons et al., *History of Labor in the United States,* vol. I (New York: Macmillan, 1918), p. 68; copyright 1918 by the Macmillan Publishing Company; extracts reprinted by permission.

3. Frank Duffy, "History of the United Brotherhood of Carpenters and Joiners of America" (unpublished manuscript in the files of the United Brotherhood of Carpenters, no date; hereafter cited as Duffy, "History"), p. 30; Commons, *History of Labor,* I, 66–71.

4. James C. Button, "The Industrial Organizations of America" (unpublished manuscript in the files of the United Brotherhood of Carpenters, no date), pp. 16–17.

5. Duffy, "History," p. 7.

6. Ibid., p. 10.

7. Ibid., p. 11.

8. Ibid., pp. 12–13.

9. Commons, *History of Labor,* I, 111.

10. Duffy "History," pp. 14–17.

11. John R. Commons et al., *A Documentary History of American Industrial Society,* vol. V (1909–1911; reissued, New York: Russell and Russsell, 1958), p. 80.

12. Ibid., p. 81.

13. Ibid., pp. 82–83. This is an interesting early statement of the idea that a reduction in working hours could be compensated for by an increase in productivity, so that output would not fall.

14. Commons, *History of Labor,* I, 189.

15. Commons, *Documentary History,* VI, 52–54.

16. U.S. Bureau of the Census, *Historical Statistics of the United States* (Washington, D.C.: Government Printing Office, 1960), p. 115.

17. Commons, *Documentary History,* VI, 55–56.

18. Ibid., pp. 336–340.

19. Ibid., p. 204.

20. Commons, *History of Labor,* I, 397.

21. Commons, *Documentary History,* VI, 76–77.

22. Ibid., p. 78.

23. Ibid., p. 82.

24. Ibid., pp. 83–86.

25. Commons, *History of Labor,* I, 319.

26. Commons, *Documentary History,* VI, 94–99.

27. This action was taken by President Van Buren on the eve of a presidential election. It had been preceded by strikes at navy yards in Washington, D.C., and Philadelphia. Sterling Spero, *Government as Employer* (New York: Remsen Press, 1948), pp. 77–83.

28. Duffy, "History," p. 33.

29. *Historical Statistics of the United States,* p. 115.

30. Commons, *History of Labor,* I, 487.

31. Duffy, "History," p. 35.

32. Ibid., pp. 39–40.

33. Ibid., pp. 39–41.

34. Commons, *Documentary History,* VIII, 314–315.

35. Duffy, "History," pp. 49–55.

36. Ibid., pp. 36–38.

37. Commons, *History of Labor,* I, 614.

38. Duffy, "History," p. 40.

39. *Historical Statistics of the United States,* pp. 90, 127.

40. Commons, *History of Labor,* II, 20.

41. Duffy, "History," pp. 56–70.

42. Ibid., pp. 68–69.

43. Ibid., p. 66.

44. Ibid., p. 71.

45. Ibid., p. 73.

46. Ibid., p. 83.

47. Ibid., p. 74.

48. Commons, *History of Labor,* II, 175–181.

49. Duffy, "History," p. 88.

50. Ibid., pp. 89–90.

51. Commons, *History of Labor,* II, 195.

52. Duffy, "History," pp. 97–99; *The Carpenter,* July 1881, p. 1, and June 1882, p. 3. (*The Carpenter* is the official monthly organ of the United Brotherhood of Carpenters.)

53. Duffy, "History," pp. 99–100. David N. Lyon, *The World of P. J. McGuire: A Study of the American Labor Movement 1870–1890* (Ann Arbor, Mich.: University Microfilms, 1972), pp. 143–149.

54. *Historical Statistics of the United States,* p. 90.

55. Duffy, "History," pp. 101–107.

56. *The Carpenter,* October 1904, p. 7.

57. Duffy, "History," p. 108.

58. See Commons, *History of Labor,* II, chap. 7.

59. John T. Dunlop, "The Development of Labor Organization," in *Insights into Labor Issues,* ed. Richard A. Lester and Joseph Shister (New York: Macmillan, 1948), pp. 190–191.

2. THE EARLY YEARS OF THE UNITED BROTHERHOOD

1. *The Carpenter,* April 1881, p. 1.

2. Typewritten copy of the diary of Peter J. McGuire, in the files of the United Brotherhood of Carpenters.

3. Samuel Gompers, *Seventy Years of Life and Labor* (New York: Dutton, 1957), pp. 82–83.

4. David N. Lyon, *The World of P. J. McGuire* (Ann Arbor, Mich.: University Microfilms, 1972), p. 28.

5. Ibid., p. 31.

6. Ibid., p. 71.

7. McGuire to Gabriel Edmonston, February 7, 1884. Unless otherwise noted, all letters cited are in the files of the United Brotherhood of Carpenters.

8. Lyon, *World of P. J. McGuire,* pp. 146–147.

9. P. J. McGuire, "A Chapter of Our History," six-page undated typescript in the files of the United Brotherhood of Carpenters.

10. Ibid.

11. *The Carpenter,* May 1881, p. 2. A century later, the "terrible ogre" is still alive and thriving in the vocabulary of economics.

12. *The Carpenter,* June 1881, p. 3.

13. Brotherhood of Carpenters and Joiners of America, *Proceedings,* First Annual Convention, August 1881, p. 1. This was the original name of the union; the preface "United" was not added until later.

14. *The Carpenter,* August 1881, p. 2.

15. Luebkert to Edmonston, August 28, 1881, in Microfilm Corporation of America, *The American Federation of Labor: The Samuel Gompers Era,* reel 1 (hereafter cited as AFL Records).

16. Ibid.

17. McGuire to Edmonston, September 24, 1881, ibid.

18. Ibid.

19. McGuire to Edmonston, November 16, 1881, ibid. The hostility between Smith and McGuire was based largely on conflicting ideologies. Smith wrote Edmonston that there was considerable opposition to the Brotherhood in Cincinnati. "The members of our Union, and especially the Germans, are much opposed to Socialism, and the pronounced Socialistic views of McGuire and Luebkert have been the great bone of contention. A majority of the members are satisfied with McGuire, but they are determined that if Luebkert has anything to do with the affairs of the Brotherhood, that they will not. They were not at all satisfied that the paper was left in his charge [during McGuire's trip to Europe]." J. R. Smith to Edmonston, November 6, 1881, ibid.

20. J. P. Sverdurn to Edmonston, November 11, 1881, ibid.

21. McGuire to Edmonston, November 27, 1881, ibid.

22. McGuire to Edmonston, January 3, 1882, ibid.

23. McGuire to Edmonston, February 21, 1882, ibid.

24. Ibid. McGuire and Ritter did not hit it off. At one of the first meetings of the general executive board, a dispute arose as to where *The Carpenter* was to be printed. Ritter argued that the decision of the New York members of the board should be decisive, but when McGuire insisted that the entire board would have to be consulted, "Ritter got angry and quit the room." United Brotherhood of Carpenters, General Executive Board, *Minutes,* December 23, 1881. (These minutes, which are variously handwritten and typed, are available in the General Office of the United Brotherhood of Carpenters. They will be referred to hereafter as GEB *Minutes.*) At the next meeting Ritter questioned McGuire's authority to purchase supplies and printing without consulting the board. McGuire replied that "he deemed it impracticable to consult the Executive Board every time he wanted a postage stamp." GEB *Minutes,* February 6, 1882.

25. McGuire to Edmonston, March 11, 1882, AFL Records, reel 1.

26. McGuire to Edmonston, March 2, 1882, ibid.

27. Ibid.

28. McGuire to Edmonston, March 11, 1882, ibid.; Thos. Bunting, "History of the Carpenters of New York City" (undated typescript in the files of the United Brotherhood of Carpenters.)

29. *The Carpenter,* September 1881, p. 2.

30. *The Carpenter,* March 1882, p. 4. Although this was an unsigned editorial, it was undoubtedly written by McGuire.

31. *The Carpenter,* May 1882, p. 4.

32. *The Carpenter,* February 1882, p. 4.

33. *The Carpenter,* April 1882, p. 1.

34. *The Carpenter,* May 1882, p. 5.

35. Ibid., p. 3.

36. *The Carpenter,* August 1882, p. 4.

37. McGuire to Edmonston, June 4, 1882, AFL Records, reel 1.

38. McGuire to Edmonston, July 18, 1882, ibid.

39. McGuire to Edmonston, June 18 1882, ibid.

40. McGuire to Edmonston, July 18, 1882, ibid.

41. "On May 8, 1882, this writer, present General Secretary-Treasurer of the United Brotherhood of Carpenters, made the proposition. He urged the propriety of setting aside one day in the year to be designated as 'Labor Day.'" *The Carpenter,* September 1897, p. 8.

42. *The Carpenter,* September 1882, p. 1.

43. Jonathan Grossman, "Who Is the Father of Labor Day?" *Labor History,* Fall 1973, p. 616.

44. Ibid., p. 616.

45. Ibid., pp. 622–623.

46. Gompers to McGuire, *The Carpenter,* August 1974, p. 3.

47. American Federation of Labor, *Report of Proceedings,* Thirty-Ninth Annual Convention, 1919, p. 331. (Cited hereafter as AFL *Proceedings,.*)

48. Brotherhood of Carpenters and Joiners of America, Convention *Proceedings,* 1882.

49. McGuire to Edmonston, May 14, 1883, AFL Records, reel 1.

50. McGuire to Edmonston, May 19, 1882, ibid.

51. "In regard to the future president of the Brotherhood, I care not who is the choice so long as he does the work. But no matter who he may be he can never have any more interest nor do any more than you have done. Let the Convention determine. Still I know not why you cannot serve." McGuire to Edmonston, July 18, 1882, ibid.

52. Duffy, "History," p. 130.

53. *The Carpenter,* November 1882, p. 4.

54. *The Carpenter,* December 1882, p. 4.

55. McGuire to Edmonston, September 15, 1883, AFL Records, reel 1.

56. *The Carpenter,* May 1884, p. 4.

57. John R. Commons et al., *History of Labor in the United States,* vol. II (New York: Macmillan, 1918), p. 375.

58. United Brotherhood of Carpenters, Convention *Proceedings,* 1884, p. 6.

59. Ibid.

60. Ibid., p. 9.

61. Ibid., p. 61.

62. *The Carpenter,* September 1884, p. 4.

63. Ibid.

64. *The Carpenter,* November 1884, p. 4.

65. *The Carpenter,* January 1885, p. 4.

66. *The Carpenter,* September 1885, p. 4.

67. *The Carpenter,* December 1885, p. 4.

68. *The Carpenter,* November 1885, p. 4. In reply to a local inquiry, the GEB noted that in fourteen months, only $25 had been appropriated for this purpose. McGuire had paid $150 out of his own pocket. GEB *Minutes,* January 1886.

69. GEB *Minutes,* March 1885.

70. Ibid.

3. THE FIRST GROWTH YEARS

1. United Brotherhood of Carpenters, Convention *Proceedings,* 1886, p. 4. Conditions in Philadelphia at the beginning of 1886 were described as follows:

"Trade dull, many members out of work. Some miserable piece workers have driven away some of our members from a job at $2.75 per day, the piece workers took two bay windows for $2 each, and three men earned $4 between them in a day and a half . . . Those are the kind of 'cattle' we have to deal with here." *The Carpenter,* January 1886, p. 5.

2. Samuel Gompers, *Seventy Years of Life and Labor* (New York: Dutton, 1957), p. 258.

3. John R. Commons et al., *History of Labor in the United States,* vol. II (New York: Macmillan, 1918), p. 377.

4. *The Carpenter,* January 1886, p. 2.

5. The Knights of Labor refused to support the strike in the belief that political rather than economic action offered a better method of attaining the end of shorter hours. The Knights were at their peak in 1886, and their action was a major factor in the outcome of the strike.

6. *The Carpenter,* February 1886, p. 4.

7. McGuire to Edmonston, February 11, 1886. AFL Records, reel 1.

8. McGuire to Edmonston, April 28, 1886, ibid.

9. Speech delivered in Boston, *The Carpenter,* April 1886, p. 3.

10. For more detail about the history of the Knights and their relations with the craft unions, see Norman J. Ware, *The Labor Movement in the United States* (New York: Appleton, 1929).

11. McGuire to Edmonston, July 8, 1882, AFL Records, reel 1.

12. *The Carpenter,* March 1886, p. 4.

13. For a detailed account of the Washington situation, see Elizabeth Fones-Wolf and Kenneth Fones-Wolf, "Knights versus the Trade Unionists: The Case of the Washington, D.C. Carpenters," unpublished manuscript.

14. McGuire to Edmonston, February 8, 1886, AFL Records, reel 1.

15. McGuire to Edmonston, February 25, 1886, ibid.

16. McGuire to Edmonston, July 24, 1886, ibid.

17. *The Carpenter,* May 1886, p. 2.

18. Commons, *History of Labor,* II, 403–409.

19. *The Carpenter,* November 1886, p. 3.

20. United Brotherhood of Carpenters, Convention *Proceedings,* August 3–7, 1886, pp. 4–7.

21. *The Carpenter,* March 1886, p. 4. The idea of equalizing funds among locals by a process of redistribution is discussed in greater detail below. McGuire undoubtedly borrowed the notion from Great Britain, where it was practiced by a number of unions, including the Amalgamated Society of Carpenters. See Sidney Webb and Beatrice Webb, *The History of Trade Unionism* (London: Longmans, Green, 1984), p. 220.

22. Philip Taft, *The A.F. of L. in the time of Gompers* (New York: Harper, 1957), p. 39. Copyright 1957 by Philip Taft. Extracts reprinted by permission of Harper & Row, Publishers.

23. McGuire to Edmonston, March 18, 1887, AFL Records, reel 1.

24. Ibid.

25. McGuire to Edmonston, March 21, 1887, ibid.

26. McGuire to Edmonston, April 18, 1887, ibid.

27. Gompers, *Seventy Years of Life and Labor,* p. 332.

28. *The Carpenter,* May 1887, p. 2.

29. *The Carpenter,* February 1887, p. 5.

30. *The Carpenter,* March 1887, p. 4.

31. GEB *Minutes,* May 1887.

32. GEB *Minutes,* April and June, 1887. Because the strike had not received

the sanction of the GEB, the locals could not have access to the protective fund. The local officers were required to sign notes payable in 60 days, and if they failed to pay, the locals would be cut off from all benefits.

33. GEB *Minutes,* July and August, 1887.

34. David N. Lyon, *The World of P. J. McGuire,* (Ann Arbor, Mich.: University Microfilms, 1972), p. 299.

35. Ibid., p. 300.

36. GEB *Minutes,* June 1887.

37. *The Carpenter,* August 1887, p. 4.

38. GEB *Minutes,* July 1887.

39. McGuire to J. B. Clark (clerk of the U.S. House of Representatives), December 9, 1887, AFL Records, reel 1.

40. *The Carpenter,* October 1887, p. 4.

41. *The Carpenter,* January 1888, p. 4.

42. *The Carpenter,* October 1888, p. 4.

43. *The Carpenter,* June 1887, p. 4.

44. United Brotherhood of Carpenters, Convention *Proceedings,* August 6–11, 1888. The same year, the AFL convention declared it "unwise for two local, national, or international organizations of any one trade to exist in the same jurisdiction." AFL *Proceedings,* Third Annual Convention, 1888, p. 21. McGuire's remarks reflected a general attitude among the craft unionists of the country.

45. McGuire to J. S. Murchie, July 18, 1887, in Duffy, "History" p. 163.

46. J. S. Murchie to McGuire, August 24, 1887, ibid., p. 164.

47. *The Carpenter,* October 1887, p. 4.

48. AFL *Proceedings,* Third Annual Convention, 1888, p. 19.

49. United Brotherhood of Carpenters, Convention *Proceedings,* 1888, p. 21.

50. GEB *Minutes,* May 1888.

51. Fones-Wolf and Fones-Wolf, "Knights versus the Trade Unionists," pp. 18–19.

52. GEB *Minutes,* June 1888.

53. Ibid.

54. Ibid., September 1888.

55. Ibid., July 1888.

56. *The Carpenter,* October 1888, p. 1.

57. United Brotherhood of Carpenters, Convention *Proceedings,* 1888, pp. 11–20.

58. Duffy, "History," p. 182.

59. *The Carpenter,* February 1889, p. 4.

60. *The Carpenter,* March 1889, p. 4.

61. GEB *Minutes,* April 1889.

62. Ibid., May 1889.

63. Ibid., June 1889.

64. *The Carpenter,* April 1889, p. 6.

65. GEB *Minutes,* August 1889.

66. United Brotherhood of Carpenters, Convention *Proceedings,* 1890, p. 11.

67. GEB *Minutes,* May 1890.

68. Ibid., February 1890.

69. Ibid., March 1890.

70. AFL *Proceedings,* Tenth Annual Convention, 1890, p. 13.

71. Gompers to McGuire, March 20, 1890, quoted in Lyon, *World of P. J. McGuire,* p. 340.

72. United Brotherhood of Carpenters, Convention *Proceedings,* 1890, p. 11; *The Carpenter,* June 1890, p. 4.

73. GEB *Minutes,* June 1890.

74. Ibid., July 26, 1890.

75. United Brotherhood of Carpenters, Convention *Proceedings,* 1890, p. 11.

76. Ibid., p. 15

77. *The Carpenter,* November 1890, p. 4.

78. GEB *Minutes,* November 26, 1890.

79. *The Carpenter,* January 1891, p. 4.

80. Ibid., February 1891, p. 4.

81. Duffy, "History," p. 197.

82. Fones-Wolf and Fones-Wolf, "Knights versus the Trade Unionists," pp. 22–25.

4. THE END OF AN ERA

1. United Brotherhood of Carpenters, Convention *Proceedings,* 1892, p. 15.

2. Robert A. Christie, *Empire in Wood* (Ithaca, N.Y.: Cornell University, 1956), p. 69.

3. United Brotherhood of Carpenters, Convention *Proceedings,* 1892, p. 15.

4. Robert A. Christie, "Empire in Wood," Ph.D. dissertation, Cornell University, 1954, p. 134. This typescript version differs in some respects from the published version cited in note 2 above.

5. This section is based in part on Jules Tygiel, "Tramping Artisans: The Case of the Carpenters in Industrial America," *Labor History,* Summer 1981, p. 348. The subject of mobility is treated more generally in Lloyd Ulman, *The Rise of the National Trade Union* (Cambridge, Mass.: Harvard University Press, 1955).

6. United Brotherhood of Carpenters, *Constitution and Laws,* January 1, 1979, section 46.

7. United Brotherhood of Carpenters, Convention *Proceedings,* 1892, p. 15.

8. Ibid.

9. Ibid., p. 53.

10. GEB *Minutes,* January 10, 1893.

11. Ibid.

12. Ibid., October 4, 1893.

13. Ibid., October 5, 1893.

14. Ibid., October 4, 1893.

15. Ibid., January 10, 1894.

16. Ibid., April 9, 1894.

17. United Brotherhood of Carpenters, Convention *Proceedings,* 1894, p. 10.

18. Ibid., p. 4.

19. *The Carpenter,* February 1895, p. 2.

20. United Brotherhood of Carpenters, Convention *Proceedings,* 1894, p. 69.

21. AFL *Proceedings,* Thirteenth Annual Convention, 1893, p. 59. Gompers won by the fairly narrow margin of 1,314 to 1,222. McGuire declined the nomination in favor of McBride. He probably could have been elected to the AFL presidency if he had agreed to run.

22. AFL *Proceedings,* Fourteenth Annual Convention, 1894, p. 41.

23. *The Carpenter,* November 1894, p. 8.

24. Ibid., February 1895, p. 2.

25. Christie, *Empire in Wood,* p. 94.

26. AFL *Proceedings,* Eighteenth Annual Convention, 1898, p. 105.

27. Christie, *Empire in Wood,* p. 94.

28. United Brotherhood of Carpenters, Convention *Proceedings,* 1894, p. 19.

29. GEB *Minutes,* January 12, 1895.

30. A recommendation to this effect was brought in by the Committee on

Organization, and concurred in by the convention. United Brotherhood of Carpenters, Convention *Proceedings,* 1894, p. 42.

31. *The Carpenter,* June 1895, p. 8.

32. Ibid., July 1895, p. 8.

33. GEB *Minutes,* July 19, 1895.

34. *The Carpenter,* July 1896, p. 8.

35. GEB *Minutes,* July 21, 1894.

36. Ibid., January 11, 1895.

37. The number of locals in good standing in 1894 was reported to be 716, compared with 440 in 1896. There is an unexplained discrepancy of 276 between these figures and the totals reported in 1896.

38. United Brotherhood of Carpenters, Convention *Proceedings,* 1896, p. 24.

39. Ibid., p. 43.

40. Christie, *Empire in Wood,* p. 78.

41. *The Carpenter,* February 1897, p. 8.

42. GEB *Minutes,* January 8, 1896.

43. *The Carpenter,* April 1897, p. 8.

44. Duffy, "History," p. 260.

45. This appears in a typed manuscript in the files of the United Brotherhood of Carpenters, which purports to be a transcript of the testimony of McGuire before the U.S. Industrial Commission in 1899. It is not included in the printed volumes issued by the Commission. This anomaly is discussed further in Appendix A.

46. United Brotherhood of Carpenters, Convention *Proceedings,* 1890, p. 16.

47. Gabriel Edmonston wrote later that he could only account for McGuire's action by the hypnotic diplomacy of Thomas Kidd. The charter gave the new union jurisdiction over machine woodworkers in factories. *The Carpenter,* October 1904, p. 9.

48. Duffy, "History," p. 211.

49. Ibid., p. 214.

50. United Brotherhood of Carpenters, Convention *Proceedings,* 1894, p. 38.

51. Duffy, "History," p. 215.

52. McGuire to B. Koenen, March 3, 1891, quoted in Duffy, "History," p. 206.

53. GEB *Minutes,* July 13, 1897.

54. Ibid., October 5, 1897.

55. Frederick S. Deibler, *The Amalgamated Woodworkers' International Union of America* (Madison: University of Wisconsin Press, 1912), p. 169.

56. GEB *Minutes,* April 5, 1898.

57. United Brotherhood of Carpenters, Convention *Proceedings,* 1894, p. 15.

58. Duffy, "History," p. 227.

59. United Brotherhood of Carpenters, Convention *Proceedings,* 1898, p. 32.

60. Ibid., p. 22.

61. Ibid., p. 37; *The Carpenter,* July 1898, p. 1.

62. United Brotherhood of Carpenters, Convention *Proceedings,* 1898, pp. 7–12.

63. *The Carpenter,* August 1915, p. 17.

64. Philip Taft, *Organized Labor in American History* (New York: Harper & Row, 1964), pp. 161–162.

65. Up to 1912, neither the general president nor the general secretary was a voting member of the GEB, although McGuire and Duffy often met with the board to discuss particular issues.

66. *The Carpenter,* August 1915, p. 7.

67. United Brotherhood of Carpenters, Convention *Proceedings,* 1898, p. 63.

68. Ibid., p. 65.

69. Ibid., p. 21.

70. United Brotherhood of Carpenters, Convention *Proceedings,* 1900, p. 46.

71. GEB *Minutes,* April 14, 1899.

72. Ibid., April 18, July 16, July 18, July 22, 1900.

73. *The Carpenter,* November 1899, p. 14. The 1886 convention had authorized the GEB to appoint "one or more suitable persons in each state and territory as Deputy Organizers." In 1888 the vice-presidents, of whom there were seven at the time, were authorized to act as organizers for their districts. Two years later the number of vice-presidents was reduced to two, and the former vice-presidents were made members of the GEB. Cattermull was elected to the GEB in 1894, and also served as a general organizer.

74. GEB *Minutes,* January 16, April 13, 1899; January 8, 13, April 9, 11, 25, July 18, 22, 23, September 15, 21, 1890.

75. GEB *Minutes,* July 15, October 13, 1899.

76. AFL, *Executive Council Minutes,* October 1902, p. 109.

77. United Brotherhood of Carpenters, Convention *Proceedings,* 1900, passim; *The Carpenter,* November 1900, p. 5, and January 1901, p. 2.

78. United Brotherhood of Carpenters, Convention *Proceedings,* 1900, p. 70.

79. Ibid., p. 85.

80. Christie, *Empire in Wood,* p. 95.

81. United Brotherhood of Carpenters, Convention *Proceedings,* 1902, p. 141.

82. GEB *Minutes,* January 6 and 21, 1901.

83. Ibid., July 15, 1901.

84. Ibid., July 18, 19, 20, 1901. Huber claimed that a preliminary audit of the books revealed a shortage of $6,000. McGuire thereupon gave the GEB a check for $6,300 and asked for more time to look over his records. It was discovered that this check was covered by a check drawn on a Brotherhood account in another bank. McGuire did come to the office and found some of the papers demanded by the board, but he refused to go before it. United Brotherhood of Carpenters, Convention *Proceedings,* 1902, p. 20.

85. GEB *Minutes,* July 23, 1901.

86. Ibid., August 7, 1901.

87. United Brotherhood of Carpenters, Convention *Proceedings,* 1902, p. 20.

88. Ibid., p. 32.

89. *The Carpenter,* December 1901, p. 2.

90. AFL Records, reel 139, November 15, 1901.

91. United Brotherhood of Carpenters, Convention *Proceedings,* 1902, p. 20.

92. *The Carpenter,* January 1902, p. 3.

93. United Brotherhood of Carpenters, Convention *Proceedings,* 1902, p. 20.

94. Morrison to Gompers, May 14, 1902, AFL Records, reel 139.

95. McGuire to Gompers, June 16, 1902, AFL Records, reel 35.

96. United Brotherhood of Carpenters, Convention *Proceedings,* 1902, Appendix, p. 12.

97. Ibid., p. 167.

98. Christie, *Empire in Wood,* p. 103.

99. Ibid., p. 104.

100. Ibid., p. 92.

101. AFL, *Executive Council Minutes,* March 19–24, 1906, p. 1.

102. Duffy to Gompers, June 21, 1906.

103. United Brotherhood of Carpenters, Convention *Proceedings,* 1906, p. 175.

104. Gompers to Duffy, June 28, 1906.

105. GEB *Minutes,* November 25, 1914.

106. Ibid., September 1, 1950.

5. ROUNDING OUT THE JURISDICTION

1. United Brotherhood of Carpenters, Convention *Proceedings,* 1906, p. 71.

2. Duffy, "History," pp. 294–295.

3. Ibid.

4. Duffy to Gompers, May 6, 1912.

5. Duffy, "History," p. 294.

6. For a detailed history of this organization and of its relationships with the Brotherhood, see Frederick S. Deibler, *The Amalgamated Wood Workers International Union of America* (Madison: University of Wisconsin Press, 1912).

7. AFL *Proceedings,* Twenty-First Annual Convention, 1901, p. 252.

8. AFL *Proceedings,* Twenty-Second Annual Convention, 1902, p. 165.

9. It should be pointed out that the Brotherhood delegates had agreed to his appointment as umpire.

10. *The Carpenter,* May 1903, p. 5.

11. Deibler, *The Amalgamated,* pp. 305–306.

12. AFL *Executive Council Minutes,* June 19–20, 1903, p. 26.

13. *The Carpenter,* October 1903, p. 4.

14. Ibid., January 1904, p. 6.

15. GEB *Minutes,* January 21, 1904.

16. United Brotherhood of Carpenters, Convention *Proceedings,* 1904, p. 205.

17. AFL *Proceedings,* Twenty-Fourth Annual Convention, 1904, p. 219.

18. AFL *Executive Council Minutes,* June 12–17, 1905, p. 29.

19. United Brotherhood of Carpenters, *January Password and Quarterly Circular,* 1907.

20. John Meiler to Gompers, June 26, 1907.

21. Huber to Gompers, July 5, 1907.

22. Duffy to Gompers, July 19, 1907.

23. AFL *Proceedings,* Twenty-Seventh Annual Convention, 1907, p. 269.

24. AFL *Proceedings,* Twenty-Eighth Annual Convention, 1908, p. 206.

25. United Brotherhood of Carpenters, Convention *Proceedings,* 1908, p. 14; *The Carpenter,* March 1910, p. 14.

26. Deibler, *The Amalgamated,* p. 187.

27. Ibid., p. 186.

28. AFL *Proceedings,* Twenty-Ninth Annual Convention, 1909, p. 289. R. A. McKee was a delegate from the International Union of Steam Engineers, while T. L. (not John L.) Lewis represented the United Mine Workers.

29. AFL *Proceedings,* Thirtieth Annual Convention, 1910, p. 105.

30. AFL *Proceedings,* Thirty-First Annual Convention, 1911, p. 320.

31. Duffy, "History," p. 224.

32. Deibler, *The Amalgamated,* p. 188.

33. Ibid., p. 190.

34. Robert A. Christie, *Empire in Wood,* (Ithaca, N.Y.: Cornell University, 1956), p. 119

35. *The Carpenter,* June 1904, p. 6.

36. Ibid., June 1906, p. 12.

37. Ibid., August 1907, p. 12.

38. Ibid., October 1907, p. 20

39. Selig Perlman, *A Theory of the Labor Movement* (New York: Macmillan, 1928), pp. 188–189.

40. United Brotherhood of Carpenters, Convention *Proceedings,* 1904, p. 32.

41. *The Carpenter,* August 1904, p. 5.

42. Christie, *Empire in Wood,* p. 252.

43. Ibid., p. 119.

44. United Brotherhood of Carpenters, Convention *Proceedings,* 1912, p. 96.

45. GEB *Minutes,* October 17, 1901.

46. United Brotherhood of Carpenters, Convention *Proceedings,* 1902, p. 203.

47. AFL *Executive Council Minutes,* November 12, 1902, p. 146.

48. Ibid., January 19–24, 1903, p. 50.

49. *The Carpenter,* March 1903, p. 6.

50. Ibid., May 1903, p. 4.

51. Ibid., July 1903, p. 4.

52. Gompers to Frank Chandler, January 13, 1904.

53. *The Carpenter,* June 1906, p. 15.

54. AFL, *Executive Council Minutes,* January 11–16, 1909, p. 33; March 23–26, 1910, p. 25.

55. United Brotherhood of Carpenters, Convention *Proceedings,* 1910, p. 517.

56. AFL *Proceedings,* Thirty-First Annual Convention, 1911, p. 310.

57. AFL, *Executive Council Minutes,* August 12–17, 1912, addendum, p. 32.

58. The full text of this plan is contained in Duffy, "History," pp. 236–237.

59. United Brotherhood of Carpenters, Convention *Proceedings,* 1914, p. 26.

60. *The Carpenter,* October 1923, p. 31.

61. United Brotherhood of Carpenters, Convention *Proceedings,* 1920, p. 21.

62. Duffy, "History," p. 300.

63. GEB *Minutes,* January 20, 1902.

64. Ibid., April 12, 1904.

65. Ibid., April 19, 1904.

66. AFL, *Executive Council Minutes,* September 12–15, 1904, p. 6.

67. Duffy, "History," pp. 303–304.

68. Nathaniel R. Whitney, *Jurisdiction in American Building Trades Union* (Baltimore: Johns Hopkins Press, 1914), p. 57.

69. United Brotherhood of Carpenters, Convention *Proceedings,* 1910, p. 61.

70. GEB *Minutes,* February 24, 1915.

71. Ibid.

72. Duffy to Gompers, June 18, 1903.

73. Gompers to Duffy, June 10, 1903.

74. United Brotherhood of Carpenters, Convention *Proceedings,* 1904, p. 44.

75. Ibid., 1906, p. 58. A Brotherhood referendum on affiliation with the Alliance carried by a margin of better than 3 to 1.

76. See Christie, *Empire in Wood,* p. 133, for details.

77. *The Carpenter,* March 1905, p. 5.

78. Duffy to Morrison, August 29, 1907.

79. Morrison to Duffy, September 11, 1907.

80. Building Trades Department, Official Report of the General Conference: Building Trades Affiliated with the American Federation of Labor, Washington, February 10–15, 1908.

81. Building Trades Department, Second Annual Convention, 1909.

82. GEB *Minutes,* January 15, 1903.

83. *The Carpenter,* July 1909, p. 28.

84. United Brotherhood of Carpenters, Convention *Proceedings,* 1910, p. 319.

85. AFL *Proceedings,* Thirtieth Annual Convention, 1910, p. 319.

86. Building Trades Department, Second Annual Convention, 1909, p. 37.

87. Duffy to Kirby, February 8, 1910.

88. Building Trades Department, Fourth Annual Convention, 1910, p. 110.

89. *The Carpenter,* January 1911, p. 20.

90. AFL, *Executive Council Minutes,* June 12–17, 1911, p. 53.

91. AFL *Proceedings,* Thirty-First Annual Convention, 1911, p. 329.

92. *The Carpenter,* May 1912, p. 23.

93. GEB *Minutes,* April 16, 1909.

94. Duffy to George Wolfe, December 21, 1907.

95. Duffy to Morrison, May 19, 1908.

96. GEB *Minutes,* April 7, 1904; April 11, 1904; January 29, 1906; July 20, 1907; October 19, 1907.

97. See Selig Perlman and Philip Taft, *History of Labor in the United States* (New York: Macmillan, 1935), p. 155.

98. GEB *Minutes,* January 11, 1909; January 19, 1909; April 15, 1910; May 1, 1911; April 9, 1913.

99. Spencer to Duffy, December 27, 1909; Duffy to Spencer, December 29, 1909.

100. Duffy to Morrison, January 19, 1910.

101. GEB *Minutes,* October 9, 1907; January 21, 1910; July 24, 1912.

102. United Brotherhood of Carpenters, Convention *Proceedings,* 1908, p. 14.

6. THE EMPLOYER OFFENSIVE AND THE UNION RESPONSE

1. Selig Perlman and Philip Taft, *History of Labor in the United States* (New York: Macmillan, 1935), pp. 84–87.

2. *The Carpenter,* June 1903, p. 4.

3. GEB *Minutes,* April 30, 1903.

4. Perlman and Taft, *History of Labor,* pp. 88–93.

5. *The Carpenter,* September 1904, p. 2.

6. Ibid., December 1904, p. 2.

7. GEB *Minutes,* February 24, 1905.

8. Ibid., April 25, 1905.

9. United Brotherhood of Carpenters, Convention *Proceedings,* 1904, p. 47.

10. Perlman and Taft, *History of Labor,* pp. 131–137.

11. *The Carpenter,* July 1906, p. 27.

12. Perlman and Taft, *History of Labor,* pp. 136–137.

13. GEB *Minutes,* January 14, 1901; May 1, 1901; October 15, 1901; January 17, 1902; April 18, 1902.

14. GEB *Minutes,* April 30, 1902. This rule had to be modified when the Box Makers were taken into the union.

15. Ibid., April 27, 1905.

16. Ibid., February 12, 1906; February 5, 1907.

17. Ibid., January 29, 1908; April 13, 1908.

18. Ibid., October 14, 1909.

19. Among the other grants of authority were the right to rule on all points of law, to pass on appeals and grievances, and to sign charters, all of which had been within the province of the general secretary.

20. United Brotherhood of Carpenters, Convention *Proceedings,* 1902, pp. 101, 192.

21. *The Carpenter,* January 1903, p. 3.

22. Ibid.
23. Ibid., March and September, 1903.
24. Ibid., April 1904, p. 6.
25. Ibid., July 1904, p. 6.
26. Ibid., October 1904, p. 5; September 1905, p. 23.
27. Ibid., February 1907, p. 31; August 1908, p. 41.
28. Ibid., January 1910, p. 21; May 1911, p. 23; May 1912, p. 38.
29. United Brotherhood of Carpenters, Convention *Proceedings,* 1904, p. 45.
30. The following account is based on a verbatim report contained as an appendix to the 1904 convention proceedings.
31. GEB *Minutes,* February 6, 1904.
32. Ibid., July 13, 1905; *The Carpenter,* October 1905, p. 14.
33. United Brotherhood of Carpenters, Convention *Proceedings,* 1906, p. 446.
34. Ibid., p. 444.
35. *The Carpenter,* February 1907, p. 33.
36. United Brotherhood of Carpenters, Convention *Proceedings,* 1910, p. 239.
37. Frank Duffy, "A History of the Union Label of the Brotherhood," typescript in the files of the United Brotherhood of Carpenters.
38. United Brotherhood of Carpenters, Convention *Proceedings,* 1906, p. 70.
39. William Haber, "The United Brotherhood of Carpenters and Joiners: A Study of Conservative Trade Unionism," Ph.D. dissertation, University of Wisconsin, 1923, p. 207.
40. Christie, *Empire in Wood,* pp. 161–163.
41. *The Carpenter,* November 1907, p. 25.
42. Ibid., January 1906, p. 19.
43. Ibid., September 1906, p. 39; December 1906, p. 22.
44. Ibid., January 1908, p. 10; February 1909, p. 30.
45. Ibid., January 1910, p. 13.
46. United Brotherhood of Carpenters, Convention *Proceedings,* 1906, p. 94.
47. Ibid., 1908, p. 72.
48. Ibid., p. 14.
49. Ibid., p. 72.
50. Ibid., p. 14.
51. Ibid., p. 207.
52. Ibid., p. 329.
53. *The Carpenter,* February 1909, p. 19.
54. Federick L. Ryan, *Industrial Relations in the San Francisco Building Trades* (Norman, Okla.: University of Oklahoma Press, 1936), p. 33. Copyright 1936 by the University of Oklahoma Press. Extracts reprinted by permission.
55. Ibid., p. 44.
56. GEB *Minutes,* October 14, 1901.
57. Ibid., January 20, 1902.
58. Ibid., July 15, 1902; *The Carpenter,* May 1902, p. 2.
59. AFL, *Executive Council Minutes,* July 22–26, 1902, p. 20.
60. Ryan, *Industrial Relations,* p. 48.
61. United Brotherhood of Carpenters, Convention *Proceedings,* 1902, p. 133.
62. Ryan, *Industrial Relations,* p. 109.
63. *The Carpenter,* July 1904, p. 5.
64. Ibid., September 1904, p. 6.
65. Christie, *Empire in Wood,* pp. 105, 157, 234.
66. *The Carpenter,* August 1908, p. 41.
67. Perlman and Taft, *History of Labor,* p. 78.
68. *The Carpenter,* February 1909, p. 30.

69. Organizer James A. Gray of the Carpenters wrote in 1908: "Our G.P. and myself left San Francisco for Portland, Oregon, with many pleasant recollections of our stay in central California, which I consider to be the best organized section of the United States, due entirely to the able manner in which the State B.T.C. conducts its affairs under the able leadership of Brother P. H. McCarthy." *The Carpenter,* September 1908, p. 32.

70. GEB *Minutes,* April 12–16, 1909.

71. United Brotherhood of Carpenters, Convention *Proceedings,* 1912.

72. GEB *Minutes,* April 14, 1910.

73. *The Carpenter,* July 1910, p. 14.

74. United Brotherhood of Carpenters, Convention *Proceedings,* 1910, p. 61.

75. Ibid., pp. 425, 451.

76. Ibid., p. 528.

77. Ibid., p. 215

78. Ibid., p. 495.

79. Ibid., pp. 503–507.

80. Ibid., p. 559.

81. *The Carpenter,* February 1911, p. III.

82. Duffy to Morrison, February 13, 1911.

83. The full report of the auditors, Lybrand, Ross Brothers, and Montgomery, one of the leading accounting firms in the United States, was printed in the *The Carpenter,* February 1911, pp. 31–48.

84. Duffy to Morrison, February 27, 1911.

85. *The Carpenter,* February 1911, p. 30.

86. At the time, the Carpenters' constitution provided that penalties imposed by locals upon members for violation of working rules could be appealed to the general president, and from his decision to the GEB. Alterations in this procedure are discussed below.

87. United Brotherhood of Carpenters, Convention *Proceedings,* 1908, p. 42.

88. GEB *Minutes,* January 17, 1912.

89. The Complaint and supporting affidavits were printed verbatim in *The Carpenter,* June 1910, pp. 2–33.

90. *The Carpenter,* September 1910, p. 24.

91. Ibid., p. 23.

92. Ibid., June 1911, p. 11.

93. United Brotherhood of Carpenters, Convention *Proceedings,* 1914. p. 405.

94. Ibid., p. 410.

95. *Paine Lumber Co.* v. *Neal,* 244 U.S. 459 (1917).

96. *The Carpenter,* August 1917, p. 30.

97. In particular, *Bossert* v. *Dhuy,* 221 N.Y. 342 (1917).

98. United Brotherhood of Carpenters, Convention *Proceedings,* 1912, p. 6.

99. Ibid., pp. 299, 96.

100. Ibid., p. 53.

101. Ibid., pp. 466, 521, 548, 542, 554.

102. Ibid., pp. 625, 633.

103. From 1910 to 1912, in 15 percent of the benefit cases allowed by the Brotherhood, the cause of death was tuberculosis. Other major causes were accidents (19 percent), heart disease (12 percent), nephritis (13 percent), cancer (8 percent), pneumonia (7 percent), and typhoid fever (3 percent). Ibid., p. 312.

104. Ibid., p. 698.

105. GEB *Minutes,* October 9, 1907.

106. Duffy to Gompers, July 15, 1912.

107. See Philip Taft, *Organized Labor in American History* (New York: Harper & Row, 1964), pp. 324–325.

108. *The Carpenter,* February 1913, p. 24.

7. CONSOLIDATING THE NATIONAL UNION

1. James M. Motley, *Apprenticeship in American Trade Unions* (Baltimore: Johns Hopkins Press, 1907), p. 75. See also Paul H. Douglas, *American Apprenticeship and Industrial* Education (New York: Columbia University Press, 1921) for data on early apprenticeship ratios.

2. *The Carpenter,* March 1888, p. 4.

3. United Brotherhood of Carpenters, Convention *Proceedings,* 1888, p. 38.

4. GEB *Minutes,* October 11, 1909.

5. United Brotherhood of Carpenters, Convention *Proceedings,* 1910, p. 559.

6. Ibid., 1912, p. 609.

7. William Haber, "The United Brotherhood of Carpenters and Joiners: A Study of Conservative Trade Unionism," Ph.D. dissertation, University of Wisconsin, 1923, pp. 213–217.

8. GEB *Minutes,* April 13, 1915.

9. United Brotherhood of Carpenters, Convention *Proceedings,* 1916, p. 6.

10. *The Carpenter,* September 1913, p. 2.

11. Ibid., July 1915, p. 17.

12. United Brotherhood of Carpenters, Convention *Proceedings,* 1914, p. 26.

13. Lloyd Ulman, *The Rise of the National Trade Union,* (Cambridge, Mass.: Harvard University Press, 1955), pp. 489–490.

14. United Brotherhood of Carpenters, Convention *Proceedings,* 1914, pp. 687, 640.

15. Ibid., p. 651.

16. See below for a discussion of this practice.

17. United Brotherhood of Carpenters, Convention *Proceedings,* 1914, pp. 578, 782, 779.

18. Ibid., p. 47.

19. Ibid., p. 533.

20. Ibid., p. 529.

21. Robert A. Christie, *Empire in Wood* (Ithaca, N.Y.: Cornell University, 1956), p. 196.

22. Ulman, *Rise of the National Trade Union,* chap. 5, deals more generally with the problem posed for American trade unions by traveling members.

23. *The Carpenter,* April 1914, p. 15.

24. Ibid., April 1915, p. 26.

25. GEB *Minutes,* April 9, 1915.

26. *The Carpenter,* March 1916, p. 17.

27. Ibid., February 1915, p. 116.

28. Duffy to Gompers, February 21, 1914.

29. Gompers to Duffy, February 25, 1914.

30. Duffy to Gompers, September 17, 1914.

31. Charles Gorman et al. to Gompers, September 11, 1915.

32. Duffy to Gompers, October 1, 1915; Gompers to Duffy, October 2, 1915.

33. Gompers to Duffy, January 7, 1916.

34. Duffy to Gompers, February 19, 1916.

35. Duffy to Gompers, May 26, 1916.

36. Maxwell C. Raddock, *Portrait of an American Labor Leader* (New York: American Institute of Social Science, 1955), p. 69.

37. GEB *Minutes,* January 18, 1916.

38. Ibid., April 10, 1916.

39. *The Carpenter,* August 1916, p. 32.

40. United Brotherhood of Carpenters, Convention *Proceedings,* 1916, p. 42.

41. GEB *Minutes,* May 12, 1916.

42. United Brotherhood of Carpenters, Convention *Proceedings,* 1916, p. 42.

43. *The Carpenter,* August 1916, p. 32.

44. Hirsh Reif, "The Carpenters' Revolution," typescript in the files of the United Brotherhood of Carpenters.

45. United Brotherhood of Carpenters, Convention *Proceedings,* 1916, p. 372.

46. GEB *Minutes,* July 13, 15, and 28, 1916.

47. United Brotherhood of Carpenters, Convention *Proceedings,* 1916, p. 398.

48. Selig Perlman and Philip Taft, *History of Labor in the United States* (New York: Macmillan, 1935), p. 476.

49. Hugh Frayne to Frank Morrison, February 26, 1913.

50. AFL, *Executive Council Minutes,* July 13–18, 1914, p. 4.

51. Ibid., April 19–24, 1915, p. 124; September 20–25, 1915, p. 5; *The Carpenter,* January 1916, p. 18.

52. AFL, *Executive Council Minutes,* February 21–26, 1916, p. 45.

53. AFL *Proceedings,* Thirty-Sixth Annual Convention, 1916, p. 130.

54. *The Bridgemen's Magazine,* January 1917, p. 11.

55. Duffy to Gompers, April 17, 1917.

56. J. M. McClory to Gompers, May 25, 1917.

57. Duffy to Morrison, June 1, 1917.

58. Gompers to McClory, June 5, 1917.

59. *The Carpenter,* January 1918, p. 26

60. State of New York, Intermediate Report of the Joint Legislative Committee on Housing, Legislative Document (1922) no. 60, Charles G. Lockwood, Chairman, pp. 38–42.

61. Ibid., pp. 43, 45.

62. Ibid., p. 48.

63. Perlman and Taft, *History of Labor,* p. 505.

64. Christie, *Empire in Wood,* p. 215.

65. Building Trades Department, *Proceedings,* Sixth Annual Convention, 1912, p. 119.

66. AFL, *Executive Council Minutes,* January 20–25, 1913, p. 14.

67. Ibid., p. 51.

68. *The Carpenter,* January 1913, p. 22.

69. Duffy to William J. Spencer, March 21, 1913.

70. Duffy to Spencer, August 7, 1913.

71. Duffy to Spencer, September 3, 1913; Spencer to Duffy, September 5, 1913.

72. Gompers to Kirby, October 16, 1913.

73. Building Trades Department, *Proceedings,* Seventh Annual Convention, 1913, pp. 82, 88, 110.

74. Duffy to Gompers, February 17, 1914.

75. AFL, *Executive Council Minutes,* May 11–16, 1914, p. 16; September 13–18, 1914, p. 7.

76. The failure of the Bricklayers to join the AFL stemmed from the fear on the part of some of their small locals that they might by drawn into strikes in sup-

port of other unions. See Ulman, *Rise of the National Trade Union,* pp. 418–421.

77. *The Carpenter,* September 1914, p. 7.

78. AFL *Proceedings,* Thirty-Fourth Annual Convention, 1914, pp. 139, 349.

79. Ibid., Thirty-Fifth Annual Convention, 1915, p. 122.

80. Ibid, pp. 404 ff.

81. Ibid. This referred to a long statement of jurisdiction that appeared in the October 1915 issue of *The Carpenter.*

82. Building Trades Department, *Proceedings,* Ninth Annual Convention, 1915, passim.

83. Brotherhood membership in 1915 was about 260,000. The Machinists had only 75,000 members at the time.

84. *The Carpenter,* March 1916, p. 4.

85. Duffy to Gompers, February 1, 1916.

86. William H. Johnston to Gompers, February 23, 1916.

87. AFL, *Executive Council Minutes,* June 26–July 4, 1916, p. 29.

88. AFL *Proceedings,* Thirty-Sixth Annual Convention, 1916, pp. 124, 374, 376.

89. AFL, *Executive Council Minutes,* October 18–27, 1917, p. 21.

90. GEB *Minutes,* February 6, 1913; AFL, *Executive Council Minutes,* July 13–18, 1914, p. 31.

91. AFL, *Executive Council Minutes,* April 19–24, 1915, p. 52.

92. GEB *Minutes,* July 12, 1917,

93. *The Carpenter,* July 1916, p. 19.

94. United Brotherhood of Carpenters, Convention *Proceedings,* 1916, p. 233.

95. Building Trades Department, *Proceedings,* Twelfth Annual Convention, 1918, p. 46.

96. AFL *Proceedings,* Thirty-Sixth Annual Convention, 1916, p. 315.

97. Duffy to Gompers, October 19, 1916.

98. Gompers to McClory, October 19, 1916; McClory to Gompers, February 12, 1917.

99. Duffy to Gompers, February 21, 1917; GEB *Minutes,* July 23, 1919.

100. GEB *Minutes,* April 8, 1915.

101. Duffy to Morrison, April 18, 1916.

102. *The Carpenter,* May 1918, p. 26.

103. GEB *Minutes,* April 5, 1917.

104. United Brotherhood of Carpenters, Convention *Proceedings,* 1916, p. 40.

105. I have discussed these issues at some length in "Why the American Labor Movement Is Not Socialist," *American Review,* Winter 1961, p. 1.

8. WORLD WAR I AND ITS AFTERMATH

1. GEB *Minutes,* April 7, 1917.

2. *The Carpenter,* August 1917, p. 37.

3. Philip Taft, *The A.F. of L. in the Time of Gompers* (New York: Harper, 1957), p. 346.

4. Quoted in Maxwell C. Raddock, *Portrait of an American Labor Leader* (New York: American Institute of Social Science, 1955), p. 91.

5. Philip Taft, *Organized Labor in American History* (New York: Harper & Row, 1964), pp. 311–312.

6. Hutcheson to Gompers, July 12, 1917, quoted in Taft, *A.F. of L. in the Time of Gompers,* p. 350.

7. Hutcheson to Gompers, September 19, 1917, quoted ibid.

8. Gompers to Hutcheson, October 21, 1917, quoted ibid., p. 350.

9. Gompers to Hutcheson, October 5, 1917, quoted ibid., p. 351.

10. *The Carpenter,* November 1917, p. 25.

11. Ibid., February 1918, p. 18.

12. Ibid., May 1918, p. 4.

13. Ibid.

14. Macy to Gompers, February 1918, ibid.

15. Hurley to Hutcheson, February 14, 1918, reproduced ibid.

16. *The Carpenter,* April 1918, p. 4.

17. Hurley to Hutcheson, February 15, 1918, reproduced in *The Carpenter,* May 1918, p. 8.

18. Woodrow Wilson to Hutcheson, February 17, 1918, ibid.

19. Ibid., p. 12.

20. *The Carpenter,* August 1918, p. 44.

21. The text of this agreement appeared in the *The Carpenter,* May 1918, p. 16.

22. *The Carpenter,* May 1918, pp. 13–14.

23. GEB *Minutes,* April 13, 1918.

24. GEB *Minutes,* October 9, 1918.

25. Taft, *A.F. of L. in the Time of Gompers,* pp. 357–358.

26. Duffy, "History," pp. 415–416.

27. *The Carpenter,* January 1919, p. 41.

28. United Brotherhood of Carpenters, Convention *Proceedings,* 1920, p. 21.

29. Ibid., p. 174.

30. *The Carpenter,* May 1919, p. 20. These sentiments were to be repeated after World War II and after the demise of the Construction Industry Stabilization Committee in 1974.

31. AFL *Proceedings,* Thirty-Ninth Annual Convention, 1919, p. 348.

32. Duffy to Gompers, September 30, 1919.

33. Hutcheson to Gompers, October 2, 1919, in AFL, *Executive Council Minutes,* October 5–22, 1919.

34. Gompers to Hutcheson, October 12, 1919, ibid.

35. Hutcheson to Gompers, October 15, 1919, ibid.

36. Philip Taft, *The A.F. of L. in the Time of Gompers,* pp. 399–406.

37. Circular letter, GEB *Minutes,* January 9, 1920.

38. The details of this episode, contained in a report by the president of the Louisiana State Federation of Labor, are given in Appendix B.

39. *The Carpenter,* May 1919, p. 33.

40. Ibid., June 1919, p. 25.

41. Duffy to Gompers, October 1, 1920; Gompers to Duffy, October 20, 1920; Duffy to S. Misiah, October 25, 1920.

42. GEB *Minutes,* April 5, 1920.

43. Duffy to Gompers, August 31, 1920.

44. Raddock, *Portrait,* pp. 132–133.

45. United Brotherhood of Carpenters, Convention *Proceedings,* 1920, p. 167.

46. Ibid., pp. 21–22.

47. Ibid., p. 462.

48. Ibid., p. 542.

49. Ibid., p. 426.

50. Morrison to Duffy, December 1, 1919; Duffy to Morrison, December 3, 1919.

9. THE NINETEEN-TWENTIES

1. The data are from U.S. Bureau of Labor Statistics, *Handbook of Labor Statistics* (Washington, D.C.: Government Printing Office, 1975), table 39.

2. Selig Perlman and Philip Taft, *History of Labor in the United States* (New York: Macmillan, 1935), pp. 489–514.

3. *The Carpenter,* February 1921, p. 29.

4. United Brotherhood of Carpenters, Convention *Proceedings,* 1924, p. 3.

5. For a detailed account of this episode, see Royal E. Montgomery, *Industrial Relations in the Chicago Building Trades* (Chicago: University of Chicago Press, 1927), chaps. 12 and 13.

6. Perlman and Taft, *History of Labor,* pp. 506–507.

7. Ibid., p. 408.

8. *The Carpenter,* October 1923, p. 28.

9. William Haber, *Industrial Relations in the Building Industry* (Cambridge: Harvard University Press, 1930), pp. 393–396.

10. Frederick L. Ryan, *Industrial Relations in the San Francisco Building Trades* (Norman, Okla.: University of Oklahoma Press, 1936), pp. 143–154.

11. Ibid., pp. 154–162.

12. GEB *Minutes,* October 20, 1921.

13. Ryan, *Industrial Relations,* pp. 164–165.

14. Ibid., p. 192.

15. United Brotherhood of Carpenters, Convention *Proceedings,* 1928, p. 47.

16. Ibid.

17. Ryan, *Industrial Relations,* pp. 195–196.

18. United Brotherhood of Carpenters, Convention *Proceedings,* 1928, pp. 105, 46.

19. GEB *Minutes,* January 1918.

20. AFL *Proceedings,* Thirty-Seventh Annual Convention, 1917, p. 401.

21. Plan of the National Board of Jurisdictional Awards of the Building Industry, 1919.

22. United Brotherhood of Carpenters, Convention *Proceedings,* 1920, p. 450.

23. Ibid., pp. 453–454.

24. GEB *Minutes,* January 14, 1921.

25. Building Trades Department, *Proceedings,* Fifteenth Annual Convention, 1921, p. 109.

26. GEB *Minutes,* July 30, 1921.

27. AFL, *Executive Council Minutes,* August 22–30, 1921, p. 67; May 10–18, 1922, p. 40.

28. Building Trades Department, *Proceedings,* Sixteenth Annual Convention, 1922, p. 86.

29. AFL, *Executive Council Minutes,* September 9–16, 1922, p. 73.

30. Ibid., November 18–25, 1922, p. 44.

31. *The Carpenter,* July 1923, p. 24.

32. Gompers to Donlin, February 15, 1924.

33. Donlin to Gompers, February 16, 1924.

34. Building Trades Department, *Proceedings,* Eighteenth Annual Convention, 1924, p. 105.

35. AFL, *Executive Council Minutes,* February 4–11, 1925, pp. 31, 56.

36. Building Trades Department, *Proceedings,* Nineteenth Annual Convention, 1925, p. 77.

37. For the text of this agreement, see *Agreements and Decisions Rendered Af-*

fecting the Building Industry, Building and Construction Trades Department, AFL-CIO, June 1, 1977. This publication is widely known as the *Green Book.* The effective date of the metal trim agreement was March 21, 1928.

38. United Brotherhood of Carpenters, Convention *Proceedings,* 1928, p. 51. The Sheet Metal Workers showed some reluctance to ratify this agreement, but finally did so after a minor change.

39. Building Trades Department, *Proceedings,* Twentieth Annual Convention, 1926, p. 50.

40. E. E. Cummins, "The National Board for Jurisdictional Awards and the Carpenters' Union," *American Economic Review,* September 1929, pp. 374–375.

41. Ibid., p. 377.

42. John T. Dunlop, "Jurisdictional Disputes," New York University, *Second Annual Conference on Labor,* 1949, p. 486.

43. Joseph Cusack to Morrison, March 3, 1920; Morrison to Hutcheson, March 10, 1920; Hutcheson to Morrison, March 16, 1920; Cusak to Morrison, March 13, 1920.

44. Hutcheson to W. H. Johnston, July 22, 1920, in AFL, *Executive Council Minutes,* August 2–11, 1920, p. 38.

45. Ibid., November 11–19, 1920, p. 40.

46. Ibid., October 5–22, 1919, p. 65; February 22–March 4, 1921, p. 11; August 22–30, 1921, p. 51; May 10–18, 1922, p. 51. Here again, one of the keys to the problem was whether the railroads wanted to do the work themselves, in which case the Maintenance of Way Employees would probably have held on to the disputed work, or whether they chose to contract it out. Apparently they opted for the latter, permitting employers of carpenters to come in.

47. GEB *Minutes,* July 15, 1919.

48. AFL, *Executive Council Minutes,* June 6–27, 1921, p. 18; GEB *Minutes,* July 26, 1921.

49. GEB *Minutes,* January 10, 1921.

50. AFL, *Executive Council Minutes,* November 14–19, 1921, p. 9.

51. United Brotherhood of Carpenters, Convention *Proceedings,* 1928, p. 52.

52. E. E. Cummins, "Jurisdictional Disputes of the Carpenters' Union," *Quarterly Journal of Economics,* May 1926, pp. 486–494.

53. United Brotherhood of Carpenters, Convention *Proceedings,* 1928, p. 53.

54. For a history of the TUEL, see David J. Saposs, *Left Wing Unionism* (New York: International Publishers, 1926).

55. William Z. Foster, *Misleaders of Labor* (New York: Trade Union Educational League, 1927), pp. 184–185. Foster has equally little sympathy for John L. Lewis: "The present head of the U.M.W.A. deserves to rank with John Mitchell as one of the most powerful and reactionary leaders in the history of the Miners' Union . . . Lewis' regime is a curse to the Miners . . . Lewis, in cooperation with the employers, rules like a despot." Ibid., pp. 132–133.

56. United Brotherhood of Carpenters, Convention *Proceedings,* 1924, p. 3.

57. Ibid., p. 4.

58. Duffy, "History," pp. 456–458.

59. GEB *Minutes,* September 27, 1924.

60. Ibid., April 6, 1925.

61. Ibid.

62. Duffy, "History," pp. 460–461.

63. GEB *Minutes,* June 10, 1926.

64. United Brotherhood of Carpenters, Convention *Proceedings,* 1928, pp. 240–245.

65. Ibid., pp. 235–247.

66. Ibid., pp. 255–256, 332–335.

67. Julius Hochman to Duffy, March 17, 1928.

68. See Walter Galenson, *The CIO Challenge to the AFL* (Cambridge, Mass.: Harvard University Press, 1960), pp. 255, 361, 631.

69. United Brotherhood of Carpenters, Convention *Proceedings,* 1924, p. 30; Duffy to William Green, February 15, 1926.

70. United Brotherhood of Carpenters, Special Circular, March 26, 1923.

71. United Brotherhood of Carpenters, Convention *Proceedings,* 1924, pp. 342–361.

72. Duffy, "History," pp. 441–442.

73. Address of C. Woudenberg, United Brotherhood of Carpenters, Convention *Proceedings,* 1928, p. 29.

74. Philip Taft, *The A.F. of L. in the Time of Gompers* (New York: Harper, 1957), pp. 483–486.

75. Duffy to Gompers, July 23, 1924.

76. Maxwell C. Raddock, *Portrait of an American Labor Leader* (New York: American Institute of Social Science, 1955), p. 158.

77. Duffy to Morrison, January 8, 1925; Morrison to Duffy, March 21, 1925.

78. Duffy to Green, May 12, 1926.

79. Green to Duffy, March 2, 1928; Duffy to Green, March 5, 1928; Green to Duffy, March 17, 1928; Duffy to Green, March 27, 1928.

80. Woll to Duffy, November 1, 1927; Woll to Duffy, December 7, 1927; Woll to Hutcheson, December 16, 1927; Duffy to Green, January 3, 1928; Green to Duffy, January 6, 1928; Duffy to Green, January 9, 1928.

81. Duffy to Green, September 11, 1929.

82. AFL, *Executive Council Minutes,* October 6–18, 1929, p. 5.

83. Lewis to Duffy, July 23, 1926.

84. Sumner H. Slichter to Duffy, quoted in *The Carpenter,* December 1926, p. 53.

85. United Brotherhood of Carpenters, Convention *Proceedings,* 1928, pp. 187–188.

86. Duffy, "History," pp. 476–477.

87. United Brotherhood of Carpenters, Convention *Proceedings,* 1928, p. 187.

88. Ibid., p. 271.

89. *The Carpenter,* March 1929, p. 38.

90. Robert A. Christie, *Empire in Wood* (Ithaca, N.Y.: Cornell University, 1956), pp. 250–251.

91. Data on Construction values are from U.S. Bureau of the Census, *Historical Statistics of the United States, Colonial Times to 1957* (Washington, D.C.: Government Printing Office, 1965), p. 381; for the construction labor force, ibid., p. 73; for wages, *Handbook of Labor Statistics,* 1975, pp. 227–228.

10. THE GREAT DEPRESSION AND THE NEW DEAL

1. It may be noted, however, that trade union wage rates were already exhibiting a marked downward inflexibility, even in the face of heavy unemployment. See John T. Dunlop, *Wage Determination under Trade Unions* (New York: Macmillan, 1944), chap. 8.

2. United Brotherhood of Carpenters, Convention *Proceedings,* 1936, p. 8.

3. *The Carpenter,* March 1930, p. 48.

4. Ibid., February 1933, p. 33.

5. Ibid., November 1932, p. 4.

6. GEB *Minutes,* March 24, 1931.

7. *The Carpenter,* November 1933, p. 14.

8. GEB *Minutes,* September 15, 1930.

9. Ibid., April 20, 1932; January 18, 1933.

10. Ibid., September 29, 1932; United Brotherhood of Carpenters, Convention *Proceedings* 1936, p. 56.

11. United Brotherhood of Carpenters, Convention *Proceedings,* 1936, p. 236

12. Ibid., p. 8.

13. Duffy, "History," p. 480.

14. Mark Perlman, *The Machinists* (Cambridge, Mass.: Harvard University Press, 1961), p. 83.

15. Duffy to Green, June 21, 1932.

16. Green to Duffy, June 22, 1932.

17. For a good account of this episode, see Philip Taft, *The AF of L from the Death of Gompers to the Merger* (New York: Harper, 1959), chap. 3.

18. United Brotherhood of Carpenters, Official Circular, May 16, 1932.

19. Open Statement to the Membership of the Brotherhood of Carpenters and Joiners, June 6, 1932; Duffy to Morris Helf, June 24, 1932.

20. GEB *Minutes,* April 26, 1932; January 9, 1934.

21. *The Carpenter,* May 1934, p. 12.

22. Duffy to Green, February 24, 1930.

23. Green to Duffy, March 12, 1930; Duffy to Green, May 26, 1930; Green to Duffy, June 19, 1930.

24. Taft, *AF of L from the Death of Gompers,* p. 25.

25. For the full text, see Maxwell C. Raddock, *Portrait of an American Labor Leader* (New York: American Institute of Social Science, 1955), pp. 177–178.

26. AFL, Executive Council Minutes, July 15, 1932.

27. GEB *Minutes,* September 30, 1932.

28. Green to Duffy, March 3, 1933; Duffy to Green, March 6, 1933; Duffy to Green, March 8, 1933.

29. Address by Frank Duffy at the University of Notre Dame, November 23, 1933, files of the United Brotherhood of Carpenters.

30. GEB *Minutes,* September 9, 1935.

31. *The Carpenter,* September 1933, p. 14; August 1933, p. 14.

32. Green to Duffy, November 14, 1935.

33. *The Carpenter,* April 1934, p. 2.

34. Lewis L. Lorwin and Arthur Wubnig, *Labor Relations Boards* (Washington, D.C.: Brookings Institution, 1935), pp. 440–442.

35. Building Trades Department, *Proceedings,* Twenty-Ninth Annual Convention, 1935, p. 127; AFL, *Executive Council Minutes,* August 5–16, 1935, p. 56.

36. *The Carpenter,* January 1929, p. 43.

37. Kenneth T. Strand, *Jurisdictional Disputes in Construction* (Pullman, Wash.: Washington State University Press, 1961), p. 66.

38. AFL *Proceedings,* Fiftieth Annual Convention, 1930, p. 133.

39. AFL, *Executive Council Minutes,* January 13–23, 1931, p. 37.

40. Duffy to Morrison, August 12, 1930.

41. Duffy to Martin Ryan, July 23, 1930.

42. Transcript, Hutcheson testimonial dinner, May 3, 1952, pp. 5–11.

43. Building Trades Department, *Proceedings,* Twenty-Fifth Annual Convention, 1931, p. 24.

44. GEB *Minutes,* June 19, 1931.

45. Building Trades Department, *Proceedings,* Twenty-Fifth Annual Convention, 1931, pp. 24, 98.

46. *The Carpenter,* July 1934, p. 20.

47. Building Trades Department, *Proceedings,* Twenty-Eighth Annual Convention, 1934, p. 52.

48. Ibid., p. 104.

49. Ibid., p. 122.

50. AFL *Proceedings,* Fifty-Fourth Annual Convention, 1934, pp. 346, 488.

51. AFL *Proceedings,* Fifty-Fifth Annual Convention, 1935, p. 107; Building Trades Department, *Proceedings,* Twenty-Ninth Annual Convention, 1935, p. 54.

52. The members of the committee were T. A. Rickert of the United Garment Workers, chairman; George M. Harrison of the Railway Clerks; and George L. Berry of the Printing Pressmen.

53. AFL, *Executive Council Minutes,* June 6–7, 1935, p. 23.

54. AFL *Proceedings,* Fifty-Fifth Annual Convention, 1935, p. 327.

55. Strand, *Jurisdictional Disputes,* pp. 67–68; Building Trades Department, *Proceedings,* Thirtieth Annual Convention, 1936, pp. 64, 70, 72, 80; *The Carpenter,* February 1937, p. 16.

56. Strand, *Jurisdictional Disputes,* p. 68.

57. GEB *Minutes,* July 28, 1937.

58. Building Trades Department, *Proceedings,* Thirty-Second Annual Convention, 1938, pp. 120, 146.

59. AFL *Proceedings,* Fifty-First Annual Convention, 1931, p. 214.

60. AFL, *Executive Council Minutes,* August 6–19, 1931, p. 37.

61. GEB *Minutes,* December 10, 1931.

62. AFL, *Executive Council Minutes,* July 12–22, 1932, p. 40.

63. AFL *Proceedings,* Fifty-Second Annual Convention, 1932, p. 81.

64. Ibid., p. 409.

65. AFL, *Executive Council Minutes,* November 25–December 3, 1933, pp. 52, 73, 101.

66. Ibid., September 6–15, 1933, p. 21.

67. GEB *Minutes,* February 20, 1936; September 22, 1936; October 27, 1936.

68. United Brotherhood of Carpenters, Convention *Proceedings,* 1936, p. 99.

69. Ibid., p. 70.

70. Ibid., p. 346.

71. Ibid., p. 353.

72. Ibid., p. 247.

73. Ibid., p. 286.

74. Ibid., p. 75.

11. THE CHALLENGE OF INDUSTRIAL UNIONISM

1. Duffy to Morrison, November 9, 1934.

2. Green to Duffy, December 18, 1934.

3. Quoted in Philip Taft, *Organized Labor in American History* (New York: Harper, 1964), p. 465.

4. Ibid., p. 468. The resolution was deliberately vague with respect to craft jurisdiction in order to avoid a floor fight.

5. AFL, *Executive Council Minutes,* January 29–February 14, 1935, pp. 56, 67, 211–221.

6. Ibid., p. 238.

7. Ibid., April 30–May 7, 1935, p. 138.

8. AFL *Proceedings,* Fifty-Fifth Annual Convention, 1935, p. 615.

9. Ibid., p. 726.

10. Robert A. Christie, *Empire in Wood* (Ithaca, N.Y.: Cornell University, 1956), p. 292.

11. Maxwell C. Raddock, *Portrait of an American Labor Leader* (New York: American Institute of Social Science, 1955), p. 204.

12. Philip Taft, *The AF of L from the Death of Gompers to the Merger* (New York: Harper, 1959), p. 144.

13. AFL, *Executive Council Minutes,* July 8–15, 1936, p. 115.

14. United Brotherhood of Carpenters, Convention *Proceedings,* 1936, p. 43.

15. Ibid., p. 365.

16. GEB *Minutes,* April 9, 1937.

17. AFL *Proceedings,* Fifty-Seventh Annual Convention, 1937, p. 457.

18. Hutcheson to Green, November 17, 1938, and Green to Hutcheson, November 23, 1938, in AFL, *Executive Council Minutes,* January 30–February 14, 1939, p. 293.

19. Much of this section is based on Walter Galenson, *The CIO Challenge to the AFL* (Cambridge: Harvard University Press, 1960), chap 11.

20. Harry Call to Green, December 5, 1933, AFL Records, reel 36.

21. Duffy to Morrison, August 29, 1934, ibid.

22. Duffy to Morrison, November 9, 1934.

23. Green to Duffy, November 13, 1934; Duffy to Green, November 20, 1934; Green to Duffy, December 18, 1934.

24. AFL, *Executive Council Minutes,* January 29–February 14, 1935, p. 83.

25. Duffy to Green, February 19, 1935; Green to Duffy, March 4, 1935; AFL circular letter to local unions of Shingle Weavers, February 20, 1935, in files of United Brotherhood of Carpenters.

26. Morrison to Duffy, March 19, 1935.

27. Vernon H. Jensen, *Lumber and Labor* (New York: Farrar & Rinehart, 1945), p. 185.

28. AFL, *Executive Council Minutes,* January 15–29, 1936, p. 4.

29. United Brotherhood of Carpenters, Convention *Proceedings,* 1936, p. 43.

30. Ibid., p. 21.

31. Ibid., p. 37.

32. Ibid., pp. 315, 317.

33. Ibid., p. 318.

34. Federation of Woodworkers, *Proceedings,* Second Semi-Annual Convention, 1937, p. 4.

35. *The Timberworker,* July 9, 1937, p. 4.

36. GEB *Minutes,* August 6, 1937.

37. Jensen, *Lumber,* pp. 215–217.

38. International Woodworkers of America, *Proceedings,* Third Constitutional Convention, 1939, pp. 287–288.

39. Ibid.

40. GEB *Minutes,* April 22, 1940.

41. Jensen, *Lumber,* p. 235.

42. Galenson, *CIO Challenge,* pp. 403–404.

43. United Brotherhood of Carpenters, Convention *Proceedings,* 1940, p. 42.

44. U.S. Bureau of Labor Statistics, Bulletin no. 840, 1945, p. 8.

45. Margaret S. Glock, *Collective Bargaining in the Pacific Northwest Lumber Industry* (Berkeley: Institute of Industrial Relations, 1955), pp. 60–61.

46. AFL *Proceedings,* Fifty-Sixth Annual Convention, 1936, p. 719.

47. AFL, *Executive Council Minutes,* October 8–21, 1936.

48. GEB *Minutes,* September 22, 1936.

49. AFL, *Executive Council Minutes,* February 8–19, 1937, pp. 147, 169, 213.

50. *The Carpenter,* January 1940, p. 2.

51. Ibid., July 1940, p. 28.

52. Ibid., October 1940, p. 2.

53. Raddock, *Portrait,* p. 237.

54. AFL, *Executive Council Minutes,* January 24–February 8, 1938, p. 154.

55. Statement of F. D. Laudemann to AFL executive council, February 6, 1939, in files of United Brotherhood of Carpenters.

56. AFL, *Executive Council Minutes,* May 13–21, 1940, pp. 79, 105.

57. Ibid., September 30–October 10, 1940, p. 27.

58. Ibid., January 10–20, 1941, p. 27.

59. Ibid., January 12–17, 1942, p. 49.

60. AFL *Proceedings,* Sixty-First Annual Convention, 1941, p. 246.

61. AFL, *Executive Council Minutes,* August 4–13, 1942, p. 46.

62. Ibid., May 17–23, 1943, p. 14.

63. Green to Duffy, June 29, 1933; Duffy to Green, July 14, 1933; Morrison to Duffy, July 18, 1933.

64. GEB *Minutes,* October 25, 1937.

65. *The Carpenter,* August 1946, p. 17.

66. AFL, *Executive Council Minutes,* January 24–February 8, 1938, p. 222.

67. Hutcheson to Green, July 28, 1938.

68. Minutes of a conference, November 30, 1938, in files of United Brotherhood of Carpenters.

69. Christie, *Empire in Wood,* p. 311.

70. *The Carpenter,* January 1940, pp. 16–18.

71. *United States* v. *Hutcheson* 312 U.S. 219 (1941).

72. United Brotherhood of Carpenters, Convention *Proceedings,* 1946, p. 251.

73. GEB *Minutes,* January 10, 1938.

74. Ibid., January 10, 1938; May 27, 1938.

75. Ibid., April 13, 1938.

76. United Brotherhood of Carpenters, Convention *Proceedings,* 1940, p. 234.

77. William L. Leiserson, *American Trade Union Democracy* (New York: Columbia University Press, 1959), p. 232.

78. United Brotherhood of Carpenters, Convention *Proceedings,* 1940, p. 236.

12. WAR AND RECONVERSION

1. GEB *Minutes,* April 1, 1941.

2. Maxwell C. Raddock, *Portrait of an American Labor Leader* (New York: American Institute of Social Science, 1955), pp. 257–258.

3. Philip Taft, *The AF of L from the Death of Gompers to the Merger* (New York: Harper, 1959), pp. 209–210.

4. *The Carpenter,* November 1941, p. 6.

5. For the full text of the agreement, see John T. Dunlop and Arthur D. Hill, *The Wage Adjustment Board* (Cambridge, Mass.: Harvard University Press, 1950), pp. 138–140.

6. *The Carpenter,* December 1941, p. 18; Taft, *AF of L from the Death of Gompers,* p. 210.

7. *The Carpenter,* January 1942, p. 5.

8. United Brotherhood of Carpenters, Circular Letter, December 22, 1941.

9. United Brotherhood of Carpenters, Convention *Proceedings,* 1946, p. 156.

10. *The Carpenter,* February 1942, p. 21.

11. Ibid., July 1942, p. 3.

12. The board was created by Executive Order 9017 on January 12, 1942. For details see the *Termination Report of the National War Labor Board* (Washington, D.C.: Government Printing Office, no date), vol. I.

13. For the text of the May 22 agreement, see Dunlop and Hill, *Wage Adjustment Board,* pp. 141–142.

14. See ibid. for a general history of the board. I was an alternate public member of the board and can attest to the efficiency and cordiality with which it operated.

15. GEB *Minutes,* August 27, 1942.

16. Taft, *AF of L from the Death of Gompers,* p. 223.

17. *The Carpenter,* October 1942, p. 46.

18. AFL, *Executive Council Minutes,* May 17–23, 1943, pp. 5, 40.

19. Ibid., August 9–16, 1943, p. 20.

20. Ibid., January 17–27, 1944, p. 88.

21. Lewis to Green, May 8, 1944, ibid., May 1–9, 1944, p. 131.

22. Ibid., February 5–15, 1945, pp. 15, 65, 106.

23. Ibid., January 21–28, 1946, pp. 61, 83.

24. AFL *Proceedings,* Sixty-Sixth Annual Convention, 1947, p. 567.

25. AFL, *Executive Council Minutes,* October 4–17, 1947, passim.

26. GEB *Minutes,* June 15, 1942; October 14, 1944; November 16, 1945.

27. Ibid., February 8, 1943.

28. *The Carpenter,* January 1943, p. 3.

29. Ibid., October 1943, p. 32.

30. Ibid., March 1944, p. 3.

31. Ibid., March 1945, p. 16.

32. Ibid., May 1945, p. 4.

33. United Brotherhood of Carpenters, Convention *Proceedings,* 1946, pp. 343 ff.

34. Ibid., pp. 261–268.

35. Ibid., p. 281.

36. Hugh Lovell and Tasile Carter, *Collective Bargaining in the Motion Picture Industry* (Los Angeles: Institute of Industrial Relations, 1955), p. 21.

37. AFL, *Executive Council Minutes,* April 30–May 8, 1945, pp. 96–99.

38. Ibid., October 15–24, 1945, pp. 9–14.

39. Ibid., pp. 132–143.

40. Ibid., January 21–31, 1946, pp. 90–132.

41. Ibid., May 15–22, 1946, pp. 71–78.

42. Ibid., August 12–20, 1946, pp. 56–67, 91–93.

43. Ronald Reagan and Richard G. Hubler, *Where's the Rest of Me?* (New York: Dell, 1965), p. 168.

44. Ibid., p. 173. In addition to Reagan, the group included Walter Pidgeon, Richard Powell, Jane Wyman, Alexis Smith, Gene Kelly, Robert Taylor, George Murphy, and Robert Montgomery.

45. AFL, *Executive Council Minutes,* April 21–25, 1947, pp. 51–57.

46. Ibid., September 8–13, 1947, pp. 83–85.

47. Ibid., October 4–17, 1947, pp. 18–19.

48. Lovell and Carter, *Collective Bargaining,* p. 24.

49. GEB *Minutes,* January 15, 1947.

50. Ibid., January 30, 1950.

51. AFL, *Executive Council Minutes,* May 10–14, 1948, p. 86; August 23–27, 1948, p. 29; November 12–23, 1948, p. 7; January 31–February 8, 1949, p. 86; May 16–20, 1949, p. 8.

52. Sal Hoffman to William Green, September 1, 1949, in AFL, *Executive Council Minutes,* October 2–11, 1949, p. 20.

53. *The Carpenter,* May 1946, p. 10.

54. Ibid., April 1947, p. 6.

55. Ibid., August 1947, p. 16.

56. Ibid., September 1947, pp. 6, 16.

57. Ibid., December 1947, p. 5.

58. Ibid., February 1948, p. 5; June 1948, p. 10.

59. Ibid., July 1948, p. 7; GEB *Minutes,* October 2, 1948.

13. ADJUSTING TO A CHANGING ENVIRONMENT

1. United Brotherhood of Carpenters, Convention *Proceedings,* 1950, p. 15.

2. Ibid., p. 247.

3. Ibid., p. 321.

4. GEB *Minutes,* September 7, 1950.

5. United Brotherhood of Carpenters, Convention *Proceedings,* 1950, pp. 321, 323.

6. *The Carpenter,* January 1951, p. 12

7. AFL, *Executive Council Minutes,* January 22–29, 1951, pp. 3, 12.

8. Ibid., January 22–February 5, 1952, p. 38; May 19–22, 1952, p. 15.

9. Ibid., May 19–22, 1952, p. 17; August 11–15, 1952, p. 18.

10. United Brotherhood of Carpenters, Convention *Proceedings,* 1954, pp. 117–119.

11. *The Carpenter,* October 1954, p. 24.

12. United Brotherhood of Carpenters, Convention *Proceedings,* 1970, p. 125.

13. See the *Green Book* for the text of the agreement and the list of signatories.

14. Building Trades Department, *Proceedings,* Forty-First Annual Convention, 1948, p. 117; ibid., Forty-Second Annual Convention, 1949, p. 142.

15. *The Carpenter,* December 1951, p. 20.

16. United Brotherhood of Carpenters, Convention *Proceedings,* 1954, p. 260.

17. *The Carpenter,* August 1957, p. 11.

18. GEB *Minutes,* February 12, 1958.

19. United Brotherhood of Carpenters, Convention *Proceedings,* 1958, p. 109. The locus of disputes tended to shift over time; the 1957 distribution was not necessarily similar for all years. However, the Carpenters were always a major client.

20. Ibid., 1954, p. 22.

21. *The Carpenter,* November 1951, p. 13.

22. Ibid., May 1952, p. 25.

23. GEB *Minutes,* September 13, 1952.

24. *The Carpenter,* December 1952, p. 8.

25. United Brotherhood of Carpenters, Convention *Proceedings,* 1958, p. 401.

26. GEB *Minutes,* October 15, 1958.

27. Ibid., February 2, 1959.

28. Arthur Goldberg, *AFL-CIO Labor United* (New York: McGraw-Hill, 1956), p. 83.

29. GEB *Minutes,* September 22, 1953.

30. Ibid.

31. *The Carpenter,* August 1953, p. 25.

32. *New York Times,* August 13, 1953, p. 1; August 14, 1953, p. 15.

33. GEB *Minutes,* September 22, 1953; *New York Times,* September 9, 1953, p. 1.

34. United Brotherhood of Carpenters, Convention *Proceedings,* 1954, p. 124.

35. GEB *Minutes,* February 10, 1955.

36. *The Carpenter,* March 1958, p. 28.

37. GEB *Minutes,* October 15, 1958.

38. This matter is discussed later in this chapter.

39. Meany to Hutcheson, August 21, 1958.

40. United Brotherhood of Carpenters, Convention *Proceedings,* 1958, p. 383.

41. GEB *Minutes,* August 12, 1957.

42. Building Trades Department, *Proceedings,* Fiftieth Anniversary Convention, 1957, p. 56; Fiftieth Regular Convention, 1959, p. 7.

43. *The Carpenter,* October 1960, p. 20.

44. Ibid., November 1953, p. 8.

45. United Brotherhood of Carpenters, Convention *Proceedings,* 1954, pp. 356, 383.

46. Ibid., p. 391.

47. 85th Congress, Senate, Select Committee on Improper Activities in the Labor or Management Field, 1958–1959. This committee will be referred to as the McClellan Committee, after its chairman, Senator McClellan of Arkansas. These widely publicized hearings helped raise to national prominence another member, Senator John F. Kennedy, and its chief counsel, Robert F. Kennedy.

48. McClellan Committee Hearings, part 31, May 1958.

49. AFL, *Executive Council Minutes,* January 30–February 7, 1950, p. 64. The local bodies were told that the paper "is not in any way connected with the American Federation of Labor and does not speak for the American Federation of Labor; and that in substance, we believe, it is not helpful to the American Federation of Labor."

50. McClellan Committee Hearings, p. 11828.

51. Ibid.

52. Robert A. Christie, *Empire in Wood* (Ithaca, N.Y.: Cornell University, 1956).

53. McClellan Committee Hearings, p. 12013.

54. United Brotherhood of Carpenters, Convention *Proceedings,* 1958, p. 131.

55. Ibid., pp. 247–249.

56. McClellan Committee Hearings, p. 12065.

57. Ibid., pp. 12018–12031.

58. Ibid., p. 12115.

59. Robert F. Kennedy, *The Enemy Within* (New York: Harper, 1960), pp. 200–201.

60. *U.S.* v. *Hutcheson,* 369 U.S. 599 (1962).

61. *The Carpenter,* December 1960, p. 3.

62. *Hutcheson* v. *State of Indiana,* 244 Indiana 345, October 3, 1963.

63. McClellan Committee, Second Interim Report, p. 591.

64. Kennedy, *The Enemy Within,* p. 203.

65. *The Carpenter,* December 1960, p. 3.

66. Philip Taft, *The Structure and Government of Labor Unions* (Cambridge, Mass.: Harvard University Press, 1954), p. 142.

67. Ibid., p. 143.

68. Morris A. Horowitz, *The Structure and Government of the Carpenters' Union* (New York: Wiley, 1962), pp. 91, 93.

69. Ibid., pp. 97–99.

70. Taft, *Structure and Government,* p. 149.

71. GEB *Minutes,* January 11, 1952.

72. Ibid., June 22, 1953; August 12, 1957; February 10, 1955.

73. Horowitz, *Structure and Government,* p. 62.
74. McClellan Committee, Interim Report, 1958, pp. 371, 448.
75. Taft, *Structure and Government,* p. 82.
76. Ibid., p. 107.
77. Ibid., pp. 104–107.
78. *The Carpenter,* October 1957, p. 19.
79. GEB *Minutes,* February 12, 1958; October 15, 1958.
80. United Brotherhood of Carpenters, Convention *Proceedings,* 1958, p. 109.
81. Ibid.
82. Ibid., p. 379.
83. Ibid., p. 362.
84. Ibid., p. 241.
85. Ibid., p. 60.
86. *The Carpenter,* November 1959, p. 8.

14. A DECADE OF PROGRESS

1. United Brotherhood of Carpenters, *Password and Quarterly Circular,* September 23, 1959.
2. Ibid., March 22, 1960.
3. United Brotherhood of Carpenters, Convention *Proceedings,* 1960, pp. 14–15.
4. Ibid., pp. 24–25.
5. *The Carpenter,* October 1960, p. 24.
6. United Brotherhood of Carpenters, Convention *Proceedings,* 1960, p. 159. In Britain, the mail ballot has been advocated as a means of defeating left-wing candidates who often gain election because of low turnouts at local meetings.
7. Ibid., p. 164.
8. Ibid., p. 251.
9. Ibid., p. 296.
10. For a good general account of the growing tensions between Canadian and American unions, see John Crispo, *International Unionism* (Toronto: McGraw-Hill, 1967).
11. This section is based on Ontario Provincial Council, "Fifty Years," manuscript in the files of the United Brotherhood of Carpenters.
12. *The Carpenter,* October 1961, p. 3.
13. United Brotherhood of Carpenters, Convention *Proceedings,* 1960, p. 183.
14. *The Carpenter,* December 1961, p. 15; GEB *Minutes,* November 1, 1961.
15. *The Carpenter,* April 1962, p. 18; May 1962, p. 18.
16. United Brotherhood of Carpenters, Convention *Proceedings,* 1966, p. 8.
17. *The Carpenter,* October 1964, p. 19.
18. United Brotherhood of Carpenters, Convention *Proceedings,* 1962, p. 451.
19. Ibid., 1970, pp. 301–302.
20. Ibid., p. 302.
21. GEB *Minutes,* November 14, 1960.
22. Ibid., February 1, 1961.
23. Ibid., July 23, 1962; November 8, 1962; November 4, 1963.
24. Ibid., July 18, 1966.
25. Ibid., February 1, 1961.
26. Ibid., July 23, 1962.

27. United Brotherhood of Carpenters, *Password and Quarterly Circular,* September 23, 1963.

28. GEB *Minutes,* May 4, 1965; November 29, 1967; November 10, 1969.

29. Ibid., June 15, 1970.

30. United Brotherhood of Carpenters, Convention *Proceedings,* 1962, p. 99.

31. Ibid.

32. Ibid., pp. 106, 450.

33. Ibid., pp. 404–407.

34. Ibid., p. 225.

35. Ibid., p. 354.

36. Ibid., p. 19

37. *The Carpenter,* March 1965, p. 5.

38. United Brotherhood of Carpenters, Convention *Proceedings,* 1966, p. 423.

39. Ibid., 1970, p. 117.

40. Ibid., 1966, p. 241. For the text of the decision, see the *Green Book,* p. 168.

41. *The Carpenter,* September 1964, p. 2.

42. United Brotherhood of Carpenters, *Password and Quarterly Circular,* September 23, 1964.

43. United Brotherhood of Carpenters, Official Circular, December 1964.

44. GEB *Minutes,* March 23, 1966; *The Carpenter,* July 1966, p. 2.

45. *The Carpenter,* November 1968, p. 3.

46. United Brotherhood of Carpenters, Convention *Proceedings,* 1966, p. 462.

47. United Brotherhood of Carpenters, *Password and Quarterly Circular,* June 24, 1969.

48. *The Carpenter,* June 1969, p. 40.

49. United Brotherhood of Carpenters, Convention *Proceedings,* 1966, pp. 8, 107.

50. Ibid., p. 45.

51. Ibid., p. 50.

52. *The Carpenter,* January 1967, p. 2; April 1968, p. 2.

53. United Brotherhood of Carpenters, Convention *Proceedings,* 1966, p. 298.

54. GEB *Minutes,* June 27, 1963.

55. *New York Times,* July 2, 1963.

56. Hutcheson to Willard Wirtz, June 11, 1963.

57. Interview with William Sidell, April 6, 1981. The Baltimore local is still largely black. It has its own business agent, and its members work on jobs throughout the city.

58. United Brotherhood of Carpenters, Convention *Proceedings,* 1970, p. 105.

15. BATTLING INFLATION AND RECESSION

1. Daniel Quinn Mills, *Government, Labor, and Inflation* (Chicago: University of Chicago Press, 1975), p. 139.

2. GEB *Minutes,* August 19, 1972; November 13, 1972.

3. United Brotherhood of Carpenters, Official Circular, December 26, 1972; February 1, 1973; March 27, 1973.

4. United Brotherhood of Carpenters, Convention *Proceedings,* 1970, p. 48.

5. Ibid., pp. 11, 51.

6. Ibid., p. 247. ABC refers to the Associated Builders and Contractors, the largest open-shop contractors' association in the country.

7. Ibid., p. 529.

8. GEB *Minutes,* January 26, 1970.

9. *The Carpenter,* February 1972, p. 7.

10. Ibid., March 1972, p. 6.

11. United Brotherhood of Carpenters, Official Circular, April 28, 1976.

12. GEB *Minutes,* November 29, 1977.

13. United Brotherhood of Carpenters, *Current Information,* June 15, 1972.

14. GEB *Minutes,* June 23, 1979.

15. *The Carpenter,* May 1971, p. 6.

16. United Brotherhood of Carpenters, Convention *Proceedings,* 1974, p. 147.

17. GEB *Minutes,* October 21, 1974.

18. United Brotherhood of Carpenters, Official Circular, September 10, 1973.

19. United Brotherhood of Carpenters, Convention *Proceedings,* 1974, p. 174.

20. Ibid., p. 67.

21. Ibid., p. 166

22. Ibid., p. 514.

23. Ibid., p. 519.

24. Paul T. Hartman and Walter H. Franke, "The Changing Bargaining Structure in Construction," *Industrial and Labor Relations Review,* January 1980, p. 176.

25. Ibid., p. 175.

26. *The Carpenter,* February 1976, p. 40.

27. GEB *Minutes,* June 22, 1976; *The Carpenter,* October 1976, p. 32.

28. United Brotherhood of Carpenters, *Password and Quarterly Circular,* September 29, 1980.

29. Ibid., December 23, 1980.

30. Public letter signed by Archibald Cox (no date).

31. GEB *Minutes,* February 1, 1981.

32. United Brotherhood of Carpenters, Convention *Proceedings,* Report of General Officers, 1981, pp. 137–148.

33. Because of general dissatisfaction with its operation, the Impartial Disputes Board was suspended on May 31, 1981, and a new format is being developed. United Brotherhood of Carpenters, Convention *Proceedings,* 1978, p. 195.

34. United Brotherhood of Carpenters, Convention *Proceedings,* 1978, p. 195.

35. *The Carpenter,* July 1973, p. 4.

36. United Brotherhood of Carpenters, Convention *Proceedings,* 1978, p. 90.

37. GEB *Minutes,* October 19, 1976.

38. United Brotherhood of Carpenters, Convention *Proceedings,* Report of General Officers, 1981, pp. 19–22.

39. United Brotherhood of Carpenters, Convention *Proceedings,* 1978, pp. 375–394.

40. Ibid., p. 292.

41. GEB *Minutes,* April 15, 1972.

42. United Brotherhood of Carpenters, Convention *Proceedings,* 1978, p. 488.

43. Ibid., p. 663.

44. *Wall Street Journal,* March 10, 1981, p. 12.

45. United Brotherhood of Carpenters, Circular Letter, April 13, 1981.

46. United Brotherhood of Carpenters, Convention *Proceedings,* 1981, pp. 258–260.

47. *The Carpenter,* May 1981, p. 9.

48. United Brotherhood of Carpenters, Convention *Proceedings,* 1981, pp. 390–409.

49. Ibid., 1978, p. 18.

50. Ibid., p. 28.

51. Ibid., p. 213.

52. Ibid., p. 479.

53. Ibid., pp. 509, 540, 595.

54. GEB *Minutes,* June 24, 1980.

55. Ibid., October 31, 1980.

56. U.S. Department of Labor, *Annual Construction Industry Report,* April 1980, p. 6.

57. United Brotherhood of Carpenters, Convention *Proceedings,* Report of General Officers, 1981, p. 17.

58. Thomas R. Brooks, *The Road to Dignity* (New York: Atheneum, 1981). The play was written by Arnold Sundgaard and featured E. G. Marshall as Peter J. McGuire, supported by members of the Goodman Theater of Chicago.

59. United Brotherhood of Carpenters, Convention *Proceedings,* 1981, p. 230.

60. Ibid., pp. 218–224.

61. Ibid., p. 312.

62. Ibid., pp. 315–316.

63. Ibid., pp. 453–458.

64. Ibid., p. 286.

16. A LOOK BACKWARD AND A LOOK AHEAD

1. U.S. Bureau of the Census, *Historical Statistics of the United States,* 1960, tables D573–577 and D589–602.

2. It is rather remarkable that the cost of living in 1907 was almost identical with that in 1881, so that no correction need be made to translate money into real wages. U.S. Bureau of Labor Statistics, *Handbook of Labor Statistics,* 1975, table 122.

3. Statements for various cities appearing in *The Carpenter* during 1881 and 1882 bear out this assertion. Washington, D.C., with a standard nine-hour day, appears to have been an exception.

4. U.S. Department of Labor, *Annual Construction Industry Report,* April 1980, p. 51.

5. *The Carpenter,* January 1882, p. 2.

6. *Annual Construction Industry Report,* 1980, p. 74.

7. *The Carpenter,* May 1881, p. 2; November 1882, p. 4.

8. In construction, the package of benefits is usually bargained as a whole, and the union is often free to divert part to fringe benefits. In this sense, workers are paying for at least part of the fringes by foregoing money wage increases.

9. Clinton C. Bourdon and Raymond E. Levitt, *Union and Open-Shop Construction* (Lexington, Mass.: Lexington Books, 1980), pp. 45–46.

10. Ray Marshall, William S. Franklin, and Robert W. Glover, *A Comparison of Union Construction Workers Who Have Achieved Journeyman Status through Apprenticeship and Other Means,* U.S. Department of Labor (processed), 1974, pp. 206–207.

11. *Monthly Labor Review,* August 1981, p. 51. According to this estimate, the outlook for millwright employment is less favorable: an increase of from 15 to 22 percent from 1978 to 1990.

12. Bourdon and Levitt, *Union and Open-Shop Construction,* pp. 45–46.

13. *The Carpenter,* December 1881, p. 4.

APPENDIX A. PETER J. McGUIRE'S VALEDICTORY

1. The source is a typewritten document in the files of the Brotherhood of Carpenters clearly labeled McGuire's testimony. It contains a statement followed by questions from members of the commission and McGuire's replies. No record of this testimony appears in the voluminous reports of the Industrial Commission, nor is there any reference to McGuire in the master list of witnesses appearing before the commission, despite the fact that other Brotherhood officers were listed. The historian of the U.S. Department of Labor made a search at my request, and reported that "Peter McGuire's testimony is not printed—it should have been listed on page 1149 in the final volume's index." Jonathan Grossman to the author, December 9, 1981. The reason for this omission is not known.

2. United States Industrial Commission, *Final Report and Hearings,* 19 vols., Washington, D.C., 1900–1902.

APPENDIX B. REPORT ON THE SITUATION AT BOGALUSA, LOUISIANA

1. The original of this report is in the files of the Brotherhood of Carpenters.

INDEX

WERTHEIM PUBLICATIONS IN INDUSTRIAL RELATIONS

Published by Harvard University Press

J. D. Houser, *What the Employer Thinks*, 1927
Wertheim Lectures on Industrial Relations, 1929
William Haber, *Industrial Relations in the Building Industry*, 1930
Johnson O'Connor, *Psychometrics*, 1934
Paul H. Norgren, *The Swedish Collective Bargaining System*, 1941
Leo C. Brown, S.J., *Union Policies in the Leather Industry*, 1947
Walter Galenson, *Labor in Norway*, 1949
Dorothea de Schweinitz, *Labor and Management in a Common Enterprise*, 1949
Ralph Altman, *Availability for Work: A Study in Unemployment Compensation*, 1950
John T. Dunlop and Arthur D. Hill, *The Wage Adjustment Board: Wartime Stabilization in the Building and Construction Industry*, 1950
Walter Galenson, *The Danish System of Labor Relations: A Study in Industrial Peace*, 1952
Lloyd H. Fisher, *The Harvest Labor Market in California*, 1953
Donald J. White, *The New England Fishing Industry: A Study in Price and Wage Setting*, 1954
Val R. Lorwin, *The French Labor Movement*, 1954
Philip Taft, *The Structure and Government of Labor Unions*, 1954
George B. Baldwin, *Beyond Nationalization: The Labor Problems of British Coal*, 1955
Kenneth F. Walker, *Industrial Relations in Australia*, 1956
Charles A. Myers, *Labor Problems in the Industrialization of India*, 1958
Herbert J. Spiro, *The Politics of German Codetermination*, 1958
Mark W. Leiserson, *Wages and Economic Control in Norway, 1945–1957*, 1959
J. Pen, *The Wage Rate under Collective Bargaining*, 1959
Jack Stieber, *The Steel Industry Wage Structure: A Study of the Joint Union-Management Job Evaluation Program in the Basic Steel Industry*, 1959
Theodore V. Purcell, S.J., *Blue Collar Man: Patterns of Dual Allegiance, in Industry*, 1960
Carl Erik Knoellinger, *Labor in Finland*, 1960
Sumner H. Slichter, *Potentials of the American Economy, Selected Essays*, edited by John T. Dunlop, 1961
C. L. Christensón, *Economic Redevelopment in Bituminous Coal: The Special Case of Technological Advance in the United States Coal Mines, 1930-1960*, 1962
Daniel L. Horowitz, *The Italian Labor Movement*, 1963
Adolf Sturmthal, *Workers Councils: A Study of Workplace Organization on Both Sides of the Iron Curtain*, 1964

Vernon H. Jensen, *Hiring of Dock Workers and Employment Practices in the Ports of New York, Liverpool, London, Rotterdam, and Marseilles*, 1964

John L. Blackman, Jr., *Presidential Seizure in Labor Disputes*, 1957

Mary Lee Ingbar and Lester D. Taylor, *Hospital Costs in Massachusetts: An Economic Study*, 1968

Kenneth F. Walker, *Australian Industrial Relations Systems*, 1970

David Kuechle, *The Story of the Savannah: An Episode in Maritime Labor-Management Relations*, 1971

Studies in Labor-Management History

Lloyd Ulman, *The Rise of the National Trade Union: The Development and Significance of Its Structure, Governing Institutions, and Economic Policies*, second edition, 1955

Joseph P. Goldberg, *The Maritime Story: A Study in Labor-Management Relations, 1957,* 1958

Walter Galenson, *The CIO Challenge to the AFL: A History of the American Labor Movement, 1935–1941,* 1960

Morris A. Horowitz, *The New York Hotel Industry: A Labor Relations Study*, 1960

Mark Perlman, *The Machinists: A New Study in American Trade Unionism*, 1961

Fred C. Munson, *Labor Relations in the Lithographic Industry*, 1963

Garth L. Mangum, *The Operating Engineers: The Economic History of a Trade Union*, 1964

David Brody, *The Butcher Workmen: A Study of Unionization*, 1964

F. Ray Marshall, *Labor in the South*, 1967

Philip Taft, *Labor Politics American Style: The California State Federation of Labor*, 1968

Walter Galenson, *The United Brotherhood of Carpenters: The First Hundred Years*, 1983

Distributed by Harvard University Press

Martin Segal, *The Rise of the United Association: National Unionism in the Pipe Trades, 1884–1924,* 1969

Arch Fredric Blakey, *The Florida Phosphate Industry: A History of the Development and Use of a Vital Mineral*, 1973.